CHEMISTRY RESEARCH AND APPLICATIONS

THE PHYSICOCHEMICAL ASPECTS OF NITRIC OXIDE IN CHEMISTRY AND BIOLOGY

FUNDAMENTALS AND RECENT DEVELOPMENT

CHEMISTRY RESEARCH AND APPLICATIONS

Additional books and e-books in this series can be found on Nova's website under the Series tab.

CHEMISTRY RESEARCH AND APPLICATIONS

THE PHYSICOCHEMICAL ASPECTS OF NITRIC OXIDE IN CHEMISTRY AND BIOLOGY

FUNDAMENTALS AND RECENT DEVELOPMENT

GERTZ I. LIKHTENSHTEIN

nova
science publishers
New York

Library of Congress Cataloging-in-Publication Data

Names: Likhtenshtein, G. I. (Gerktls Il'ich), author.
Title: The physicochemical aspects of nitric oxide in chemistry and biology: fundamentals and recent developments / Gertz I. Likhtenshtein (Department of Chemistry, Ben-Gurion University of the Negev, Beer-Sheva, Israel and Institute of Problems of Chemical Physics, Russian Academy of Science, Chernogolovka, Moscow region, Russian Federation).
Description: Hauppauge, New York: Nova Science Publishers, Inc., 2018. | Series: Chemistry research and applications | Includes bibliographical references.
Identifiers: LCCN 2018032694 (print) | LCCN 2018038516 (ebook) | ISBN 9781536139594 (ebook) | ISBN 9781536139587 (hardcover)
Subjects: LCSH: Nitric oxide--Physiological effect. | Nitric oxide--Metabolism. | Nitric oxide.
Classification: LCC QP535.N1 (ebook) | LCC QP535.N1 L55 2018 (print) | DDC 572/.53--dc23
LC record available at https://lccn.loc.gov/2018032694

Published by Nova Science Publishers, Inc. † New York

CONTENTS

PREFACE

Since discovery in 1772, for more than almost 220 years, nitric oxide (NO) remained to be in the "shadow" of its three atomic "sisters," O_2, N_2, and CO and was almost unknown in biology. It was only after the outstanding research of the Nobel Laureates Robert F. Furchgott, Louis J. Ignarro, and Ferid Murad's triumphal march of this tiny molecule with a huge impact in biology, biochemistry, physiology, biomedicine, and clinical practice occurred.

Nitric oxide, as a signaling molecule, directly or nondirectly significantly contributes in regulation of transcriptional, translational, and post-translational processes, aging, inflammation, and major human pathologies, including cancer, diabetes, cardiovascular and gastrointestinal tract disorders, atherosclerosis, arteriosclerosis, thrombosis, Alzheimer's, Parkinson's, Huntington's disease, neurodegenerative diseases, etc.

The processes involving NO embrace practically all the aspects of modern biochemistry, biophysics, and molecular biology. For the last decades, substantial advances have been made in understanding of molecular mechanism of these processes. A wide variety of advanced biochemical and physicochemical methods allow one to shed light on the NO activity. Nevertheless, despite the many studies conducted and many effects reported, the exact mechanisms of its still remains elusive and continue to be sources of controversy. That's why in the literature the role of NO is characterized by a cascade of epithets: double-faced Janus, dichotomy, dual role, complex and quite perplexing, controversy and confusion, double-edged sword, unanswered dilemma, conflicting findings, paradigm shift, etc.

This book represents a view on the area from physical chemist with long and broad expertise in chemical biophysics, mechanisms of enzyme catalysis, and physical methods application in biology, in particular. The main intention was not to provide the reader with an exhaustive survey of each topic of vast literature, but rather to discuss the key theoretical and experimental background and focus on recent developments. Thus, researchers working in the fields of biochemistry, molecular physiology, biomedicine, and related fields would find the fundamentals of physical properties and chemical activity of NO, information about the sources and methods of analysis of NO, structure and action mechanism of corresponding key enzymes, as well as a review of the major milestones. This gained knowledge is expected to promote some critical thinking to solve new emerging problems in their fields. Chemists, physicists, biochemists, and molecular biophysicists would benefit from an overview of current physiological and medicine achievements and problems in the NO area and may offer original ways to solve them.

With a wealth of experimental and theoretical works devoted to NO (the international program SkiFinder shows more than 40,000 references for "nitric oxide"), here we restrict the discussion to the physicochemical and biochemical aspects. The literature of the nitroxide "empire" is so vast, and many scientists have made important contribution in the area, that it is impossible in the space allowed for this book to give a representative set of references. The author apologizes to those he has not been able to include. More than a thousand references are given in the book, which should provide a key to essential relevant literature.

Chapters 1–4 form the theoretical and experimental background for the central chapters 5 and 6. Basic physical and chemical properties of NO are described in chapter 1, as the molecular basis for deep understanding of the biochemical and physiological impact of this molecule. Chapter 2 is dedicated to the endogeneous and exogeneous NO donors. Chapter 3 is a brief outline of the contemporary knowledge of structure and properties of NO metallocomplexes, playing key roles as active centers of iron-containing enzymes and nitrosyl dithiocarbamate complexes, NO traps, and donors. Experimental methods of analysis of nitroxides is a subject of chapter 4. In chapter 5, a general survey is made of data on structure, molecular dynamics, catalytic mechanism, and inhibitors of NO synthase, which produces NO in diverse biological processes of vital importance. Chapter 6 describes soluble guanylate cyclase (sGC), which, as an NO receptor, is a key metalloprotein in mediating NO signaling transduction. Chapter 7 shows mechanisms of initiation of reactive nitrogen species (RNS) and its effects on biological molecules' activity and injury, and on the normal and pathological functions of the entire organisms by radical and signaling mechanisms. Nitric oxide contribution to genetics and aging is considered in Chapter 8. Five subsequent chapters 9–13 are brief reviews of the NO involved in the occurrence of pathology and treatment of various diseases: cancer (chapter 9), arterial and bone diseases (chapter 10), central nervous system (chapter 11), neurodegenerative (chapter 12) disorders, and miscellaneous diseases (chapter 13).

This monograph is intended for scientists working in areas related to the impact of NO to biochemistry, molecular biology, biophysics, physiology, and biomedicine. The book will also be useful for doctors and pharmacologists involved in clinical practice. This text as a whole, or as separate chapters, can also be employed as subsidiary manuals for instructors and graduate and undergraduate students of university physics, biophysics, chemistry, biochemistry, and biomedicine departments, and researchers in laboratories of pharmacological companies.

Chapter 1

NITRIC OXIDE PHYSICAL CHEMISTRY

ANNOTATION

Nitric oxide (NO) is a small molecule consisted of two atoms, oxygen and nitrogen. The molecule chemical activity and corresponding impact in the biological and physiological processes are determined by its physical properties and quantum mechanical characteristics, in particular. In the NO molecular diagram, an unpaired electron occupies the forth antibonding $\pi^*\pi^*$- orbitals, is involved in a trielectronic bond between oxygen and nitrogen, and can be considered as a stable radical. Therefore, NO readily reacts with free reactive radicals but very slowly with saturated molecules. This unique property provides a fast reaction with superoxide with the subsequent formation of peroxinitrite and other reactive nitrogen species, on the one hand, and the chemical inertness with respect to biological molecules, on the other hand. The ability to be a ligand of transition metal complexes enables NO to serve as a powerful signaling molecule for vital physiological processes.

1. PHYSICAL PROPERTIES

1.1. General Characteristics

Nitric oxide of the chemical formulas

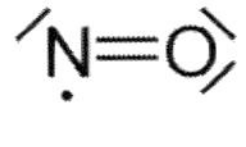

or

has a molar mass 30.01 g·mol^{-1}, density 1.3402 g dm^{-3}, melting point −164°C, boiling point −152°C, bond distance 1.15 Å (https://en.wikipedia.org/wiki/Nitric_oxide#Further_reading), bond energy 150 kcal/mol. [1], (ΔHf_{298}), 630.57 kJ/mol (https://labs.chem.ucsb.edu/), ionization energy 9.2642 ev, proton affinity 531.8 kJ/mol, and gas basicity 505.3 kJ/mol. [2]. Relative atomic mass of ^{14}N is 14.003 and isotopic composition is 0.996 for ^{15}N, atomic mass and isotopic composition are 15.0001 and 0.003, respectively (https://www.ncsu.edu/.../Isotopic Mass_NaturalAbundance). The small difference in electronegativity between nitrogen and oxygen gives NO a modest electrical dipole moment of 0.159 D. The concentration of nitrogen monoxide in a solution exposed to a partial pressure of 1 atm (101.3 kPa) is 1.93 mM at 25°C and 1.63 mM at 37°C [3]. The NO diffusivities in pure water at 25°C were found to be (2.21) x 10^{-5} cm^2 s^{-1} [4], with agreement of a recent theoretical calculation [5].

1.2. Nitric Oxide Quantum Chemistry

Nitric oxide is characterized by electron configuration $(1\sigma)^2$ $(2\sigma)^2$ $(3\sigma)^2$ $(4\sigma)^2$ $(1\pi)^4$ $(5\sigma)^2$ $(2\pi)^1$ [6, 7] and rotational symmetry around the molecular axis. According to molecular orbital (MO) theory, NO has five fully bonding orbitals with unpaired electron resulting in the fourth antibonding $\pi^*\pi^*$-orbitals (Figure 1.1).

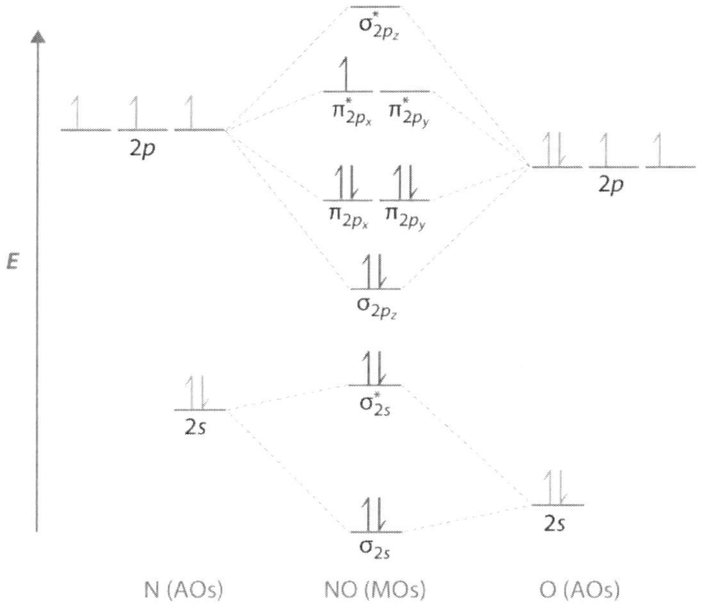

Figure 1.1. Molecular orbitals diagram of nitric oxide

Nitric oxide MOs are formed from both the nitrogen and oxygen p atomic orbitals, but it is closer in energy to the nitrogen. Therefore, it is expected that this electron has more nitrogen and less oxygen character. The nitrogen and oxygen are held together by 2.5 bonds. The NO molecular orbital constants are shown in Table 1.1.

The unpaired electron in the antibonding orbitals weakens over bonding of the nitrogen to oxygen by one-half. The transfer of additional electron density into the antibonding additionally weakens over the bond. Conversely, the transfer of the electron away from the antibonding orbital significantly stiffens the NO bond. These features of the NO molecule substantially determine its physicochemical properties and redox potential, in particular. The resulting magnetic moment is determined by the electronic spin only.

Table 1.1. Nitric Oxide Molecular Orbital Constants: National Institute of Standards and Technology (NIST) Chemistry Webbook

Mol. Orbital	B (eV)	U (eV)	N	Q
3σ	43.70	76.55	2	1
4σ	25.32	77.04	2	1
1π	18.49	55.37	4	1
5σ	15.87	62.25	2	1
2π	9.26	65.27	1	1

B: Binding Energy
U: Average Kinetic Energy
N: Electron Occupation Number
Q: Dipole Constant)

1.3. Investigations by Physical Methods

The nonparamagnetic ground state of an isolated NO in the gas phase cannot be detected by electron spin resonance (ESR) spectroscopy. The detection of NO radicals with electron paramagnetic resonance (EPR) becomes possible in thermally excited state at room temperature (kT ~ 25 meV) [8-9]. The origin of the nine-line EPR pattern of NO dilute gas at (T = 295 K) (Figure 1.2) can be explained by the following scheme: the larger splitting is caused by the zero-field splitting (ZFS) interaction and the smaller splitting by the hyperfine coupling to the magnetic moment of the ^{14}N nucleus (I = 1).

Typical NO absorption spectra were obtained using for a gas mixture of 0.5% of NO in N_2 at atmospheric pressure gas sensor based on quartz-enhanced photoacoustic spectroscopy and external cavity quantum cascade laser show wavenumbers (cm^{-1}) in areas 1900.0–1900.2 and 1900.4–1900.6 [10]. In work [11], a quantum cascade laser source for access to the strong fundamental absorption band in the mid-infrared spectrum of nitric oxide was utilized. Room temperature measurements at pressures ranging from 2–20 atm were conducted to measure the NO spectrum across a range from 1920 to 1937 cm^{-1} Absorption spectra of the NO and SO_2 molecules was used in stationary gas analyzer designed in [12]. A configuration coordinate model for the single selected mode Q was suggested in [14]. In the frame of this model, the vibrational surfaces of the single selected mode correspond to the N = O vibrational level of the ground state (g) and to special excited vibrational levels of the valence state (v). The Rydberg states were also calculated. Nitric oxide infrared spectra bands were absorbed at 1890, 1605, 1575, and 1510 cm^{-1} [13].

Potential energy, electric field gradient (EFG) at both nuclei, and electric dipole moment functions for the electronic ground states of NO^+, NO, and NO^- were calculated at the internally contracted multireference configuration interaction (icMRCI) level using augmented correlation-consistent basis sets [7]. On the base of the calculation, the ^{14}N nuclear quadrupole coupling constants were determined, the vibrational dependences of the electric properties were estimated, and the character of bonding in the triad of NO species was established. Some applied aspects of the infrared spectroscopy related to NO analysis were discussed in [15, 16].

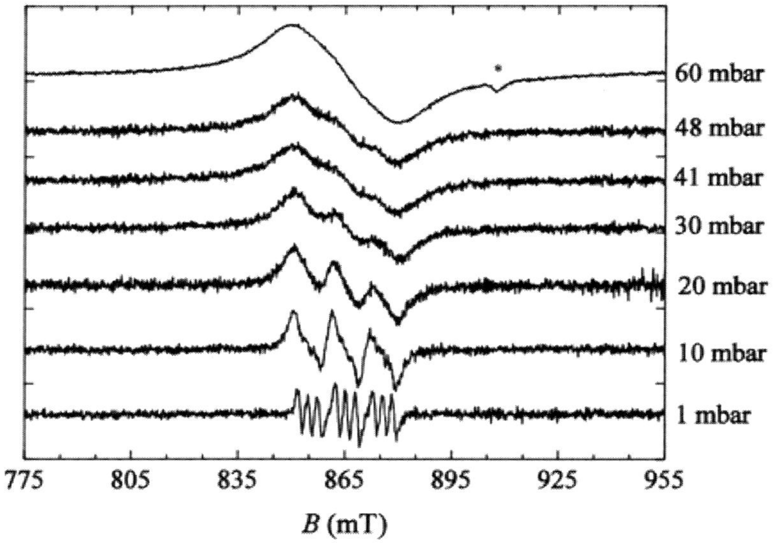

Figure 1.2. X-band EPR spectra of samples NO_p ($1 \leq p \leq 60$) with different room temperature pressures (denotation at the *right*). The spectra were measured at room temperature using a modulation amplitude of 0.5 mT and *a* mw power of 10 mW. The *asterisk* marks an artifact [9]. With permission from Springer Nature.

1.4. Chemical Reactions

1.4.1. Nitric Oxide Basic Chemical Reactions

1.4.1.1. Gas Reactions

Though NO formally is a radical, its reactivity is partially damped due to involving an unpaired electron in formation of the trielectronic bond. Therefore, NO readily reacts with free reactive radicals but slowly with saturated molecules; that is of great importance for the molecular activity in the biological condition.

For example, the rate expression for reactions

$NO + H^{\cdot} \rightarrow NOH$,
$NH_2 + NO \rightarrow NONH_2$

and

NO + Cyclohexadienyl, 6-hydroxy- → Products

were found to be k = $1.34x10^{-31}$ [cm^6/$molecule^2$ s] $(T/298\ K)^{-1.32}$ $e^{-3076\ [J/mole]/RT}$ [17], $96x10^{-11}$ [cm^3/molecule s] $(T/298\ K)^{-2.37}$ $e^{-3625\ [J/mole]/RT}$ [18], and $1.0x10^{-12}$ [±$4.98x10^{-13}$ cm^3/molecule s [19], respectively.

Presence of atmospheric NO resulting from industrial activity may be one of the causes of the gradual depletion of the ozone layer in the upper atmosphere. Under sunlight radiation, NO reacts with ozone (O_3), converting the O_3 to molecular oxygen. Ozone depletion leads to a steady decline of about 4% in the total amount of O_3 in the earth's stratosphere. Gas phase studies showed that the reaction occurs by the following the mechanism:

$NO+O_3 = NO^*_2(^2B_1) + O_2(1a)$,
$NO+O_3 = NO_2(^2A_1) + O_2(1b)$, and $NO^*_2 = NO_2+hv(2)$ $M+NO_2 =$
$NO_2 + M$ (3)

with corresponding rate constants of the primary steps are $k_{1a}=(7.6 \pm 1.5)\times 10^{11} \exp(-4180 \pm 300/RT)$ and $k_{1b}=(4.3 \pm 1.0)\times 10^{11} \exp(-2330 \pm 150/RT)$ $cm^3\ mole^{-1}\ sec^{-1}$ [20].

The special role of the reactions of NO with oxygen has caused a significant number of experimental and theoretical studies [20-29]. On the basis of experimental measurements, several kinetic models have been suggested to describe the reaction mechanism and to explain the negative temperature dependence of the rate constant. Investigation of the oxidation of NO by oxygen in the gas phase at temperatures between 226 and 758 K and at pressures ranging from about 0.2 to 30 torr has shown that the reaction is first order against oxygen and second order against nitrogen monoxide [21]. The reactions run with the rate constants:

$$\log_{10} k(\mathrm{l^2 mol^{-2} s^{-1}}) = -(5.18 \pm 1.00) + (2.70 \pm 0.25)\log_{10} T + (700 \pm 50)/T$$

or by

$$k(\mathrm{l^2 mol^{-2} s^{-1}}) = (350 \pm 100)10^{(390 \pm 50)/T} + (8000 \pm 2000)10^{-(440 \pm 50)/T}$$

In work [22] the process $2NO + O_2 = 2NO_2$ has been also established as a third-order homogeneous reaction with rate constant k (L2 x mol(-2) x s(-1)) = 1.2 x 10(3) e(530/T) (=1.2 x 10(3) x 10(230/T)), measured in a temperature range of 273 to 600 K. The two-step mechanism including NO_3 formation

$NO + O_2 \rightarrow NO_3$

$NO + NO_3 \rightarrow 2NO_2$

was suggested in [23]. The synchronous trimolecular reaction via classical transition state

$2NO + O_2 \rightarrow [ONOONO]^{\#} \rightarrow 2NO_2$.

was proposed by Gershinowitz, and Eyring, [24]. An alternative mechanism was developed by McKee [25], who considered the peroxide ONOONO ($N_2\,O_4$) as a key intermediate (not the transition state) in the reaction $2NO + O_2 \rightarrow 2NO$. In this work density functional theory with nonlocal corrections and Hartree-Fock theory with perturbative electron correlation were used to study the process. The large reorganization energy of NO_2.causes negative bond dissociation energy of -17.5 kcal/mol. Assuming O-O cleavage as the rate-determining step, the calculation gives the correct rate law ($[O_2][NO_2]$) and an overall activation barrier ($E_{overall}$), which is consistent with the scheme:

$O_2, + NO \leftrightarrow NO_3$ (1)

$NO_3 + NO \rightarrow ONOONO$ (2)

$ONOONO \rightarrow 2NO_2$

with the rate = Kk $[NO_2][O_2]$.

Quantum chemical calculation indicated that weak intermolecular interaction between two NO molecules, a *cis* conformation [26]. The natural bond orbital (NBO) analysis gives an illustration of the formation of the dimer bonding and antibonding orbitals concomitant with the breaking of the π bonds with bond order 0.5 of the monomers. The dimer binding is weak because it is counteracted by partially filling the antibonding dimer orbital and the repulsion between those fully or nearly fully occupied nonbonding dimer orbitals. The calculated overall activation energy is Ea=-0.1 kcal/mol, which is in agreement with the experimental determinations. The binding energy obtained by molecular mechanics method for the two NO molecules of a transconformation (Figure 1.3) was reported as 3.11 kJ/mol. This value is close to the thermal energy at ambient condition (~2.5 kJ/mol) [26].

Singlet and triplet potential energy surfaces of the reaction between molecular oxygen and two nitric oxide (II) molecules were calculated by quantum chemical methods (coupled cluster, CASSCF, and density functional theory: B3LYP, TPSS, VSXC, BP86, PBE, B2-PLYP, B2K-PLYP) [27]. It was taken in consideration that ONOONO may exist in different conformers [28]. Elementary steps involving various N_2O_4 isomers (*cis-cis-*, *cis-trans-*, *trans-trans-*ONOONO, *cis-* and *trans-*$ONONO_2$, O_2NNO_2) as well as Coupe-type quasi-aromatic hexagonal ring intermediate CCINT and van-der-Waals molecular clusters were also studied. A valley ridge inflection point is located on the minimum energy path connecting $NO_2{\cdot}O_2N$ and *cis-*$ONONO_2$. Figure 1.3 demonstrates results of detail calculation of the profile of the singlet potential energy surface (PES) for the reaction $2NO + O_2 \rightarrow 2NO_2$.

On the base of corresponding calculations, the main channel of reaction $2NO + O_2 \rightarrow 2NO_2$ through the following reactions was suggested (ΔrH is reaction energy *and* TSi is a corresponding transition state):

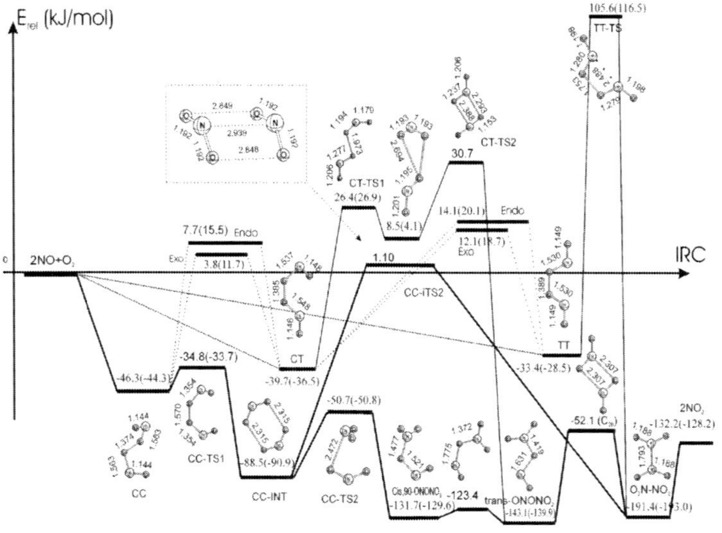

Figure 1.3. Profile of the singlet PES for the thermal reaction $2NO + O_2 \rightarrow 2NO_2$, obtained at the level UB3LYP/6-311+G(2d). The dotted and solid lines indicate the minimum energy pathways. Bond lengths are in Å; relative enthalpies are in kJ/mol. Values in parentheses taken from ref 24 are at the B3LYP/6-311G* level [27]. Copyrights 2009 American Chemical Society.

Table 1.2. Enthalpy of Reactions 1 and 2 (kJ/mol) Evaluated at Different DFT Levels and Compared to the Experimental Data from Reference [27]. Copyrights 2009 American Chemical Society

reaction	B3LYP	B3LYP	TPSS	VSXC	PBE	BP86	experiment
	aug-pc3			6-311+G(2d)			
$2NO + O_2 \rightarrow 2NO_2$	−123.82	−119.70	−167.12	−146.46	−209.28	−197.18	−105.55
$2NO_2 \rightarrow N_2O_4$	−42.81	−44.09	−62.96	−67.7	−82.76	−76.71	−53.60

$2NO + O_2 \rightarrow$ *cis-cis*-ONOONO
(ΔrH) -10.1 kJ/mol, $Ea = 0$ kJ/mol)

cis-cis-ONOONO $\rightarrow$TS1 $\rightarrow NO_2 \cdot O_2N$
(ΔrH) -42.5 kJ/mol, $Ea = 10.4$ kJ/mol)

NO_2 O_2N $\rightarrow$TS2 $\rightarrow$*cis*-$ONONO_2$
(ΔrH) -41.3 kJ/mol, $Ea = 37.8$ kJ/mol)

cis-$ONONO_2$ $\rightarrow$TS3 $\rightarrow$ *trans*-$ONONO_2$
(ΔrH) -10.4 kJ/mol, $Ea = 8.6$ kJ/mol)

trans-$ONONO_2$ $\rightarrow$ O2NNO_2
(ΔrH) -44.4 kJ/mol, Ea) 90.0 kJ/mol

Calculated values of enthalpy of reactions $2NO + O_2 \rightarrow 2NO_2$ (1) and $2NO_2 \rightarrow N_2O_4$ (2) were compared with experimental one (Table 1.2).

Overall, the activation enthalpy of the reaction $2NO + O_2 \rightarrow 2NO_2$ at absolute zero ΔHq(0 K) was evaluated as -4.5 - -6.5 kJ/mol. This agrees with the experimental value Ea of -4.5 kJ/mol at $T = 298$-600 K and $E \approx 0$ kJ/mol at $T \approx 650$ K.

The barrierless formation of cis-cis peroxide in the oxidation reaction of nitrogen oxide 2NO ($^2\Pi$) + O_2 ($^3\Sigma_g$) $\rightarrow$ O=NO-ON=O (^{1}A) (I) was theoretically investigated in work [29]. It was suggested that the reaction proceeds in the two-triplet state with total zero spin by the spin catalysis mechanism with the negative activation energy. According to the density functional theory (DFT) calculation of the process, the possible intermediate, the *cis–trans* isomer, has energy that is higher by 6.3 kJ/mol; the least stable is the *trans–trans* isomer. The following B3LYP/6-311++G(3df) calculated geometric parameters (bond lengths in A, angles in deg) of different isomers of the N2O4 molecule were reported: the structure of the ground state of D2h symmetry (a); *cis–cis* isomer O = NO– ON = O of C2 symmetry, calculated torsion angle τN1O2O3N2 = 88.9° (b); cis–trans isomer O = NO–ON = O of C1symmetry, calculated torsion angle τN1O2O3N2 = 88.2° (c); *trans–trans* isomer O = NO–ON = O of C2 symmetry [29]. The O–O bond in transition state of *cis–cis* O = NO–

ON = O was found to extend by 14% compared to its initial length in initial state with a simultaneous contraction of O–N bonds.

The interaction of two NO radicals in mutually perpendicular planes with respect to the O–O molecular axis of oxygen results in the formation of a structure of low-energy (ΔH = –30 kJ/mol). The activation energy (Ea) was obtained to be 7.4 kJ/mol. Fast dissociation of the intermediate is provided at a high rate of the whole process $2NO + O_2 \rightarrow 2NO_2$. Enthalpy of this reaction was estimated as –122 kJ/mol, in agreement with the experiment yields. Analysis of the molecular orbital interactions of two NO ($^2\Pi$) radicals with a biradical O_2 ($^3\Sigma g$) molecule showed that trimolecular collision $2NO + O_2$ in mutually perpendicular planes relative to the O–O molecular axis of oxygen results in the formation of the most stable metastable O = NO–ON = O peroxide in *cis– cis* configuration [29]. The authors concluded that in the reaction mechanism of the NO oxidation in the key stage is the activationless formation of metastable *cis–cis* peroxide: 2NO ($^2\Pi$) + O_2 ($^3\Sigma g$) $\rightarrow$ *cis–cis*-O = NO–ON = O (1A), and the driving force of the trimolecular $2NO + O_2$ interaction is singlet spin coupling of π^* electrons of NO molecules with *g π electrons of triplet oxygen.

1.4.1.2. Reactions in Solution

The direct reaction NO with superoxide with formation of peroxinitrite

$$NO + O_2^{\cdot -} \rightarrow ONOO^-$$

is one of the most important processes in the area, which causes the whole cascade of chemical elementary processes of great biological importance [30].

Kinetics of reactions relevant to superoxide with NO was studied in a number of works and the most thoroughly in [31] and references therein. Rate constants relevant to the reaction of superoxide with Nitrogen Monoxide were listed. Three independent methods were employed: (1) photolysis of alkaline peroxynitrite solutions resulting in the formation of superoxide and nitrogen monoxide, (2) photolysis of hydrogen peroxide in the presence of NO with formation of hydroxyl radicals, which in turn react with hydrogen peroxide to form superoxide, (3) and photolysis of nitrite giving rise to NO and hydroxyl radicals and after that superoxide. For the reaction of superoxide with nitrogen monoxide, a value of rate constant of (1.6) $\times$ 10^{10} M^{-1} s^{-1}, the weighted average of these three photolysis experiments was reported. The reactions of several organic peroxyl radicals with NO in aqueous solution, measured in works [32, 33], where found to be very fast, with k = 1-3 x 10^9 L $mol^{-1}s^{-1}$. These reactions lead to the formation of organic peroxynitrite, which in turn, decays with a rate constant of k = 0.1-0.3 s^{-1}. Reactive alkoxyl, nitrogen dioxide free radicals, and organic nitrate are products of the peroxynitrite decomposition.

The basic possibility of involving NO and NO reactive species in chemical reactions is defined by their thermodynamic potential. Standard Reduction Potentials of Nitrogen and Oxygen Containing Species (in V) [33] were found as 1.21 (NO^+/NO•), 0.39 (NO•/NO, triplet) and 0.39 (NO•/NO^-singlet). High ionization energy of NO is not compensated with solvation in water, and the molecule possesses high negative reduction potential (1.21 V). In ambient conditions in a biological system, oxidation of NO is thermodynamically forbidden.

Damping of radical character of NO limits its chemical reactivity in reactions with nonradical species. In biology, these reactions are carried out by products of NO oxidation by oxygen, such as $ONOO^-$, N_2O_3 and NO_2 [33-41]. Typical example of such reactions is nitrosation of thiols [34] and references therein. On the basis of kinetic observations, the following mechanism of nitrosation of glutathione, cysteine, and N-acetylcysteine was suggested [35]

$$2NO + O_2 \rightarrow 2NO_2 \; k^1 = 6.3 \times 10^6 M^{-2} s^{-1}$$
$$NO_2 + NO \rightarrow N_2O_3 \; k^2 = 1.1 \times 10^9 M^{-1} s^{-1},$$
$$N_2O_3 + H_2O \rightarrow 2H^+ + 2NO_2 \; k^3[H_2O] = 1.6 \times 10^3 s^{-1}$$
$$N_2O_3 + RSH \rightarrow H^+ + NO_2^- + RSNO \; k^4 = A \times 10^5 M^{-1} s^{-1}$$

where A = 1.5 -3 for low molecular weight thiols.

The S-nitroso species formed from a thiol or a disulfide group, and NO can react back into a thiol by reduction and consequently release NO [36]. It was suggested that this property of the S-nitroso species allows them to serve as the depot of NO in biological systems.

Nitrogen dioxide ($NO_2^•$) is an important chemically active biological oxidant that can be derived not only from peroxynitrite via the interaction of NO with superoxide, but from nitrite with peroxidases, or from autoxidation of NO as well. NO_2·can react with various compounds via addition to double bonds [38], hydrogen abstraction [37], or electron transfer mechanisms [37, 38]. Rate constants for the reaction of $NO_2^•$ with glutathione, cysteine, and urate at pH 7.4 were estimated as ~ 2×10^7, 5×10^7, and 2×10^7 $M^{-1} s^{-1}$, respectively [39]. Thiolate reacted much faster than undissociated thiol.

Peroxynitrite can also react directly with cysteine, methionine of aromatic amino acids tyrosine, tryptophan, phenylalanine, and histidine with formation of nitration species [40]. Protonated peroxynitrite (peroxynitrous acid, ONOOH; pKa = 6.8) can produce $^•NO_2$ and a hydroxyl radical [41]. One-electron oxidation by $^•NO_2$ generates the thiyl and tyrosyl radicals. Fast reactions of these radicals with NO yield nitrosothiol and nitrotyrosine, respectively. Another possible mechanism for RSNO formation involves the reaction of •NO_2 with •NO to yield N_2O_3, which reacts with thiol by transfer of the nitroso group (transnitrosation). ONOO− also has the ability to modify a variety of amino acids in proteins, including oxidation of sulfur-containing amino acids (cysteine and methionine) and nitration of aromatic amino acids (tyrosine, tryptophan, phenylalanine, and histidine).

A diagram of the important biological reactions of •NO, NO_2, and $ONOO^-$ with nonmetals and its subsequent reactive nitrogen species was suggested in [41].

The nitrosation of various thiols and morpholine by oxygenated ˙NO solutions at physiological pH was investigated [34]. The formation rates and the yields of the nitroso compounds were determined using the stopped-flow technique. The reaction run with the stoichiometry by $4{}^{\bullet}NO + O_2 + 2RSH/2RR'NH \rightarrow 2RSNO/2RR'NNO + 2NO_2^- + 2H^+$, the rate law is $-d[O_2]/dt = k_1[{}^{\bullet}NO]^2[O_2]$ with $k_1 = (2.54 \pm 0.26) \times 10^6\ M^{-2}\ s^{-1}$ and $-d[{}^{\bullet}NO]/dt = 4k_1[{}^{\bullet}NO]^2[O_2]$ with $4k_1 = (1.17 \pm 0.12) \times 10^7\ M^{-2}\ s^{-1}$, independent of the kind of substrate present. The nitrosation of thiols and amines by NO_2 and N_2O_3 is limited by the formation of ONOONO (or $ONONO_2$ or O_2NNO_2). N_2O_3 is capable of directly nitrosating thiols and amines with rate constants exceeding $6 \times 10^7\ M^{-1}\ s^{-1}$. Under physiological conditions, where $[{}^{\bullet}NO] < 1\ \mu M$ and $[O_2] < 200\ \mu M$, the rate formation of *S*-nitrosothiols was found to be less than about $10^{-1}\ min^{-1}$. Therefore, *S*-nitrosothiols can unlikely serve as carrier molecules of ˙NO *in vivo*.

High ionization energy of NO is not compensated with solvation in water, and the molecule possesses high negative reduction potential (1.21 V). Therefore, in ambient conditions in a biological system, oxidation of NO is thermodynamically forbidden. Nitric oxide can react with aromatic amines only after activation with oxygen in organic solvent [43]. In every case, *N*-nitrosamines were obtained as the main products.

Presence of a phenyl fragment in tyrosine, important antioxidant, has caused special interest to reaction of NO with phenols [44–46]. For example, in work [45] reaction of three phenols with NO, that is, a sterically hindered phenol, which produces persistent phenoxyl-free radicals upon hydrogen atom abstraction, 2,4,6-tri-tert-butylphenol(1H), 2,6-di-tertbutyl- 4-methylphenol, and a-tocopherol, was examined in sodium dodecyl sulfate micelles. Experiments have shown that reactions with aqueous solutions of NO led to mixtures of nitro and nitroso derivatives, depending upon the phenol, while nitration is the major reaction with peroxynitrite. The reaction produces phenoxyl-free radicals upon hydrogen atom abstraction. Nitrosation occurs on phenol substrates bearing a free *para*-position with respect to the OH group with the exception of 1-naphthol. The slowness of these reactions allows us to conclude that they are not realized in cells. Nevertheless, the phenolic compounds in cells can be attacked by free radicals formed from the NO species activated by oxygen.

Antioxidant Etoposide (VP-16) [44] containing a phenolic OH group in the 4′-position, is one of the most active anticancer drugs. Reactions of VP-16 with an NO donor DEANO and with •NO/•NO_2 in $CHCl_3$. were investigated by ESR (Figure 1.4) [44]. Forming a radical VP-16• and subsequently other products from the phenolic OH of VP-16 reaction with •NO-derived species was evaluated.

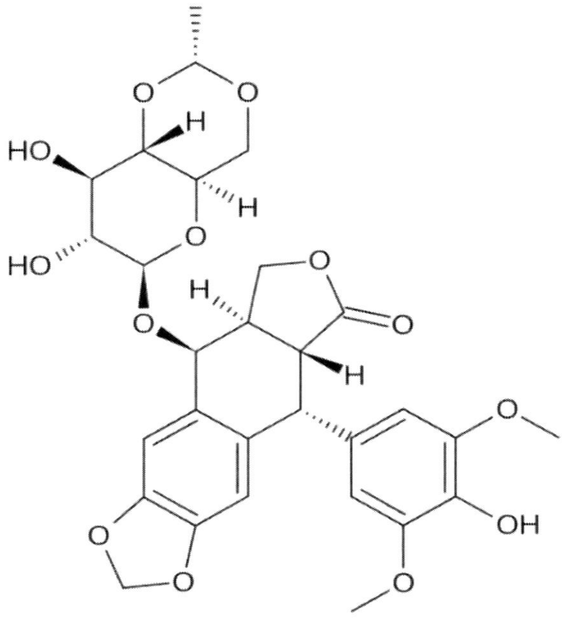

Nitrosation of tryptophane by $NO_2^{\cdot}$ was subject of several investigations. Tryptophan moiety in proteins and peptides is sensitive to nitrosation by reactive NO species. The experimental rate constant of reaction of tryptophan with NO_2 was determined as k = 2,1 10^6 M^{-1} s^{-1} [47].

Using photolysis-chemiluminescence analysis, modified Saville assay, differential UV-is spectroscopy, and bioassays, nitrosation indole nitrogen of dipeptide L-glycyl-L-tryptophan (Gly-Trp) was established [48]. N-nitrosation of N-acetyltryptophan and lysine-tryptophan-lysine by N_2O_3-generating systems in the presence of oxygen at pH 7.4 UV-visible spectroscopy, ^{15}N NMR spectrometry, and polarographically was examined [49]. The experiments indicated that the indole ring of the tryptophan derivatives was nitrosated by N_2O_3 with rate constant (4.4 – 6.4 $10^7 M^{-1} s^{-1}$) and melatonin with (9.2 $10^7 M^{-1} s^{-1}$). Ascorbate induced the release of NO from N-acetyl-N-nitrosotryptophan.

The reactions of $NO^{\cdot}$ and nitrogen dioxide ($NO_2^{\cdot}$) radicals with free zwitterionic tryptophan and N-formyl-tryptophanamide (a model for tryptophan as a protein residue) in aqueous have been studied using density functional theory and transition state theory [50]. Three possible reaction channels have been considered (Figure 1.5). The calculation showed that reaction of free zwitterionic tryptophan with NO_2 can occur in aqueous solution, at pH = 7.4 and 298 K only as single electron transfer (SET process with the Gibbs free energies $\Delta G°$, = 2.39 kcal/mol)·and Gibbs free energies of activation $\Delta G^{\neq}$ = 8.5 kcal/mol) and rate coefficients *k*app (NO_2) k = 3.4 10^6 M^{-1} s^{-1}. For direct reaction with NO, the estimated value of $\Delta G°$, = 79.64 and the reaction is thermodynamically forbidden.

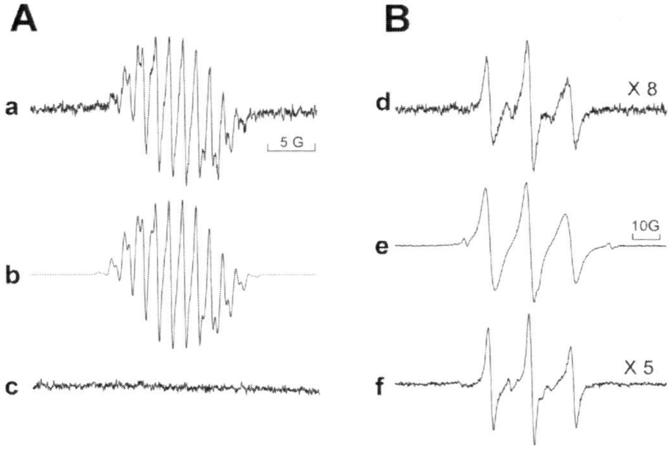

Figure 1.4. Panel A: ESR spectra of the VP-16 radical formed from (a) reaction of VP-16 (1 mM) with an NO donor DEANO (1 mM) or •NO gas (2 mM) in PBS at pH 7.4; (b) computer Figure simulation; and (c) spectrum recorded after 20 min. Panel B:(d) ESR spectra of radicals formed from the reaction of VP-16 (1 mM) with •NO/•NO_2 gas in $CHCl_3$ at a low •NO:VP-16 ratio of 2:1 recorded after 2-3 min; (e) at a high •NO:VP-16 ratio (5:1); and (f) from VP-16 quinone (1 mM) and •NO/•NO2(at a ratio of NO:VP-16-quinone 5:1) in $CHCl_3$. ESR spectra were recorded with a Bruker EMX operating in X-band. ESR settings were center field, 3486 G; scan range, 30 G; modulation amplitude, 1.0 G; modulation frequency 100 KHz; nominal microwave power, 21.49 mW; microwave frequency, 9.80 GHz and receiver gain, 5×10^4 [44]. Copyrights 2003 American Chemical Society.

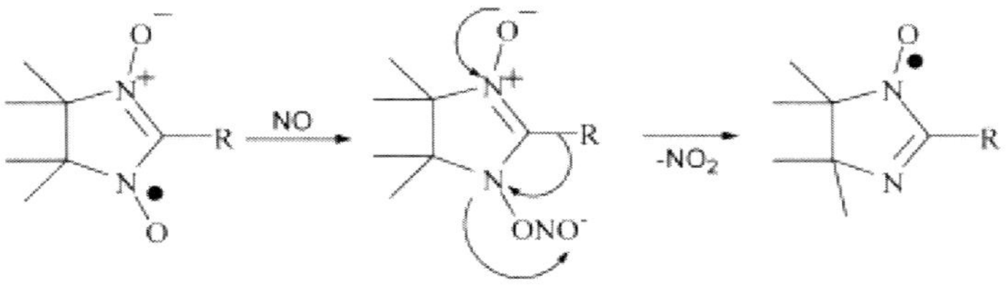

Figure 1.5. Reaction of nitronyl nitroxide radical with the NO molecule [52]. With permission from Elsevier.

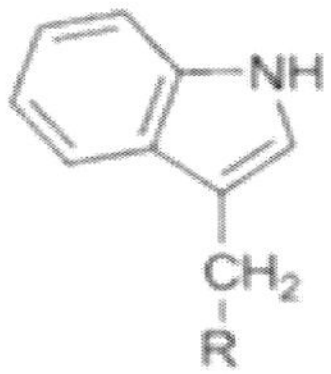

Nitronyl nitroxides radicals react with the NO molecule to yield labile intermediates, which eliminate NO_2 and form imino nitroxide radicals (Figure 1.5) [52] and references therein.

Reaction of NO generated from a NO donor SNAP with pyrene-nitronyl super molecule was investigated in [51,52].

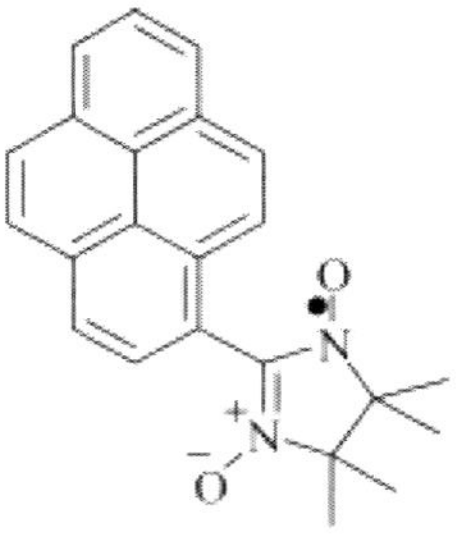

PN

An evolution of ESR spectra of the pyrene-nitronyl (PN) in the presence of (SNAP) is shown in Figure 1.6.

The reaction was monitored also by the measurement of the change in fluorescence from the pyren fragment. In the initial state, the strong intramolecular fluorescence quenching of the fluorophore fragment by the nitroxide fragment occurs. The transformation of the EPR signal is accompanied by a 32-fold increase in the fluorescence intensity since the imino nitroxide radical is a weaker quencher than the nitronyl one. Since fluorescence technique is three orders of magnitude more sensitive than ESR, this phenomenon makes it possible to detect nanomolar concentrations of NO compared to a sensitivity threshold of only several micromolar for the EPR and optical absorption techniques.

Interrelation between NO and ascorbic acid was in the focus of attention of serious of works [53–55] and references therein. Nitric oxide can react with aspartic acid only after activating in the presence of oxygen. Remarkably, *in vivo* ascorbate is effective in counteracting peroxynitrite, nitrogen dioxide, and N_2O_3 [53]. The role of ascorbate, as a nitrosation modulating agent for in a system of concomitant generation of NO and superoxide, was characterized experimentally monitoring by ESR and ^{15}N NMR spectrometry and by CBS-QB3 calculations. SIN-1 (3-morpholinosydnonimine) was used as a donor of nitric oxide/superoxide/ peroxynitrite, and the effect of ascorbate on SIN-1 mediated nitrosation kinetics was determined [54]. The rate constants of reaction of peroxynitrite with ascorbate was found as 2.35x102 M^{-1} S^{-1}. It was found that that peroxynitrite is the most predominant species in the SIN-1 system. The primary oxidized product of dehydroascorbic acid (DHA) efficiently consumes NO, and O-nitrosoascorbate is a key intermediate. Computational modeling also predicted that ascorbate may increase the availability of NO by removing superoxide. Taken all together, these results suggested that ascorbate may inhibit nitrosation by acting as a scavenger for superoxide, peroxynitrite, $\bullet NO_2$, and/or other radical species.

The NO_2 formed by autoxidation of NO reacts with ascorbate

$$ASC^- + {}^\bullet NO_2 \rightarrow ASC^{\bullet -} + H^+ + NO_2^-$$

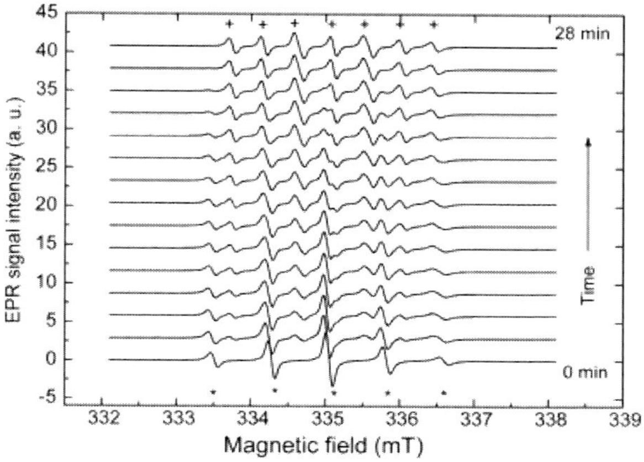

Figure 1.6. Evolution of the EPR spectrum of the pyrene-imino-nitroxide (0.1 mM) after the addition of 0.1 mM SNAP a solution of 0.1 mM PN in Ringer solution (pH 7.2); T = 25°C. Crosses indicate the EPR signals of PI after the reaction is over. Stars indicate the EPR signals of PN before addition of SNAP [52]. With permission from Elsevier.

with the rate constant $k = 3.6x10^7 M^{-1}s^{-1}$ followed by a cascade of reactions [55, 56]:

$$^{\bullet}NO_2 + {}^{\bullet}NO \rightleftharpoons N_2O_3$$

$k = 1.1x10^9 M^{-1}s^{-1}$

$$N_2O_3(+H_2O) \rightleftharpoons 2\,NO_2^- + 2\,H^+$$

$k = 2x10^3 M^{-1}s^{-1}$

$OH + NO_2^- = HONO_2^-$

$k = 1.0 \times 10^{10} M^{-1} sec^{-1}$

REFERENCES

[1] Pimentel GC, RD Spratley RD. *Chemical Bonding Clarified Through Quantum Mechanics*. Holden-Day, Inc., San Francisco, CA, 1969.

[2] National Institute of Standards and Technology. Material Measurement Laboratory.

[3] Wilhelm E., R. Battino R, and R. J. Wilcock RJ. 1977. "Low-pressure solubility of gases in liquid water." *Chem. Rev.* 77:219–262.

[4] Zacharia IG, Deen WM. Diffusivity and solubility of nitric oxide in water and saline. *Ann Biomed Eng.* (2005) 33:214-22.

[5] Pokharel S, Pantha N, Adhikari NP. Diffusion coefficients of nitric oxide in water: A molecular dynamics study. *International Journal of Modern Physics B:*

Condensed Matter Physics; Statistical Physics; Atomic, Molecular and Optical Physics (2016), 30, 1650205.

[6] Pfennig B. 2015. *Principles of Inorganic Chemistry*. Hoboken, New Jersey: John Wiley & Sons, Inc.

[7] Polak R, Fiser J. A comparative icMRCI study of some NOþ, NO and NO electronic ground state properties. *Chemical Physics* (2004) 303 73-83.

[8] Whittaker JW. Molecular paramagnetic resonance of gas-phase nitric oxide. *J. Chem. Educ*. (1991) *68,* 421.

[9] Mendt M, Pöppl A. The Line Width of the EPR Signal of Gaseous Nitric Oxide as Determined by Pressure and Temperature-Dependent X-band Continuous Wave Measurements. *Appl. Magn. Reson.* (2015) 46, 1249-1263.

[10] Spagnolo V, Kosterev AA, L. Dong L, Lewicki R, Tittel FK. NO trace gas sensor based on quartz-enhanced photoacoustic spectroscopy and external cavity quantum cascade laser. *Applied Physics B.* (2010) 100: 125-130. .

[11] Spearrin RM, Schultz IA, Jeffries JB, Hanson RK. Laser absorption of nitric oxide for thermometry in high-enthalpy air. *Meas. Sci. Technol*. (2014) 25, 125103 (7pp).

[12] Azbukin AA, Buldakov MA, Korolev BV, Korol'kov VA, Matrosov II, Tikhomirov AA. A Stationary Gas Analyzer of Nitric and Sulfur Oxides. *Instruments and Experimental Techniques* (2006), 49, 839–843. © Pleiades Publishing, Inc., 2006. Original Russian Text 2006, published in *Pribory i Tekhnika Eksperimenta*, 2006, No. 6, pp. 105-109.

[13] London JW, Bell AT. Infrared spectra of carbon monoxide, carbon dioxide, nitric oxide, nitrogen dioxide, nitrous oxide, and nitrogen adsorbed on copper oxide. *Journal of Catalysis* (1973) 31, 32-40.

[14] Suisky D, Chergui M, Schwentner N. Theoretical description of interference effects in the absorption spectra of NO in rare-gas matrices. *Chem. Phys. Lett*. (1992) 200, 325-332.

[15] Thompson BT, Mizaikoff B. Real-time Fourier transform-infrared analysis of carbon monoxide and nitric oxide in side stream cigarette smoke. *Appl Spectrosc.* (2006) 60, 272-8.

[17] Tsang W, Herron JT. Chemical kinetic data base for propellant combustion. I. Reactions involving NO, NO_2, HNO, HNO_2, HCN and N_2O. *J. Phys. Chem.* (1991)20, 609.

[18] Song S, Hanson RK, Bowman CT, Golden DM. Shock Tube Determination of the Overall Rate of NH_2 + NO -> Products in the Thermal De-NOx Temperature Window. *Int J. Chem. Kinet.* (2001). 33, 715- 721.

[19] Zellner R, Fritz B, Preidel MA. A cw UV laser absorption study of the reactions of the hydroxy-cyclohexadienyl radical with NO_2 and NO. *Chem. Phys. Lett.* (1985) 1211 985.

[20] Clough PN, BA. Mechanism of chemiluminescent reaction between nitric oxide and ozone. *Trans. Faraday Soc.* (1967) 63, 915-925.

[21] Olbregts J. Termolecular reaction of nitrogen monoxide and oxygen: A still unsolved problem. *Int. J. Chem. Kinet.* (1985) 17, 835.

[22] Tsukahara H, Ishida T, Mayumi M. Gas-phase oxidation of nitric oxide: Chemical kinetics and rate constant *Nitric Oxide: Biol. Chem.* (1999), 3, 191.

[23] Treacy JC, Daniels F. Kinetic Study of the Oxidation of Nitric Oxide with Oxygen in the Pressure Range 1 to 20 Mm. 1 *J. Am. Chem. Soc.* (1955) 77, 2033–2036 [Morris, V. R., Bhatia, S. C., Hall, J. J. H. *J. Phys. Chem.* (1990) 94, 7418.]

[24] Gershinowitz, H, Eyring H. The theory of trimolecular reactions *J. Am. Chem. Soc.* (1935), 57, 985-991.

[25] McKee ML. Ab Initio Study of the N_2O_4 Potential Energy Surface. Computational Evidence for a New N_2O_4 Isomer. *J. Am. Chem. Soc.* (1995) 177, 1629-1637.

[26] Zhang Hui, Zheng Gui-Li, Lv Gang Geng, Yi-Zhao Ji Qing. Covalent intermolecular interaction of the nitric oxide dimer $(NO)_2$. *Chinese Physics B* (2015) 24, Number 9.

[27] Gadzhiev OB, Ignatov SK, Razuvaev AG, Masunov AE. Quantum Chemical Study of Trimolecular Reaction Mechanism between Nitric Oxide and Oxygen in the Gas Phase. *J. Phys. Chem. A* (2009) 113, 9092–9101.

[28] Wang X, Qin Q-Z. Peroxide linkage N_2O_4 molecule: Prediction of new ONOONO isomers *Int. J. Quantum Chem.* 2000, 76, 77.

[29] Zakharov II, Minaev BF. A quantum chemical study of the structure of O = NO-ON = O peroxide and the reaction mechanism of no oxidation in the gas phase. *Journal of Structural Chemistry* 2012, 53, 1-11.

[30] Ohara A, Bonini MG, Amanso AM, Linares E, Santos CCX, De Menezes SL. Nitrogen dioxide and carbonate radical anion: two emerging radicals in biology. *Free Radic Biol Med.* 2002 May 1;32(9):841-59.

[31] Nauser T, Koppenol WH.
Rate Constants Relevant to the Reaction of Superoxide with Nitrogen Monoxide. The Rate Constant of the Reaction of Superoxide with Nitrogen Monoxide: Approaching the Diffusion Limit. *J. Phys. Chem. A* (2002) 106, 4084-4086.

[32] Padmaja S, Huie RE. The reaction of nitric oxide with organic peroxyl radicals. *Biochem. Biophys. Res. Commun.*, (1993) 195, 539-544.

[33] Idh J, Andersson B, Lerm M, Raffetseder J, Eklund D, Woksepp H, Werngren J, Mansjö M, Sundqvist T, Stendahl O, Schön T. Reduced susceptibility of clinical strains of Mycobacterium tuberculosis to reactive nitrogen species promotes survival in activated macrophages. *PLoS One.* (2017)12, e0181221. doi: 10.1371/journal.pone.0181221. eCollection 2017.

[34] Goldstein S, Czapski G. Mechanism of the nitrosation of thiols and amines by oxygenated NO solution: the nature of the nitrosating species. *J. Am. Chem. Soc.* (1996) 118, 3419-3425.

[35] Kharitonov AR, Sundquist ARVS. Kinetics of nitric oxide autoxidation in aqueous solution. *J. Biol. Chem.* (1995) 269, 5881-5883.

[36] Girard P., Potier P. NO, thiols and disulfides. *FEBS Letters* (1993), 320(1), 7-8.

[37] Singh RJ, Goss SPA, Joseph J. Nitration of γ-tocopherol and oxidation of α-tocopherol by copper-zinc superoxide dismutase/H_2O_2/NO_2^-: role of nitrogen dioxide free radical *Proc. Natl. Acad. Sci* USA (1998) 95, 12912.

[38] Huie RE, Neta P. Kinetics of one-electron transfer reactions involving chlorine dioxide and nitrogen dioxide *J. Phys. Chem.* (1986) 90, 1193.

[39] Ford E, Hughes MN, Wardman P Kinetics of the reactions of nitrogen dioxide with glutathione, cysteine, and uric acid at physiological. *Free Radic. Biol. Med.* (2002) 32, 1314.

[40] Alvarez B, Radi R. Peroxynitrite reactivity with amino acids and proteins. *Amino Acids.* (2003) 25, 295-311.

[41] Lancaster JR, Jr. Nitric oxide: a brief overview of chemical and physical properties relevant to therapeutic applications. *Future Science OA* (2015)1, No.

[42] Goldstein S, Czapski, G. Mechanism of the Nitrosation of Thiols and Amines by Oxygenated .bulledNO Solutions: the Nature of the Nitrosating Intermediates *J. Am. Chem. Soc.* (1996) 118, 3419-3425.

[43] Itoh, T.; Matsuya, Y.; Maeta, H.; Miyazaki, M.; Nagata, K.; Ohsawa, A Reaction of Secondary and Tertiary Amines with Nitric Oxide in the Presence of Oxygen. *Chem. Pharm. Bull.* (1999) 47, 819-823.

[44] Sinha BK, Bhattacharjee S, Chatterjee S, Jiang J , Motten AG, Kumar A, Espey MG, Mason RP. Role of *Nitric Oxide* in the Chemistry and Anticancer Activity of Etoposide (VP-16,213) *Chem. Res. Toxicol.* 2013 Mar 18; 26(3): 379-387.

[45] Yenes S, Messeguer A. A study of the reaction of different phenol substrates with nitric oxide and peroxynitrite. *Tetrahedron* (1999), 55, 14111-14122.

[46] Conforti F, Menichini F. Phenolic compounds from plants as nitric oxide production inhibitors. *Curr Med Chem.* (2011)18,1137-45.

[47] Augusto O, Bonini MG, Amanso AM, Linares E, Santos CCX, De Menezes SL. *Free Radic. Biol. Med.* (2002) 32, 841.

[48] Zhang YY, Xu AM, Nomen M, Walsh M, Keaney JF Jr, Loscalzo J. Nitrosation of Tryptophan Residue(s) in Serum Albumin and Model Dipeptides. Biochemical characterization and bioactivity. *J Biol Chem.* (1996) 1271(24):14271-14279.

[49] Kirsch M, Fuchs A, de Groot H. Regiospecific Nitrosation of N-terminal-blocked Tryptophan Derivatives by N_2O_3 at Physiological pH. *J. Biol. Chem.* (2003) 278, 11931-11936.

[50] Perez-Gonzalez A, Munoz-Rugeles L, Alvarez-Idaboy JR. Tryptophan versus nitric oxide, nitrogen dioxide and carbonate radicals: differences in reactivity and implications for oxidative damage to proteins. *Theoretical Chemistry Accounts* (2016) 135, 1-9.

[51] Joseph J, Kalyanaraman B, Hyde JS. Trapping of nitric oxide by nitronyl nitroxides: an electron spin resonance investigation. *Biochem. Biophys. Res. Commun.*, (1993) 192, 926-934.
[52] Lozinsky EM, Martina LV, Shames AI, Uzlaner N, Masarwa A, Likhtenshtein GI, Meyerstein D, Martin VV, Priel Z. Detection of NO from pig trachea by a fluorescence method, *Analytical Biochemistry* (2004) 326, 139-145.
[53] Kytzia A, Korth H-G, Sustmann R, de Groot H, Kirsch M. On the mechanism of the ascorbic acid-induced release of nitric oxide from N-nitrosated tryptophan derivatives: scavenging of NO by ascorbyl radicals. *Chemistry - A European Journal* (2006) 12, 8786-8797.
[54] Hu TM, Chen YJ. Hu Teh-Min, Chen Yu-Jen. Nitrosation-modulating effect of ascorbate in a model dynamic system of coexisting nitric oxide and superoxide. *Free Radical Research* (2010) 44, 552-556.
[55] Ross AB, Mallard WG, Helman WP, Buxton GV, Huie RE, Neta P. *NDRL/NIST Solution Kinetics Database 3.0*, NDRL/NIST, Gaithersburg, MD, 1998.
[56] Treinin A, Hayon E. Treinin A, Hayon E. Absorption spectra and reaction kinetics of NO_2, N_2O_3, and N_2O_4 in aqueous solution. *J. Am. Chem. Soc.* (1970) 92, 5821-5828.

Chapter 2

Nitric Oxide Donors

Annotation

In the past decades, hundreds of NO donors have been developed and widely used in biological and biomedical research and medicine. Nowadays, the large variety of NO donors can serve as a tool to explore the wide range of properties of NO in cancer and other diseases and as pharmacological drags. In the present chapter, we briefly discuss classic nonenzymatic NO donors and NO-containing multitarget hybrid drugs. The latter combines the benefits of both NO donors and currently used therapeutics.

2.1. Introduction

Three main mechanisms by which NO is released from potential NO donors in biological processes have been proposed: (1) spontaneous release, (2) chemical reactions with acid, alkali, metal ions and complexes, thiols, etc., and (3) enzymatic reactions [1–5].

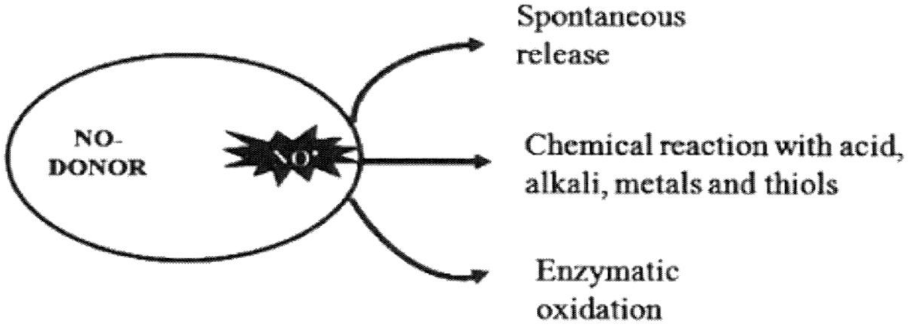

Figure 2.1. Mechanisms by which NO is released from NO donors [2]. Copyrights 2017 American Chemical Society.

Table 2.1. Current Major Classes of Nitric Oxide Donors [1]. American Chemical Society

Index, Name	Representative Compounds	Pathway of NO Generation: Non-enzymatic	Enzymatic
A. Organic nitrates	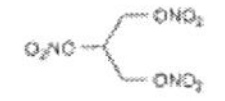	thiols	Cyt-P450, GST and a membrane-bound enzyme
B. Organic nitrites	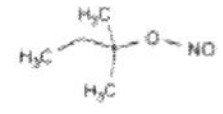	hydrolysis and trans-nitrosation; thiols; light; heat	cytosolic and microsomal enzymes; xanthine oxidase
C. Metal-NO complexes	$Na_2[Fe(CN)_5(NO)] \cdot 2H_2O$	light; thiols; reductants; nucleophiles	a membrane-bound enzyme
D. *N*-Nitrosamines	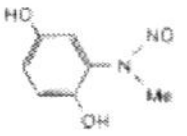	$\bar{O}H$; light	Cyt-P450 related enzymes
E. *N*-Hydroxyl nitrosamines	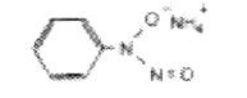	light; heat	peroxidases
F. Nitrosimines	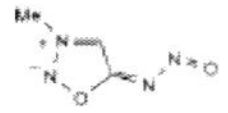	thiols; light	?
G. Nitrosothiols l	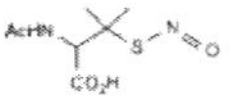	spontaneous; enhanced by thiols, light and metal ions	unknown enzymes
H. *C*-nitroso compounds	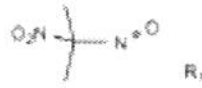	light; heat	?
I. Diazetine dioxides	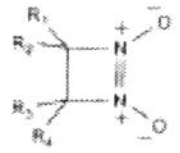	spontaneous; thiols	?
J. Furoxans and benzofuroxans	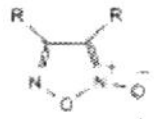	thiols	unknown enzyme
K. Oxatriazole-5-imines	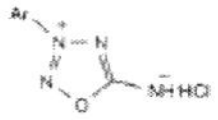	thiols	?
L. Sydnonimines		spontaneous, enhanced by light, oxidants and pH>5	prodrugs require enzymatic hydrolysis
M. Oximes	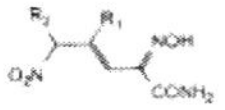	spontaneous; O_2/Fe^{III}-porphyrin	?
N. Hydroxylamines	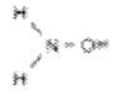	autoxidation enhanced by metal ions	catalase/H_2O_2
O. *N*-Hydroxyguanidines	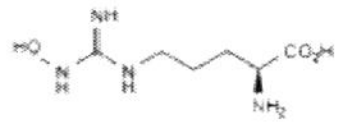	oxidants	NOSs, Cyt-P450
P. Hydroxyureas	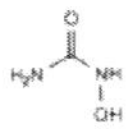	H_2O_2/CuZn-SOD or ceruloplasimin; H_2O_2/Cu^{2+}; heme-containing proteins	peroxidase

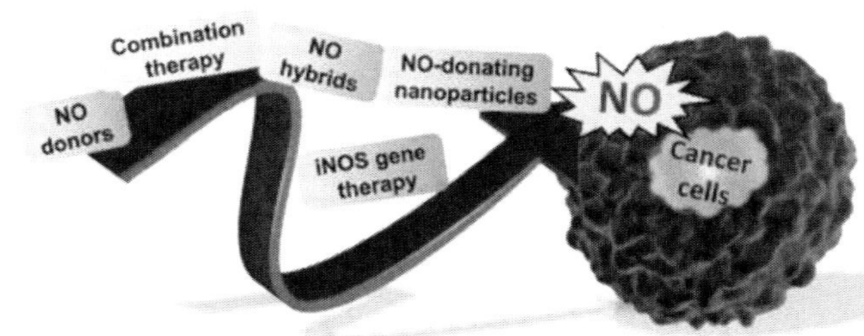

Figure 2.2. Schematic illustration key role of nitic oxide in cancer therapy [2 Copyrights 2017 American Chemical Society.

Commonly adopted classifications of non-enzymatic NO donors are presented in Figure 2.1 and Table 2.1. The following compounds are included in the NO donors list: nitrates, diazeniumdiolates, *S*-nitrosothiols, *N*-nitrosamines, metal-NO complexes, N-hydroxy-N-nitrosamines, *N*-nitrosimines, *C*-nitroso compounds, furoxans and benzofuroxans, oxatriazole-5-imines, oximes, and hydroxylamines.

An example of contribution of NO in treating diseases is presented in Figure 2.2.

The NO-containing non-steroidal anti-inflammatory drug (NO-NSAID) and correspondent NO antioxidant drug (NO-AD) constitute the most promising class of therapeutics of double action.

2.2. Monofunctional Donors

2.2.1. Organic Nitrates

The organic nitrates and sodium nitroprusside (Figure 2.3) are widely used in biological research, biomedicine, and medicine [3–8].

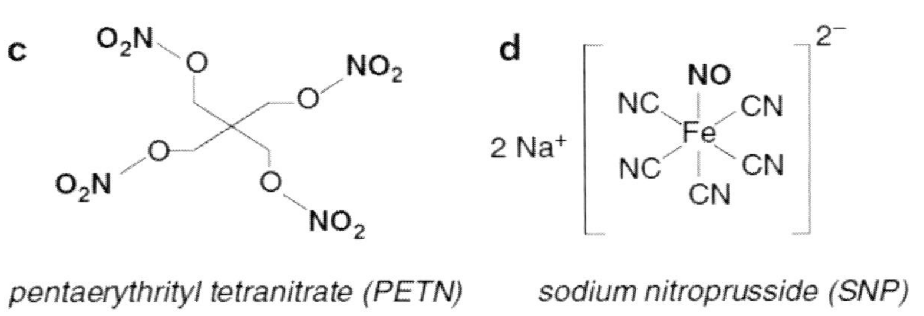

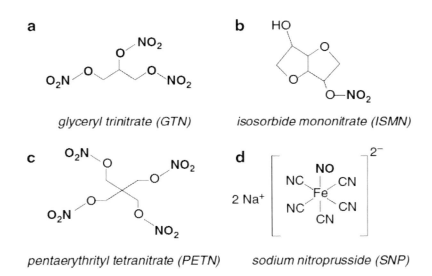

Figure 2.3. Chemical structure of NO donor drugs used clinically: the organic nitrates (a–c) and sodium nitroprusside (d). The NO-containing moiety is shown in bold [3]. With permission from John Wiley and Sons, 2007.

The major biological effects of nitrates are attributable to the formation of NO. Nevertheless, NO releases from organic nitrates and requires either enzymatic or nonenzymatic bioactivation. A three-electron reduction is involved in these processes [6]. Glyceryl trinitrate (GTN) has a biological half-life of 1 to 4 seconds, and its metabolites (1,2-glyceryl dinitrate and 1,3-glyceryl dinitrate) have a half-life of up to forty minutes [3].

Due to the high electronegativity of oxygen, nitrosyl nitrogen atoms exhibit high electron deficiency and are highly susceptible to nucleophilic attack by oxygen-, nitrogen-, and sulfur-nucleophiles. This property also provides transfer of a nitrosyl group (transnitrosation) to nitrites, which requires one-electron reduction. Key example includes formation of *S*-nitrosothiols by the transfer of organic nitrites to a sulfhydryl group [8]. Under strong alkaline conditions, organic nitrates are susceptible to hydrolysis (SN_2 nucleophilic substitution to give alcohol and nitrate), â-H elimination (forming alkene), and R-H elimination (producing aldehyde and nitrite).

2.2.2. Miscellaneous Nitric Oxide Donors

2.2.2.1. Diazeniumdiolates

The diazeniumdiolates (NONOates) structure are compounds containing the X-$[N(O)NO]^{-}$ structural unit in which X is a secondary amine group (Figure 2.4).

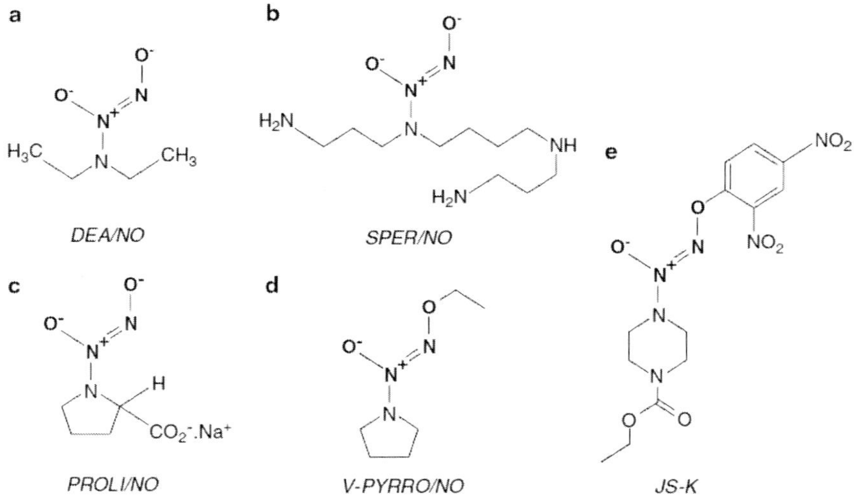

Figure 2.4. Chemical structure of five examples of the diazeniumdiolate (NONOate) class of NO donor drug. The diolate group (shown in bold) releases NO in solution, although prior cleavage of the molecule to release the terminal oxygen may be required [3]. With permission from John Wiley and Sons, 2007.

In aqueous media, diazeniumdiolates

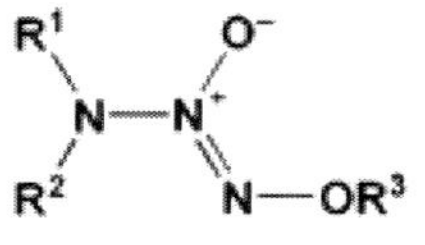

spontaneously generate up to two molecules of NO per $[N(O)NO]^-$. Their half-lives range from two seconds (for X = L-prolyl) to twenty hours [for X = $(H_2NCH_2CH_2)_2N$] at pH 7.4 and 37°C. In the first-order dissociation, Angeli's salt (X = O^-), produces NO^- and becomes an NO source having the short half-life of ~two minutes [9]. In the presence of selected oxidizing agents, diazeniumdiolate-derived NO can also be used to generate reactive nitrogen/oxygen species with higher nitrogen oxidation states (+3 and +4) [10]. The mechanism of NO release for the diazeniumdiolates suggests the following steps: (1) NO is generated spontaneously on protonation of the anionic portion's R_1R_2N nitrogen and (2) covalently bound R_3 is removed to free the anion before spontaneous NO generation.

Secondary amine-based diazeniumdiolates generate NO upon spontaneous decomposition, with half-lives of decomposition ranging from two seconds to twenty hours, depending on the amine backbone [11]. The mechanism of NO and HNO release from isopropylamine (IPA/NO IPA/NO) is given in Figure 2.5. Under physiological conditions, IPA/NO has a short half-life of 6.7 min and generates HNO via a tautomerization pathway, while at lower pH, NO is produced via protonation of the nitroso oxygen, followed by tautomerization and N-N bond cleavage [12].

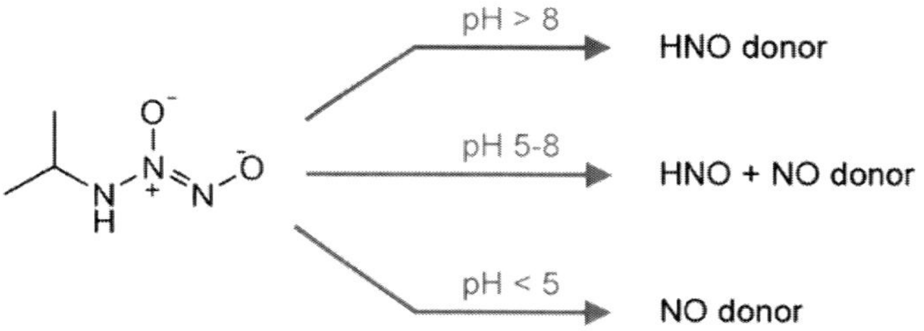

Figure 2.5. Mechanisms available for Angeli's Salt leading to release of HNO or NO [13]. Copyrights 2011 American Chemical Society.

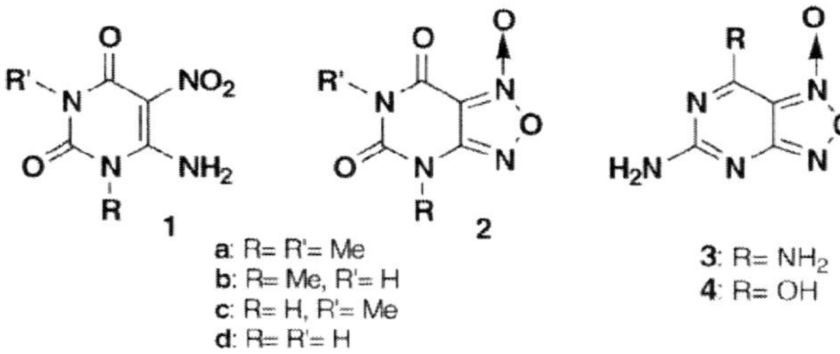

Figure 2.6. Chemical structure of NO donors described in [34]. Copyrights 2017 American Chemical Society.

The decomposition mechanism of Angeli's Salt ($Na_2N_2O_3$) is dependent on pH ([13].

2.2.2.2. S-Nitrosothiols

Nitrosothiol compounds, such as a S-nitroso-N-acetylpenicillamine (SNAP) and S-nitrosoglutathione (GSNO), yields NO, NO^+, and NO^- under action of heat, UV light, superoxide, seleno compounds, and metal ions (Cu^+, Fe^{++}, Hg^{++}, and Ag^+) [3, 13, 14]. These compounds may serve for the storage, transfer, and delivery of NO in biological systems [14]. For example, the presence of trace transition metal ions (e.g. Cu+) stimulates the catalytic breakdown of S-nitrosothiols to NO and disulfide. Decomposition of S-nitrosothiol can be also stimulated by ascorbate in the presence of reduced transition metals. At a pH of 6.0–8.0 and a temperature of 37°C, transition-metal ion chelators stabilize SNAP with the half-life of six hours. Photolysis of S-nitrosothiols results in the formation of NO and disulfide via the intermediacy of thiyl radicals.

2.2.2.3. Sydnonimines

In the presence of oxygen, 3-morpholinosydnonimine (SIN-1)

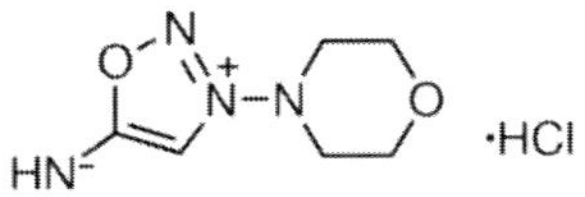

produces peroxynitrite (OONO-) and is considered a spontaneous donor of NO and frequently used without causing tolerance even during long-term treatment [15, 16]. In plasma, SIN-1 has a half-life of one to two hours. Release of NO from sydnonimines in an alkaline pH is facilitated by oxygen and irradiation from visible light [16].

2.2.2.4. Oxatriazole-5-Imines

The results of work [17] suggest that the mesoionic 3-aryl substituted oxatriazole-5-imine derivatives GEA 3162 (1,2,3,4-oxatriazolium- 5-amino-3-[3,4-dichlorophenyl] chloride) and GEA 3175 ([1,2,3,4-oxatriazolium-3-[3-chloro-2-methylphenyl]-5- [(4-methylphenyl)sulfonyl] amino-]-, hydroxide inner salt) release NO, similar to "classic" SIN-1 and SNAP. The release of NO and NO_2 by GEA 3175 was increased 140-fold in the presence of human plasma. All of the four compounds studied converted oxyhaemoglobin to methaemoglobin and formed a paramagnetic NO-haemoglobin complex.

2.2.2.5. N-Nitrosamides and Nitrosourea

N-nitrosamides

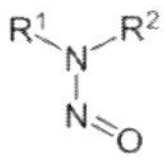

and nitrosourea,

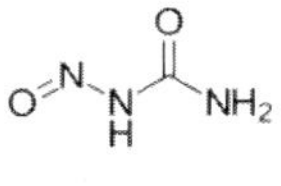

are potential NO donors. *N*-nitrosoguanidines, *N*-nitrosocarbamates, and other *N*-acyl-*N*-nitroso compounds, having an electron-withdrawing group nitroso group, release $NO^{\cdot}/NO^{+}$ through homolytic or heterolytic cleavage of the N-NO bond [18]. It was found that the homolytic cleavage of N-NO bonds, generating NO radical, is thermodynamically more favorable than the heterolytic cleavage, which generates NO^{+} after one-electron oxidation of the R-carbon atom, a highly unstable R-nitro amino radical is generated, which releases NO.

2.2.2.6. Metal-Nitric Oxide Complexes

Sodium nitroprusside (SNP) [$Na_2Fe(CN)_5NO$] (Figure 2.3), which liberates 1 mol of NO per M-NO happened to be among the most widely employed NO donor [19–21]. Its capability of producing NO depends on its interaction with sulfhydryl-containing molecules present in vivo. To establish the mechanism of interaction between SNP and sulfhydryl-containing compounds, such as cysteine and glutathione, the radical and nonradical species, produced in the reaction, were detected by EPR, UV-vis, and IR spectroscopy [21]. The results suggested that an electron-transfer process is the key step, which leads to the formation of the reduced SNP radical, the main detectable radical intermediate, and the corresponding S-nitrosothiol (scheme 2.1). The latter can be considered the real storage and transporters of NO.

Biological and chemical porphyrin nitrosyls are commonly very stable and cannot serve as NO donors. In contrast, the nitrosyl derivative of iron(II) octaethylporphyrin [Fe(oetap)(NO)] exhibits fast ligand-promoted NO dissociation in the presence of coordinating ligands, such as pyridine and N-methylimidazole [22]. According to a mechanism suggested on the base the denitrosylation kinetic research of [Fe(oetap)(NO)], the process involves rapid equilibrium binding of axial ligand followed by a rate-determining loss of NO from the six-coordinate intermediate.

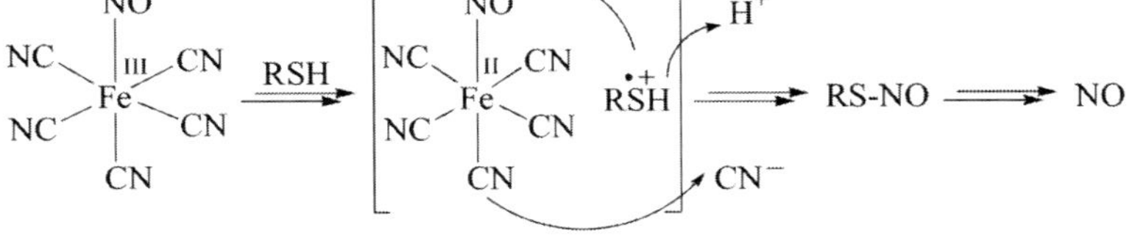

Scheme 2.1. Proposed mechanism of interaction between SNP and sulfhydryl-containing compounds [21]. Copyrights 2005 American Chemical Society.

The NO(L)- $(NH_3)_4Ru]Cl_3$ complex (L=imidazole, theophylline and caffeine) releases NO upon reduction [23]. The affinities of ruthenium ion for NO and the reduction potential of the complex can be modulated by varying ligands. For example, the imidazole ligand coordinated to the Ru(II) through a carbon atom facilitates NO release in the reduced form. Reduction of the complex *R,R,S,S-trans*-$[RuCl(NO)(cyclam)]^{2+}$ (1,4,8,11-

tetraazacyclotetradecane) ($E°$ = −0.1 V) results in the rapid loss of Cl^- by first-order kinetics with $k = 1.5\ s^{-1}$ and the slower loss of NO ($k = 6.10 \times 10^{-4}\ s^{-1}$, $\Delta H^{\ddagger} = 15.3$ kcal mol^{-1}, $\Delta S^{\ddagger} = -21.8$ cal mol^{-1} K^{-1}) [24].

Recently, iron-sulfur cluster nitrosyls have been proved to be promising NO donors (section 3.2).

2.2.2.7. N-Hydroxy-N-Nitrosamines

N-Hydroxy-*N*-nitrosamines (*N*oxy- *N*-nitrosamines), such as *N*-hydroxy-*N*-nitrosamines, Cupferron

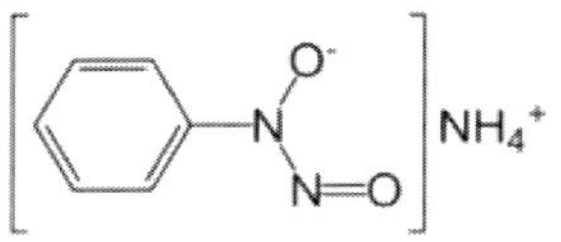

alanosine, and dopastin can decompose under physiological conditions to release NO [25–27]. Cupferron derivatives in ethanol thermally or photochemically decompose to azoxy compounds, NO and NH_3. and NO. Cupferron also releases NO under enzymatic, electrochemical, as well as chemical oxidation.

2.2.2.8. N-Nitrosamines

Derivatives of N-Nitrosamine

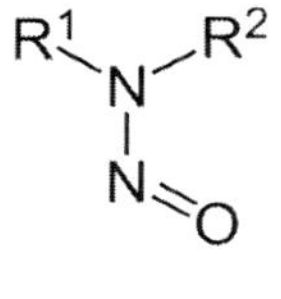

can release both NO and N_2O. Spontaneous release of a small amount of NO from the N-nitrosamines was observed on incubation in neutral buffers [28]. Nitrosodimethylamine treated with Fenton reagent generates NO as estimated by ESR technique using cysteine-Fe(II), and N-methyl-D-glucaminedithiocarbamate (MGD)-Fe(II) complexes. Results indicate that the N-nitrosamines can be decomposed accompanying concomitant release of NO on contact with reactive oxygen species (ROS) in physiological conditions.

2.2.2.9. C-Nitroso Compounds

C-Nitroso compounds contain a nitroso group attached to a carbon atom. Tertiary *C*-nitroso compounds, such as 2-methyl-2-nitrosopropane (MNP)

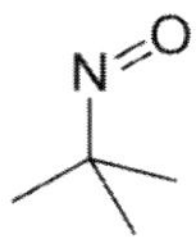

can serve as NO donors [3, 29, 30]. It was reported that substituted C-nitroso compounds act as donors of neutral NO through a first-order homolytic C-N bond scission to release up to 88% NO in DMSO at 25°C [29]. C-Nitroso compounds are sources of biologically active NO and display potent NO bioactivity in a rabbit aortic ring assay.

2.2.2.10. Diazetine Dioxides

The rate constants of diazetine dioxides (DD), 3-Bromo-3,4,4-trimethyl-3,4-dihydrodiazete 1,2-dioxide and 3-bromo-4-methyl- 3,4-hexamethylene-3,4-dihydro-diazete 1,2-dioxide, and other 3-halogeno-DD derivatives decomposition in water and in DMSO have been determined using nitronylnitroxides as spin traps for NO [31]. Rate constants of derivatives of 3,4-dihydro-1,2-diazete 1,2-dioxides decomposition were found to be in the range from 10^{-8} to 6.5×10^{-7} c^{-1} in water and between 3×10^{-7} and 1.6×10^{-5} c^{-1} in dimethylsulfoxide. Reactivity toward thiols of 3-bromo- and 3-chloro-3,4,4-trimethyl-DD, 3-bromo- and 3-chloro-4-methyl-3,4-hexamethylene-DD, 3,3,4,4-tetramethyl-DD (3) and 3-methyl-3,4-hexamethylene-DD was investigated. Two suggested pathways in the reaction have been discussed: (1) reversible nucleophilic addition of the thiolate anion at the *N*-oxide oxygen atom resulted in the formation of an unstable intermediate and (2) the spontaneous decomposition forming an intermediate, which can be reduced by thiol. In the latter reaction, intermediate undergoes spontaneously hemolytic decomposition accompanied by NO formation.

2.2.2.11. Furoxans and Benzofuroxans

Furoxan, 1,2,5-oxadiazole 2-oxide

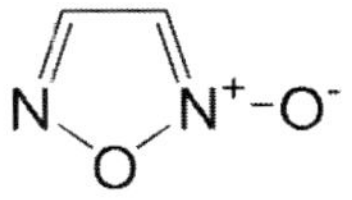

and Benzofuroxan

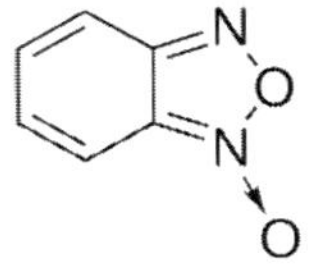

are heterocycles of the isoxazole family [32]. It was reported that for furoxan derivatives, the attack of RS- at position 3 or 4 leads to intermediates that undergo ring opening to the nitroso derivatives followed by the NO formation by oxidation of eliminated nitrosyl anions (NO^-) [33]. Chemical structure of NO donors was synthesized and investigated in [34]. The generation of NO and NO-related species from 4*H*-[1,2,5]Oxadiazolo[3,4-*d*]pyrimidine-5,7-dione 1-oxides (2) occurs under physiological conditions in the presence of thiols, such as *N*-acetylcysteamine, cysteine, and glutathione under physiological conditions (Figure 2.7, scheme 2.2) [34].

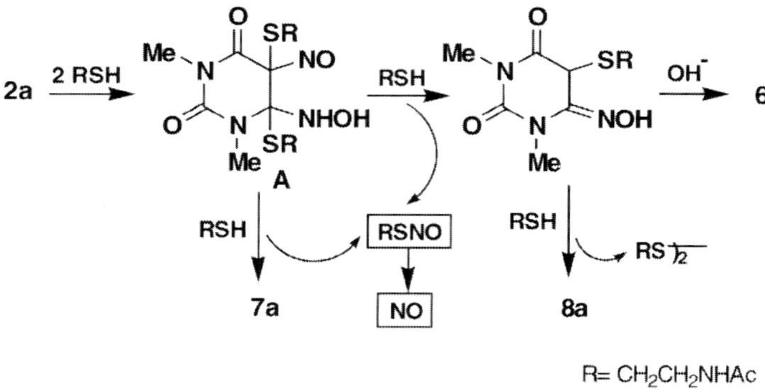

Figure 2.7. Mechanisms for the present reactions resulting in the NO generations. The attack of thiols to both of the 3a and 7a positions in 2a and the subsequent ring opening produce the key intermediary adduct A. Further attack of another thiol to the *C*-5 nitroso group of A releases thionitrite, a NO precursor, together with the formation of 6a, 7a, and 8a [34]. Copyrights 2017 American Chemical Society.

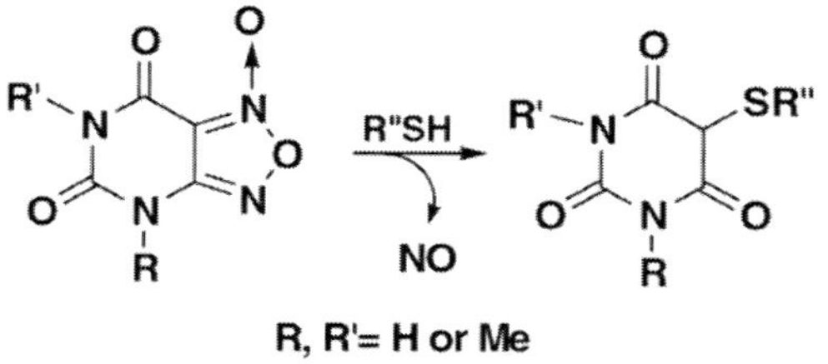

Scheme 2.2. NO release from 4H-[1,2,5]Oxadiazolo[3,4-d]pyrimidine-5,7-dione 1-oxides [34]. Copyrights 2017 American Chemical Society.

2.2.2.12. Oximes

Oximes belong to the imines, with the general formula $R^1R^2C = N O H$, where R^1 is an organic side-chain and R^2 may be hydrogen:

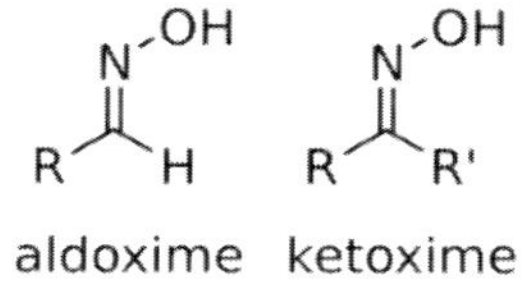

Alkyl- and aryloximes, quaternized pyridine aldoximes (2- and 4-PAM), hydroxamic acids can produce NO under mild oxidative conditions [35].

2.2.2.13. N-Hydroxyguanidines

Derivatives of hydroxyguanidines

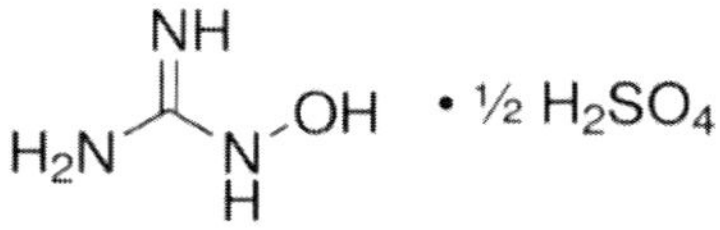

combine the imino group of guanidine with hydroxylamino group of hydroxyurea [36]. *N*-hydroxyguanidines generate NO by photosensitized oxygenation of hydroxyguanidines. *N*-Hydroxyalkylguanidine compounds, N-butyl-N-hydroxyguaindine, and N-(N-hydroxyamidino) piperidine (NHAP), N –Hydroxy-L –arginine were oxidized with lead tetra-acetate (Pb[OAc]$_4$) and potassium ferricyanide/hydrogen peroxide (K_3FeCN_6/H_2O_2). The process was accompanied, generating significant amounts of NO [37]. Oxidation with K_3FeCN_6/H_2O_2, (Pb[OAc])$_4$, lead oxide (PbO_2), and silver carbonate (Ag_2CO_3) occurs presumably through the initial release of nitroxyl (HNO). This reaction produces N^{δ}-cyanoornithine, citrulline, and NO.

2.2.2.14. Hydroxylamines

Hydroxylamine (HA), a natural product of mammalian cells, is widely used as an NO donor and exhibits a wide range of biological activities [3]. Hydroxylamine was converted to nitrite and nitrate via peroxynitrite in aqueous solution. The ability of hydroxylamine to generate NO in the presence of myoglobin (Mb) and hydrogen peroxide (H_2O_2) has been proved in [38]. ESR measurements at 77 K showed the formation of the ferrous nitrosyl myoglobin, Mb-NO, in the reaction mixtures containing Mb, H_2O_2, and HA. The following steps of the reaction mechanism was suggested: $MbFe^{3+}$ is oxidized by H_2O_2, producing Ferryl Mb ($MbFe^{4+}$=O), which oxidizes HA nitroxide radical. The nitroxide radical, in turn, reacts with FerrylMb to release NO. Recently it has been reported that the hydroxylamine oxidoreductase from the anammox bacterium, *Candidatus Kuenenia stuttgartiensis t* catalyzes the oxidation of hydroxylamine to NO by using bovine cytochrome *c* as an oxidant [39].

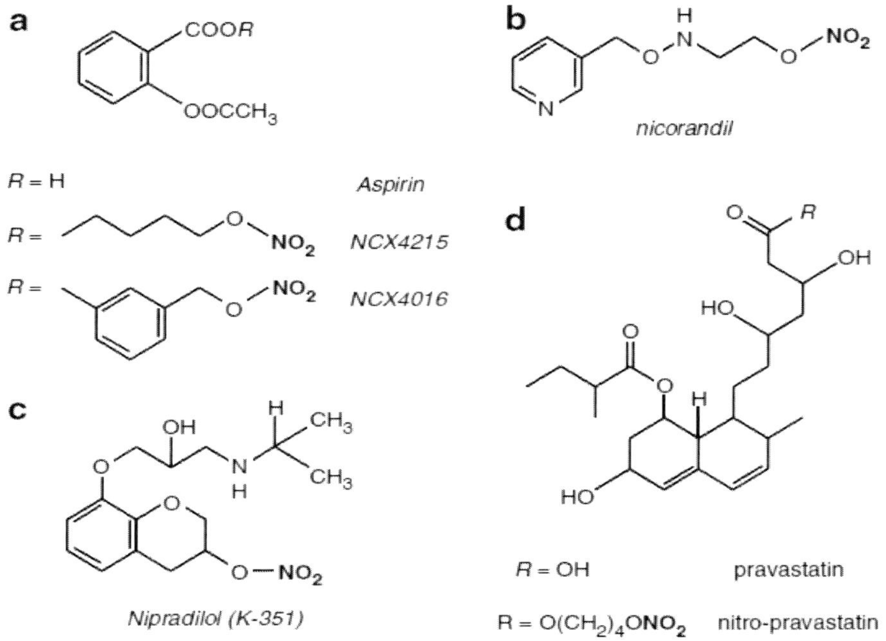

Figure 2.8. Chemical structure of NO donor hybrid drugs containing a nitro-oxy moiety. The nitro-oxy group is shown in bold [3]. With permission from John Wiley and Sons, 2007.

2.3. Nitric Oxide-Donor Hybrid Drugs

A design of multitarget hybrid drugs containing NO is one of the most promising trends in the area (Figures 2.8-2.12, 2.10–2.14) [1–3, 40–48]. This approach is carried out by joining a drug and NO through an appropriate linker (spacer). The linker can be either susceptible to metabolic cleavage or be hard. Such hybrid supermolecules can be considered as either prodrugs or codrugs. For last decade, more than a hundred target combinations used to develop multitarget drugs were investigated. The design and the synthesis of prodrugs for nonsteroidal anti-inflammatory drugs (NSAIDs) have been given much attention by medicinal chemists [42] and references therein. The organic esters of nitric acid, the organic nitrates, and glyceryl trinitrate (GTN) in particular (Figure 2.10), appeared to be the most important therapeutic group.

A class of NO-donor aspirin-like drugs derived from aspirin derivatives by linking acyl moieties bearing NO-donor *O*-nitro groups to the –OH function of salicylic acid was reported [41]. The following hemical structures of hybrid NO-releasing anti-inflammatory ester prodrugs, diazen-1-ium-1,2-diolate ester prodrugs of aspirin (1a–c), ibuprofen (2a–c) and indomethacin (3a–c), the 3-(nitrooxymethylphenyl) ester of aspirin (4), some nitrooxybutyl ester prodrugs of aspirin (5) indomethacin (6), the nitroglyceryl esters of indomethacin (7) and ibuprofen (8) [43]:

Analogues of aspirin, 2-Hydroxysulfamoylbenzoic acid
and its ethyl benzoate ester

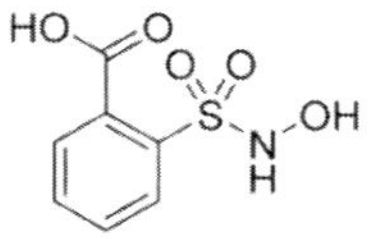

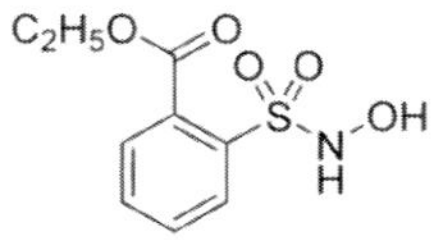

were synthesized through a one-step reaction in which the carboxyl group was replaced by an ethyl ester, and/or the acetoxy group was replaced by an N-substituted sulfonamide (SO_2NHOR_2):R_2 = H, Me, CH_2Ph (pharmacophore) [44]. The SO_2NHOH moiety present in these compounds was found to be a good NO donor upon incubation in phosphate buffer at pH 7.4. Other reported chemical structures of the NO-NSAIDs drugs:

NO–containing compounds such as JS-K

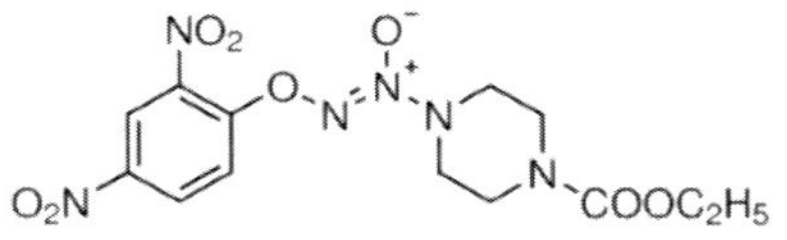

and NO-releasing hybrids such as NO– and NONO–nonsteroidal anti-inflammatory drugs can act as the s-nitrosylation agents [45]. Chemical structures of *m*-NO-aspirin, NONO-aspirin, NO-naproxen, and JS-K were also given.

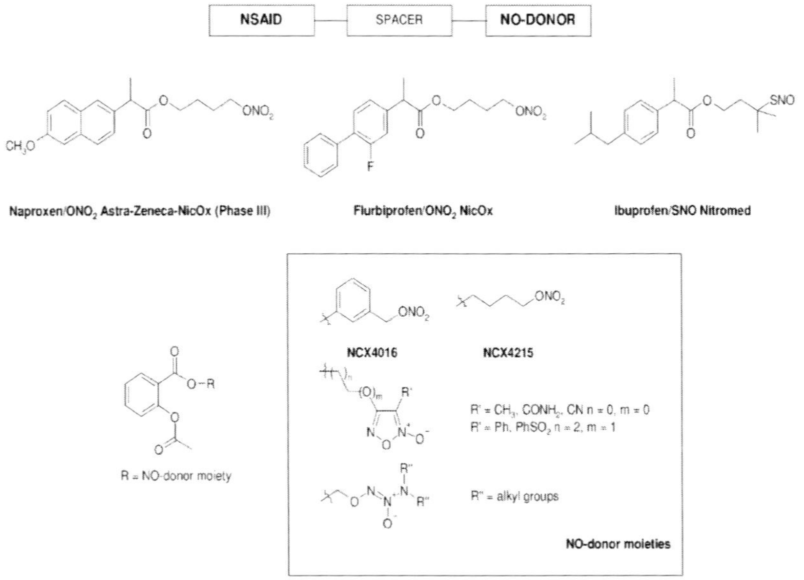

Figure 2.9. Examples of NO-NSAIDs [41]. With permission from Walter De Gruyter.

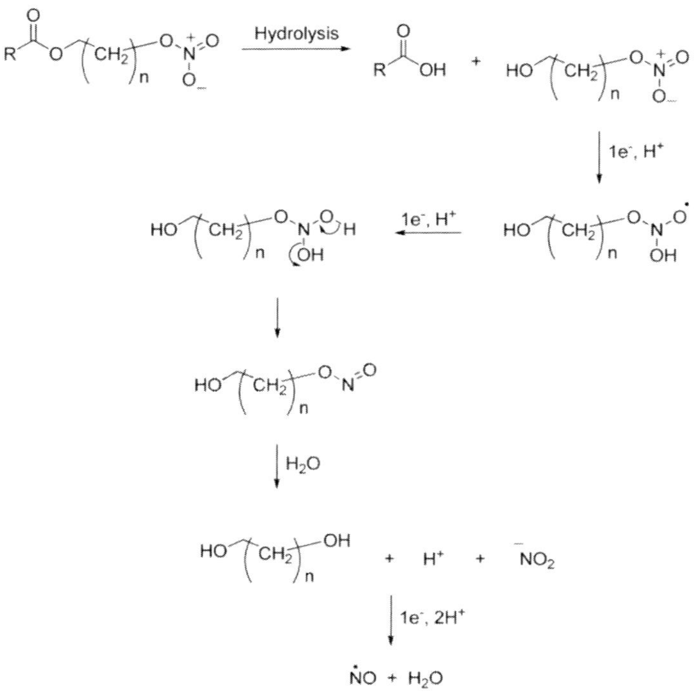

Figure 2.10. Chemical structure and hydrolysis of simple NO-NSAIDs [42]. With permission from *International Journal of Molecular Science*, 2012.

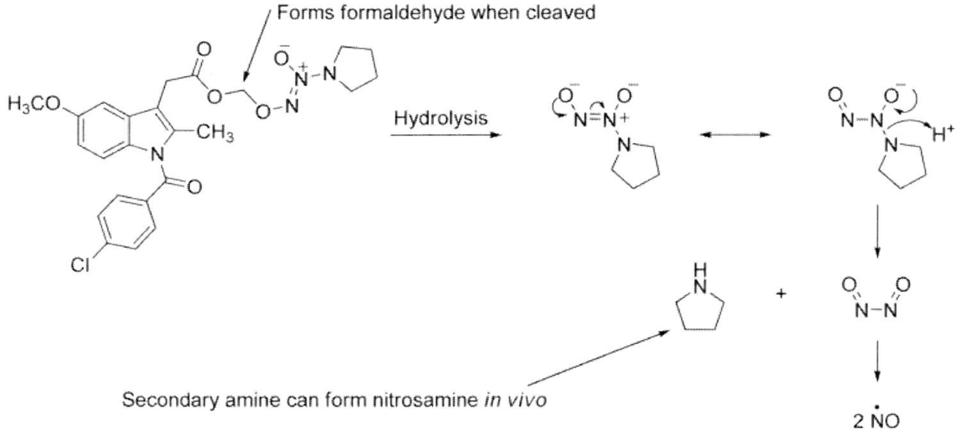

Figure 2.11. *In vivo* release of nitric oxide from NONOate-containing NO-NSAIDs [42]. With permission from *International Journal of Molecular Science*, 2012.

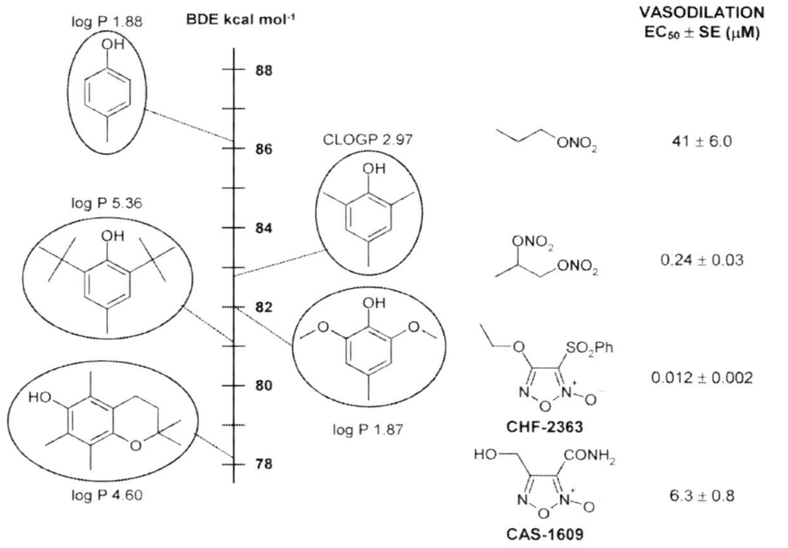

Figure 2.12. Phenols and NO-donors used in the symbiotic approach to obtain NO-donor antioxidant [41]. With permission from Walter De Gruyter.

Other examples of NO-NSAID are shown in Figure 2.9 [41].

Various hybrid compounds including a nitrate-containing prodrug of diclofenac, ethanesulfohydroxamic acid esters of indomethacin, naproxen, the methyl ether of ibuprofen ethansulfonhydroxamic acid ester, NO-releasing ester of diclofenac, 2-hydroxysulfamoylbenzoic acid, ethyl 2-hydroxysulfamoylbenzoate, furoxan containing ester of aspirin, tyrosyl-NSAID prodrugs, NO-diclofenac prodrugs, amide of indomethacin, and nitrate-containing amide of flurbiprofen were described in review [42].

The mechanisms of NO release from NO-NSAIDs, in which an NSAID molecule connected via an alkyl spacer to a nitrate group (-ONO_2), are described in Figures 2.10 and 2.11.

2.4. Nitric Oxide-Donor Antioxidant

NO-donor antioxidant is a new class of products in which selected phenols were conjugated to NO-releasing moieties [14, 46]. NO-released hybrid antioxidant drugs (NO-AO) containing phenol derivatives expose the wide antioxidant profile. The phenol moieties of (NO-AO) were characterized by having widely modulated both –OH bond dissociation energies and lipophilicities modulated [46] (Figure 2.12, Table 2.2).

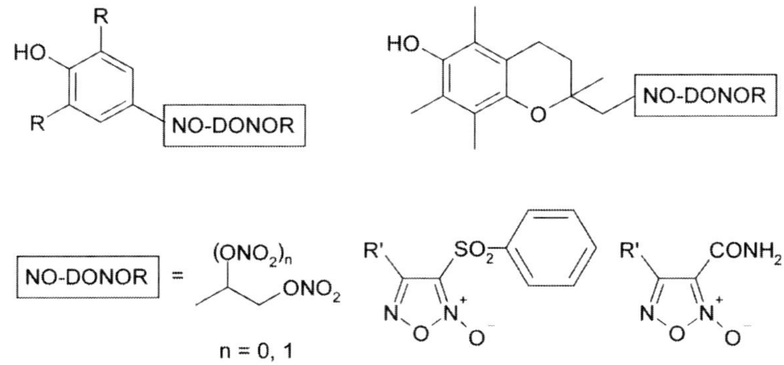

Figure 2.13. Hybrid nitric oxide donors with antioxidant activity [46]. Copyrights 2 011 American Chemical Society.

Table 2.2. Antioxidant and Vasodilating Activity of the Nitric Oxide-Donor Phenols [41]. With permission from Walter De Gruyter

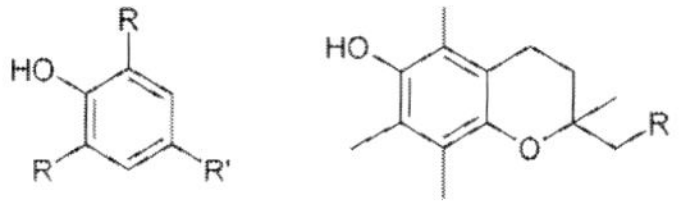

A B

	Struct.	R	R'	Antioxidant activity[a] IC_{50} (95% CL) μM	Vasodilating activity[b] $EC_{50} \pm SE$, μM	+1 μM ODQ	IC_{50}/EC_{50}
1	A	H	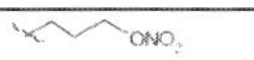	143 (133-153)	1.0 ± 0.2	>100	131
2	A	OCH_3		5.9 (5.5-6.4)	4.3 ± 0.6	>100	1.4
3	A	*t*-Bu	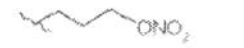	2.0 (1.9-2.1)	40 ± 1	>100	0.05
4	B		-	0.15 (0.15-0.16)	1.2 ± 0.1	10 ± 1	0.12
5	A	H	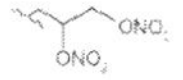	185 (176-195)	0.13 ± 0.03	65 ± 4	1423
6	A	OCH_3	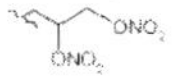	5.4 (5.0-5.8)	0.64 ± 0.09	49 ± 4	8.4
7	A	*t*-Bu	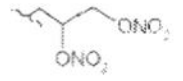	2.6 (1.9-3.5)	3.3 ± 0.4	>100	0.8
8	A	H	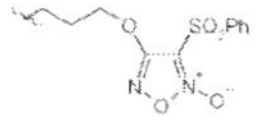	47 (45-48)	0.012 ± 0.001	0.36 ± 0.09	3917
9	A	OCH_3	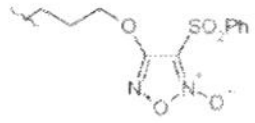	3.4 (3.2-3.5)	0.022 ± 0.003	0.50 ± 0.13	154
10	A	*t*-Bu	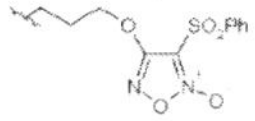	2.0 (1.9-2.0)	0.11 ± 0.03	4.8 ± 0.5	18
11	B	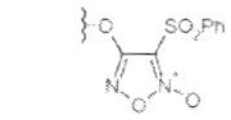	-	0.49 (0.48-0.50)	0.044 ± 0.004	0.67 ± 0.09	11
12	A	*t*-Bu	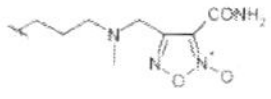	1.2 (1.1-1.2)	0.41 ± 0.08	7.4 ± 1.1	2.9
13	B	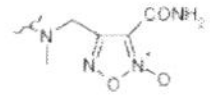	-	0.14 (0.14-0.14)	1.5 ± 0.1	19 ± 1	0.09

The 3-carbamoylfuroxan and the 3-phenylsulfonylfuroxan substructures products were tethered with NO-donor moieties [41]. The antioxidant activities of the hybrid products were assessed by detecting the 2-thiobarbituric acid reactive substances. The obtained products were proved to inhibit in a concentration-dependent manner the auto-oxidation of lipids in microsomial membranes of rat hepatocytes (Table 2.2).

The antioxidant potencies can be described by a parabolic dependence on the lipophilicity (log *P*) and a linear dependence on the calculated O–H bond dissociation energy (ΔHabs).

A class of phenols able to release NO was designed through a symbiotic approach using selected phenols and selected nitrooxy and furoxan NO-donors (Figure 2.13). The antioxidant activities of the hybrid products were assessed by detecting the 2-thiobarbituric acid reactive substances (Table 2.2).

REFERENCES

[1] Wang PG[1], Xian M, Tang X, Wu X, Wen Z, Cai T, Janczuk AJ. Nitric Oxide Donors: Chemical Activities and Biological Applications. *Chem. Rev.* (2002) 102, 1091-1134.

[2] Huang Z, Fu J, Zhang Y. Oxide Donor-Based Cancer Therapy: Advances and Prospects. *J Med Chem.* (2017) 60, 7617-7635.

[3] Miller MR, Megson IL. Recent developments in nitric oxide donor drugs. *Br J Pharmacol.* (2007) 151, 305-321.

[4] Peng GW, Tingwei Bill Cai, Naoyuki Taniguch (eds). *Nitric Oxide Donors.* WILEY VCH, 2005.

[5] Alberto G, Boschi D, Chegaev K, Cena C, Di Stilo A, Fruttero R., Lazzarato L, Rolando B, Tosco P. Multitarget drugs: Focus on the NO-donor hybrid drugs. *Pure Appl. Chem.* (2008) 80, 1693-1701.

[6] Thatcher GR, Nicolescu AC, Bennett BM, Toader V. Nitrates and NO release: contemporary aspects in biological and medicinal chemistry. *Free Radic Biol Med* (2004) 37, 1122-1143.

[7] Parker JD and Parker JO. Nitrate therapy for stable angina pectoris. *N Engl J Med* (1998) 338, 520-531.

[8] Meloche BA, O'Brien PJ. S-nitrosyl glutathione-mediated hepatocyte cytotoxicity *Xenobiotica* (1993), 23, 8687.

[9] Amatore C, Arbault S, Ducrocq C, Hu S, Tapsoba I. Angeli's salt ($Na_2N_2O_3$) is a precursor of HNO and NO: A voltammetric study of the reactive intermediates released by Angeli's salt decomposition. *ChemMedChem.* (2007) 2, 898-903.

[10] Keefer LK. Progress toward clinical application of the nitric oxide-releasing diazeniumdiolates. *Annu. Rev. Pharmacol. Toxicol.* (2003) 43, 585-607.

[11] Bharadwaj G, Benini PCZ, Basudhar D, Ramos-Colon CN, Johnson GM, Larriva MM, Keefer LK, Andrei D, Miranda KM. Analysis of the HNO and NO donating properties of alicyclic amine diazeniumdiolates. *Nitric Oxide.* 2014 Nov 15; 0: 70-78.

[12] Salmon DJ, Torres de Holding CL, Thomas L, Peterson KV, Goodman GP, Saavedra JE, Srinivasan A, Davies KM, Keefer LK, Miranda KM. HNO and NO release from a primary amine-based diazeniumdiolate as a function of pH. *Inorg. Chem.* (2011) 50, 3262-3270.

[13] Kıroğlu OE, Aydinoglu F, Oğülener N.. The effects of thiol modulators on nitrergic nerve- and S-nitrosothiols-induced relaxation in duodenum. *Journal of Basic and Clinical Physiology and Pharmacology* (2013), 24(2), 143-150.

[14] Wang PG, Xian M, Tang X, Wu X, Wen Z, Cai T, Janczuk AJ: Nitric oxide donors: Chemical activities and biological applications. *Chem. Rev.* (2002) 102, 1091-1134.

[15] Coert BA., Anderson RE, Meyer FB. Effects of the nitric oxide donor 3-morpholinosydnonimine (SIN-1) in focal cerebral ischemia dependent on intracellular brain pH. *Journal of Neurosurgery (2002)* 914-921.

[16] Ullrich T, Oberle S, Abate A and Schroder H: Photoactivation of the nitric oxide donor SIN-1. *FEBS Lett* (1997) 406: 66-68.

[17] Kankaanranta H, Rydell E, Petersson AS, Holm P, Moilanen E., Corell T, Karup G, Vuorinen P, Pedersen SB, Wennmalm A, Metsä-Ketelä T. Nitric oxide-donating properties of mesoionic 3-aryl substituted oxatriazole-5-imine derivatives. *Br. J. Pharmacol.* (1996) 117, 401-406.

[18] Zhu X-Q, He J-Q, Li Q, Xian M, Lu J, Cheng JP. N−NO Bond Dissociation Energies of *N*-Nitroso Diphenylamine Derivatives (Or Analogues) and Their Radical Anions: Implications for the Effect of Reductive Electron Transfer on N−NO Bond Activation and for the Mechanisms of NO Transfer to Nitranions. *Org. Chem.* (2000) 65, 6729.

[19] Marks GS, McLaughlin BE, Jimmo SL, Poklewska-Koziell M, Brien JF and Nakatsu K. Time-dependent increase in nitric oxide formation concurrent with vasodilation induced by sodium nitroprusside, 3-morpholinosydnonimine and S-nitroso-Nacetylpenicillamine but not by glyceryl trinitrate. *Drug Metab Dispos* (1995) 23, 1248-1252.

[20] Butler AR, Megson IL. Non-heme iron nitrosyls in biology. (*Chem Rev.* 2002) 102,1155-1166.

[21] Grossi L, D'Angelo S. Sodium nitroprusside: Mechanism of NO release mediated by sulfhydryl-containing molecules. *J Med Chem.* (2005) 48, 2622-2626.

[22] Bohle DS, Debrunner P, Fitzgerald JP, Hansert B, Hung C-H, Thomson A. Electronic origin of variable denitrosylation kinetics from isostructural $\{FeNO\}^7$ complexes: X-ray crystal structure of [Fe(oetap)(NO)]. *J. Chem. Commun.* (1997), 91-92.

[23] Lopes LGF, Wieraszko A, El-Sherif Y, Clarke MJ. The *trans*-labilization of nitric oxide in Ru^{II} complexes by C bound imidazoles *Inorg. Chim. Acta* (2001) 312, 15.

[24] Lang DR, Davis JA, Lopes LGF, Ferro AA, Vasconcellos LCG, Franco DW, Tfouni E, Wieraszko A, Clarke MJ. A Controlled NO-Releasing Compound: Synthesis, Molecular Structure, Spectro-scopy, Electrochemistry, and Chemical Reactivity of *R,R,S,S-trans*-$[RuCl(NO)(cyclam)]^{2+}$(1,4,8,11-tetraazacyclotetradecane). *Inorg. Chem.* (2000) 39, 2294.

[25] Marvel CS. Cupferron. *Organic Syntheses*, Coll. Vol. 1, p. 177; Vol. 4, p. 19.

[26] Saavedra JE, Keefer LK. NO better pharmaceuticals. *Chem. Br.* (2000) 36, 30- 33.

[27] Hwu JR, Yau CS, Tsay S-C, Ho T-I. Thermal- and photo-induced transformations of *N*-aryl-*N*-nitrosohydroxylamine ammonium salts to azoxy compounds *Tetrahedron. Lett.* (1997) 38, 9001.

[28] Hiramoto K, Ryuno Y, Kikugawa K. Decomposition of N-nitrosamines, and concomitant release of nitric oxide by Fenton reagent under physiological conditions. *Mutat. Res.* (2002) 26,103-111.

[29] Chakrapani H, Bartberger MD, Toone EJ. C-nitroso donors of nitric oxide. *J. Org. Chem.* (2009) 74,1450-1453.

[30] Calder A, Forrester AR. Hepburn SP 2-Methyl-2-nitrosopropane and its Dimer. *Organic Syntheses, Coll.* Vol. 6, p.803; Vol. 52, p.77.

[31] Kirilyuk IA, Utepbergenov DI, Mazhukin DG, Fechner K, Mertsch K, Khramtsov VV, Haseloff RF. Thiol-Induced Nitric Oxide Release from 3-Halogeno-3,4-dihydrodiazete 1,2-Dioxides. *J. Med. Chem.* (1998) 41, 1027.

[32] Paton RM. 1,2,5-Oxadiazoles. *Compr. Heterocyclic Chem. II* (1996) 4, 229.

[33] Medana C, Ermondi G, Fruttero R, Di Stilo A, Ferretti C, Gasco A. Furoxans as Nitric Oxide Donors. 4-Phenyl-3-furoxancarbonitrile: Thiol-Mediated Nitric Oxide Release and Biological Evaluation A. *J. Med. Chem.* (1994) 37, 4412-4416.

[34] Sako M, Od S, Ohara S, Hirota K, Maki Y. Facile Synthesis and NO-Generating Property of 4*H*-[1,2,5]Oxadiazolo[3,4-*d*]pyrimidine-5,7-dione 1-Oxides *J. Org. Chem.* (1998) 63, 6947.

[35] Pappport Z, Leabman JF. *The Chemistry of Hydroxylamines, Oximes and Hydroxamic Acids,* Part 1 Wiley 2008.

[36] Chern JW, Leu YL, Wang SS, Lou R, Lee CF, Tsou PC, Hsu SC, Liaw YC, Lin HW. Synthesis and Cytotoxic Evaluation of Substituted Sulfonyl-*N*-hydroxyguanidine Derivatives as Potential Antitumor Agents. *J. Med. Chem.* (1997) 40, 2276.

[37] Fukuto JM, Wallace GC, Hszieh R, Chaudhuri G. Chemical oxidation of N-hydroxyguanidine compounds: Release of nitric oxide, nitroxyl and possible relationship to the mechanism of biological nitric oxide generation. *Biochem. Pharmacol.* (1992) 43, 607.

[38] Taira J, Misik V, Riesz, P. Nitric oxide formation from hydroxyl-amine by myoglobin and hydrogen peroxide. *Biochim Biophys Acta.* (1997), 1336, 502-5O8.

[39] Irisa T, Hira D, Furukawa K, Fujii T. Reduction of nitric oxide catalyzed by hydroxylamine oxidoreductase from an anammox bacterium. *J. Biosci. Bioeng.* (2014) 118, 616-621.

[40] Morphy R, Rankovic Z. Designed multiple ligands. An emerging drug discovery paradigm. *J. Med. Chem.* (2005) 48, 6523-6543.

[41] Gasco A, Boschi D, Chegaev K, Cena C, Di Stilo A, Fruttero R, Lazzarato L, Rolando B, Tosco P. Multitarget drugs: Focus on the NO-donor hybrid drugs. *Pure Appl. Chem.* (2008). 80, 1693-1701.

[42] Qandil AM. Prodrugs of Nonsteroidal Anti-Inflammatory Drugs (NSAIDs), More Than Meets the Eye: A Critical Review. *Int J Mol Sci.* (2012) 13, 17244-17274.

[43] Abdellatif KR, Chowdhury MA, Dong Y, Das D, Yu G, Velazquez CA, Suresh MR, Knaus EE. Dinitroglyceryl and diazen-1-ium-1,2-diolated nitric oxide donor ester prodrugs of aspirin, indomethacin and ibuprofen: Synthesis, biological evaluation and nitric oxide release studies. *Bioorg. Med. Chem. Lett.* (2009) 19, 3014-3018.

[44] Kaur J, Bhardwaj A, Huang Z, Knaus EE. Aspirin analogues as dual cyclooxygenase-2/5-lipoxygenase inhibitors: Synthesis, nitric oxide release, molecular modeling, and biological evaluation as anti-inflammatory agents. *ChemMedChem* (2012) 7, 144-150.

[45] Khosrow Kashfi. Nitric Oxide-Releasing Hybrid Drugs Target Cellular Processes Through S-Nitrosylation. *Dis Therap.* (2012) 3, 97-108.

[46] Boschi D, Tron GC, Lazzarato L, Chegaev K, Cena C, Di Stilo A, Giorgis M, Bertinaria M, Fruttero R, Gasco A. NO-Donor Phenols: A New Class of Products Endowed with Antioxidant and Vasodilator Properties *J. Med. Chem.* A (2006) 2886-2897.

[47] Rigas B, Kashfi K. Nitric-oxide-donating NSAIDs as agents for cancer prevention. *Trends Mol. Med.* (2004) 10, 324-330.

[48] Thatcher GR, Nicolescu AC, Bennett BM, Toader V. Nitrates and NO release: Contemporary aspects in biological and medicinal chemistry. *Free Radic. Biol. Med.* (2004) 37, 1122-114.

Chapter 3

NITROSYL IRON COMPLEXES

ANNOTATION

This chapter highlights advances made in the area of metal nitrosyl complexes. The chapter focuses on synthesis and physicochemical properties of isolated metal nitrosyl complexes with potential biological effects and their plying role in NO storage and transport. The direct interaction of NO with the heme group of heme proteins, including myoglobin, hemoglobin nitric oxide synthase, guanylyl cyclase, cytochrome P450, catalase, and peroxidase enzymes, appears to be a primary impotence in the molecule impact in biologic and physiologic processes. Discovery of various nitrosyl dithiocarbamate complexes and detail studies of its structure and chemical reactions have paved a way for wide and effective application of these compounds for solving actual problems in biochemistry, biomedicine, and physiology.

3.1. NITRIC OXIDE HEMIN COMPLEXES

In a metalloporphyrin nitrosyl complex, charge transfer from the π* NO orbital to the metal provides the triple-bond character of the N–O bond and a linear M–NO bond (~180°) [1]. A net charge transfer from a low valent metal can give a coordinated nitroxyl anion (NO-) for which a M–N–O bond angle of ~120° is anticipated. Figures 3.1 and 3.2 show cases of NO binding to a metalloporphyrin center.

DFT calculations have elucidated fundamental aspects of the structure and bonding of a variety of metalloporphyrin-diatomic complexes, including NO [2] Results of the calculation are presented in Figure 3.3.

Association rate constants of NO with heme proteins ranges from 2 x 10^7 to 200 M^{-1} s^{-1} [1, 3, 4]. The dissociation rate constants ranging from 100 to 0.03 s^{-1} were found to be for the ferric species and at least 1000-fold slower for the ferrous form. For example, the rate constant of recombination of heme with NO was found to be 2.5 x IO^{7} M^{-1} s^{-1}' for hemoglobin and 1.7 x 10^{6} M^{-1} s^{-1}' for myoglobin, correspondingly [3]. The following values of the dissociation rate constants (in s^{-1}) have been reported: heme–NO (2.1 $\times 10^{-5}$) Mb–NO (8.0 $\times 10^{-5}$) and Hb–NO (1.8 $\times 10^{-5}$) [4]. The rate constant of irreversible trapping reaction

$$HbO_2 + NO\ Hb^+ + NO_3^-$$

was estimated as $k = (3{-}5) \times 10^7$ M^{-1} s^{-1} [4].

In view of involving nitrite in many biological processes, the reductive nitrosylation ($Fe^{III}(P) + 2NO + H_2O = Fe^{II}(P)(NO) + NO_2^- + 2H^+$) of the ferriheme models, Fe^{III}(TPPS) (TPPS = tetra[4-sulfonatophenyl] porphyrinato), for example, and met-myoglobin causes special interest [5]. This process in aqueous solutions can be catalyzed by general base (Figure 3.4) in the presence the nitrite ion, NO_2^- as an autocatalyst in the reductive nitrosylation of ferriheme compounds or in the reductive nitrosylation of ferriheme compounds [5].

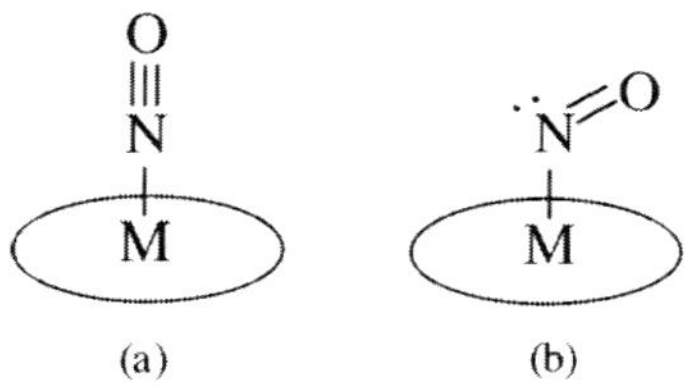

Figure 3.1. Illustration of limiting cases of NO binding to a metalloporphyrin center as (a) the nitrosyl cation (NO+) with a M–N–O bond angle of ~180° or as (b) the nitroxyl anion (NO-) with a M–N–O bond angle of ~120° [1]. Copyrights 2005 American Chemical Society.

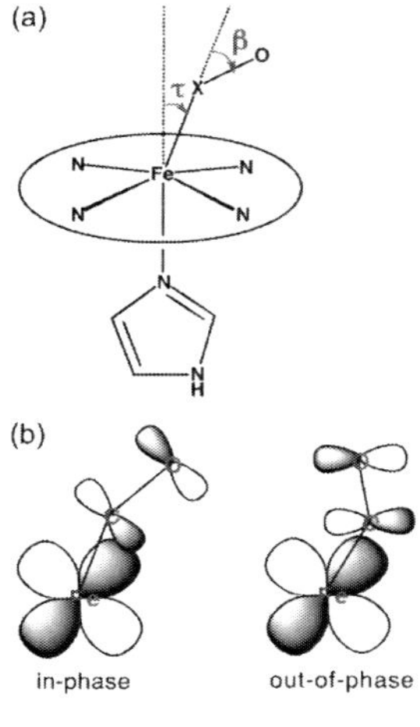

Figure 3.2. (a) Definition of the tilt and bend angles, (b) in- and out-of-phase tilting-and-bending deformations [2]. Copyrights 2005 American Chemical Society.

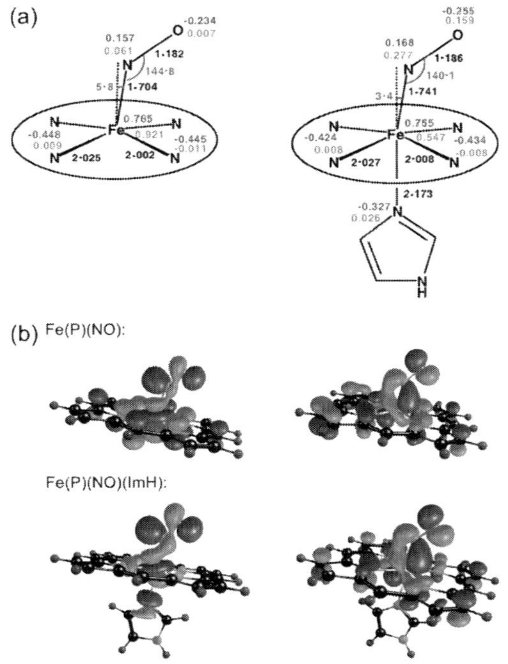

Figure 3.3. Calculated results on Fe(P)(NO) and Fe(P)(ImH)(NO). (a) Optimized distances (Å, in black), angles (deg, in red), Mulliken charges (in blue), and spin populations (in magenta). (b) Selected MOs involving Fe(d)-NO(ð*) bonding. The MOs to the left are the SOMOs, while those to the right are doubly occupied [2].

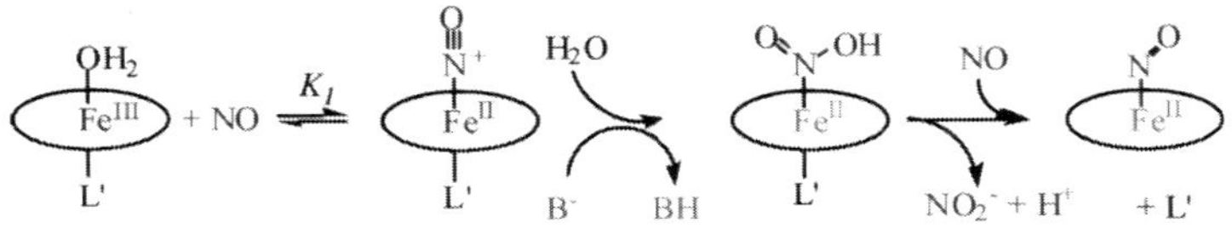

Figure 3.4. General base catalysis of the reductive nitrosylation of ferriheme compounds [5]. Copyrights 2005 American Chemical Society.

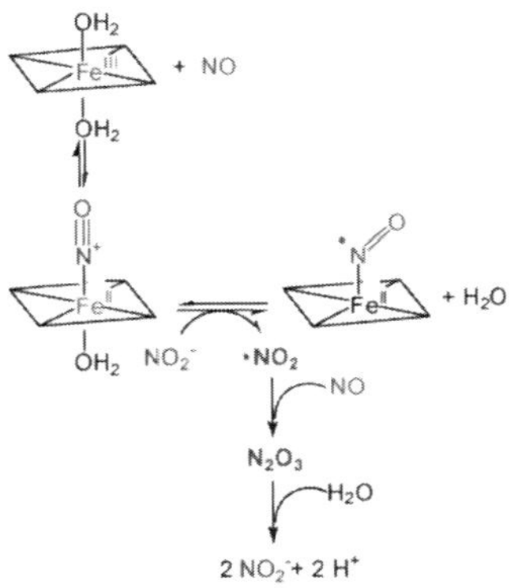

Figure 3.5. Outer-sphere electron-transfer mechanism for nitrite catalysis [5]. Copyrights 2005 American Chemical Society.

The nitrite catalysis can occur via the outer sphere (Figure 3.5) or inner (Figure 3.6) sphere electron-transfer mechanisms.

The investigation of the reaction of NO with ironheme models is also of interest, owing to their potential participation in biological processes and use as NO traps. Principle possibility of reaction of NO with the nitrato iron(III) complex $Fe^{III}(TPP)(\eta^2\text{-}O_2NO)$ (1, TPP = meso-tetraphenyl porphyrinate^{2-}) was demonstrated in [6]. The reaction of NO with ferriheme model $Fe^{III}(TPPS)(H_2O)_2$ (TPPS = tetra[4-sulfonato-phenyl]porphinato) in solution to form $Fe^{III}(TPPS)(NO)$

$$Fe^{III}(TPPS)(H_2O)_2 + NO \rightleftharpoons Fe^{III}(TPPS)(H_2O)(NO) + H_2O$$

was found to be rapid (k_{NO} = 4.5 × 10^5 M^{-1} s^{-1}) and characterized by the following thermodynamic parameters: ΔH = ~65 kJ mol^{-1}, ΔS = ~60 J mol^{-1} K^{-1}, and ΔV = ~20 cm^3 mol^{-1} [7]. For the back reaction the following values were found: ΔH = 76 kJ mol^{-1}, ΔS = ~41 J mol^{-1} K^{-1}, and ΔV. = 20 cm^3 mol^{-1}). For comparison, the reaction for the ferro-heme model $Fe^{II}(TPPS)$ is significantly faster (k_{NO} = 1.5 × 10^9 M^{-1} s^{-1}). For the anionic TPPS complex, the value of equilibrium constant K_{NO} for formation of the ferric nitrosyl complex

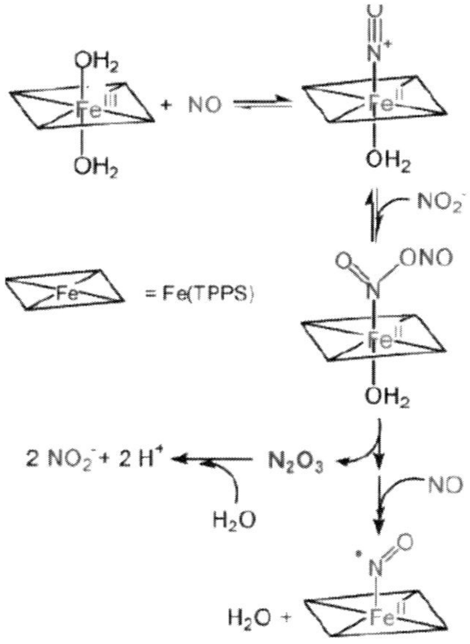

Figure 3.6. Inner-sphere mechanism for nitrite catalysis [5]. Copyrights 2005 American Chemical Society.

$$Fe^{III}(P) + NO \xrightarrow{K_{NO}} Fe^{III}(P)(NO)$$

was determined to be 1.32 × 10^3 M^{-1} (μ = 0.10 M, 298 K) [5].

The heam group of NO synthase occupied a key position in the enzyme active center. Experiments on flash photolysis and stopped-flow spectrophotometry at 23. 6 °C showed that NO reacts rapidly ($ka > 2 \times 10^7\ M^{-1}\ s^{-1}$) with neuronal NOS in both its ferric and ferrous oxidation states [8]. The dissociation rate constant was found to be less than $10^{-4}\ s^{-1}$ for the ferrous form of heam and about 50 s^{-1} for the ferric form. Protein pocket cavities trap small amounts NO in nanoseconds, a much larger in picosecond time scales. After rapid photodissociation of the nitrosyl heme complex, a molecule of NO in one of these cavities might collide many times before escaping into the solvent (geminate recombination effect). Interestingly, that role of internal cavities was also pointed out for the reaction myoglobin with NO [9]. The crystallographic model indicates two of the cavities inside the protein matrix (secondary docking sites). A molecule of NO in one of these cavities might diffuse inside the protein and, before escaping into the solvent, might collide many times per microsecond with the iron bound O_2. This increase in collision frequency by internal diffusion might enhance reactivity and lead to formation of the initial product peroxynitrite, which rapidly isomerizes to nitrate. Another example of important ferriheme proteins are nitrophorins 1–4 (NP1–4), which transport NO in biological systems [10]. For different nitrophorins, the initial association is fast ($k_a = 1.5–33\ \mu M^{-1}\ s^{-1}$), and equilibrium is strongly shifted to heme-NO complex (Keq = 1–850 nM)

The study the ultrafast reaction of movement of the heme iron induced by NO binding to hemoglobin (Hb) and myoglobin (Mb) is an excellent example of application of advance physical techniques to process of formation of nitrosylheme complexes [11]. The process was followed by the picosecond spectral evolution of absorption band III (~760 nm) and vibrational modes (iron–histidine stretching, ν_4 and ν_7 in-plane modes) in time-resolved resonance Raman spectra (Figure 3.7). The time constants of band III intensity kinetics induced by NO rebinding were found to be as 25 ps and 40 ps for Hb and Mb, respectively.

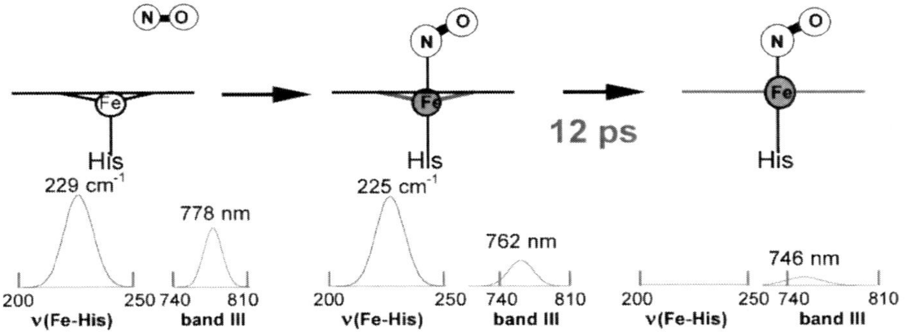

Figure 3.7. Illustration of heme iron motion triggered by nitric oxide binding to hemoglobin and myoglobin probed by picosecond spectral resonance Raman time-resolved techniques [11]. Copyrights 2012 American Chemical Society.

3.2. Iron Complexes with Thiol-Containing Ligands

3.2.1. General

Iron complexes with thiol-containing ligands complexes can be classified by the following properties: number of iron atoms, number of NO molecules, chemical nature of bonds iron thiol ligands, and solubility and nonsolubility in water. Bridging the two iron atoms by two sulfur atoms is denoted as "μ-S", while the term "μ-N-C-S" describes bridging via atoms sulfur and nitrogen. To describe the number of d-type electrons present in a complex, the Enemark-Feltham notation is used [12]. For example, the dimeric reduced Roussin's red ester (rRRE)

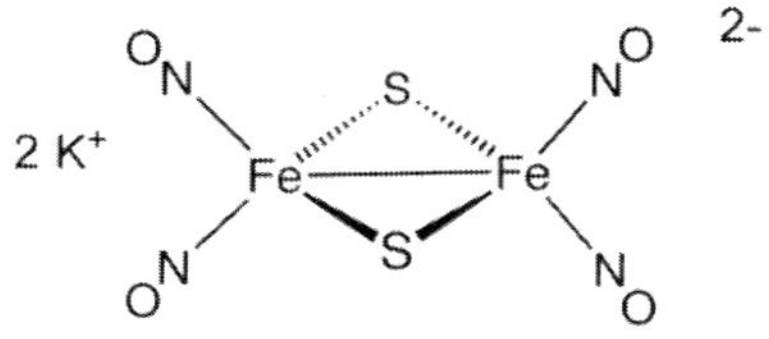

are designated as $\{Fe(NO)2\}^9$ and are considered as the homodinuclear form of a dinitrosyl iron complex (DNIC).

The coordination chemistry of these complexes has expanded rapidly involving set synthetic methods and the use of an arsenal modern, physicochemical techniques, such as X-ray crystallography, electron paramagnetic resonance (EPR), nuclear magnetic resonance (NMR), optical absorption, infrared (IR), Mössbauer spectroscopy, mass spectrometry, electrochemistry, nuclear resonance vibrational spectroscopy (NRVS), stopped-flow optical spectroscopy, chemical kinetics, and advance theoretical calculations [29–34].

In view of the large diversity of currently available nytrosyl iron sulfur complexes, in this section we confine ourselves to a few typical examples.

3.2.2. Mononitrosyl Iron Dithiocarbamate Compounds

Starting from the 1960 pioneering works of Vanin and Blumenfitld with colleagues [13, 14], iron(II)–dithiocarbamate complexes were widely employed *in vivo* and *in vitro* as traps of NO in biological samples [15–19].

The mechanism of formation of EPR-detectable mononitrosyliron complexes (MNIC) with dithiocarbamate derivatives

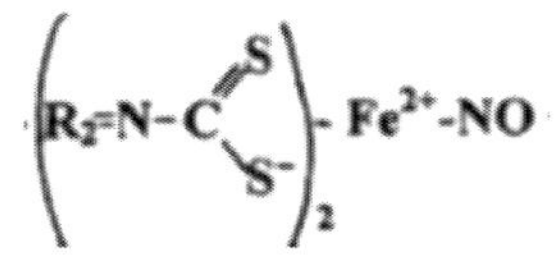

with dithiocarbamate derivatives was described in [20].

Both iron(II)– and iron(III)– -dithiocarboxy (dtc) complexes,

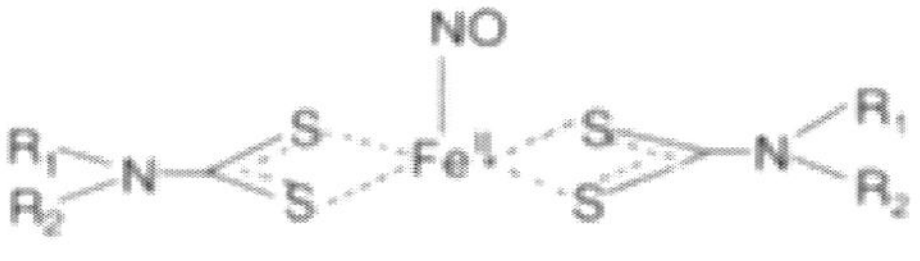

which were synthesized and investigated in [17], react rapidly (k = 10^8 M^{-1} s^{-1}) with NO to produce a corresponding paramagnetic nitrosylferrate (II,III) complex [17]. Thermodynamic and kinetic properties of the nitrosylferrate complex were intensively studied. As an example, the electrode potentials of iron complexes of N-(dithiocarboxy)sarcosine (DTCS) and N-methyl-d-glucamine (MGD) dithiocarbamate were found to be -56 and -25 mV at pH 7.4, respectively. For the FeII(dtcs)2 and FeII(mgd)2, data on the rate constant of the autoxidation reaction with hydrogen peroxide and with reductants were also presented in [17].

3.2.3. Dinitrosyl Dithiolate Iron Complexes with Thiolate Ligands

The formation of paramagnetic non-heam iron dinitrozyl complexes in biological system (mouse liver) first has been detected in [21] and found the greatest usefulness in cells and tissues investigation [22-30]. Mononuclear dinitrosyl iron complexes (M-DNIC) with thiolcontaining ligands were identified by the characteristic EPR signal at g = 2.03. The following schemes of the mechanisms of conversion of Be and M-DNIC with thiol-containing ligands into MNIC with dithiocarbamate derivatives, possessing of the iron-dinitrosyl fragment, [Fe(NO)2], were proposed and discussed in details [20–30]: (1) the ability to generate neutral NO molecules and nitrosonium ions (NO^+) [23]; (2) the reversible equilibrium between the Fe(NO)2 fragments and their constituent components [30]; (3) the Fe(NO)2 fragment formation during the interaction of iron atom with neutral molecules of NO [26]; and (4) the Fe(NO)2 fragment formation during the interaction of (1) and reduction mechanism of Fe(NO)2 fragment of M-DNIC with glutathione [30]. Reduction mechanisms of Fe(NO)2 fragments of B-DNIC with thiol-containing ligands and of Fe(NO)2 fragment of M-DNIC with thiol-containing ligands and corresponding change in its electronic configuration d^7 to d^9 was also considered [30].

In the nitrosyl iron complexes, the diamagnetic group, $Fe^{2+}(NO)_2$ with electron configuration d^8, converts into a paramagnetic $Fe^{+}(NO^{+})_2$ group as a result of disproportionation of NO ligands and substitution of newly generated NO^- for NO [23]. High nitrosylating activity of nitrosonium ions in the $\{(RS^-)_2Fe^+(NO^+)_2(^-SR)_2\}^-$ complex provides ability to induce *S*-nitrosylation of thiols. The mechanism of $Fe^{2+}(NO)_2$ transition to this configuration during the interaction of iron atoms with two molecules of NO was proposed.

The ability of mononuclear dinitrosyl iron complexes (M-DNICs) with thiolate ligands to act as NO donors and to trigger S-nitrosation of thiols was demonstrated in [26]. The distribution of unpaired electron density in M-DNIC corresponds to the low-spin (S = 1/2) state with a d^7 electron configuration of the iron atom and predominant localization of the unpaired electron on MO (d_{Z2}) was evaluated. In such a structure, the positive charge on the nitrosyl ligands diminishes, and weak binding of thiolate ligands to the iron atom can occur.

Binuclear dinitrosyl iron complexes (B-DNIC) was detected by optical spectroscopy in animal tissues in the presence glutathione using absorption and ESR spectroscopies [31]. In such a condition, the transformation of the optical absorption spectrum of B-DNIC and appearing of EPR signals of MNIC-MGD were detected. Binding of two iron-dinitrosyl [Fe (NO)] fragments in B-DNIC monitored by detection of absorption spectra MNIC caused spin pairing with antiferromagnetic interaction. Mechanisms of generation MNIC from dinitrosyl iron complexes thiolcontaining ligands and conversion of B- and M-DNIC with thiol-containing ligands into MNIC with dithiocarbamate derivatives were proposed. Both M- and B-DNIC with thiol-containing ligands have the ability to act as donors of NO molecules and nitrosonium ions (NO^+) [29].

The water-soluble neutral $\{Fe(NO)_2\}^9$ DNIC $[(S(CH_2)_2OH)\ (S(CH_2)_2NH_3)Fe(NO)_2]$ (DNIC 2) complex

2

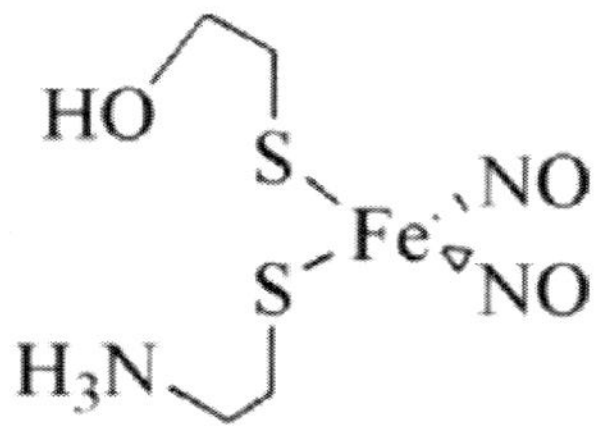

exposing ESR signal was obtained by conversion of dinuclear $\{Fe(NO)_2\}^9$-$Fe(NO)_2\}^9$. [32]. DNIC 2 is stable in tetrahydrofuran THF and was characterized by IR, UV−vis, EPR (Figure 3.8), and single-crystal X-ray diffraction.

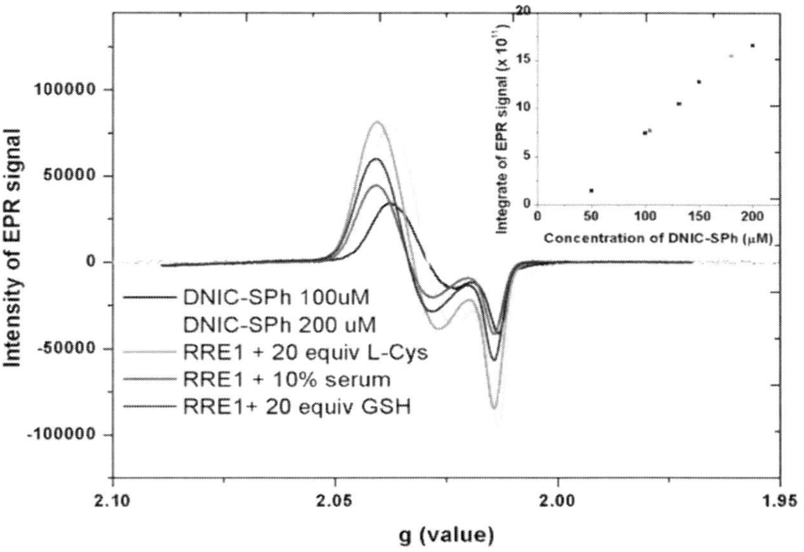

Figure 3.8. EPR spectra derived from reaction of RRE 1 and 20 equiv of l-cysteine, 20 equiv of reduced glutathione (GSH), and 10% serum in PBS, respectively, and standard DNIC [(PhS)2Fe(NO)2]− (100 μM, 200 μM [32]. Copyrights 2016 American Chemical Society.

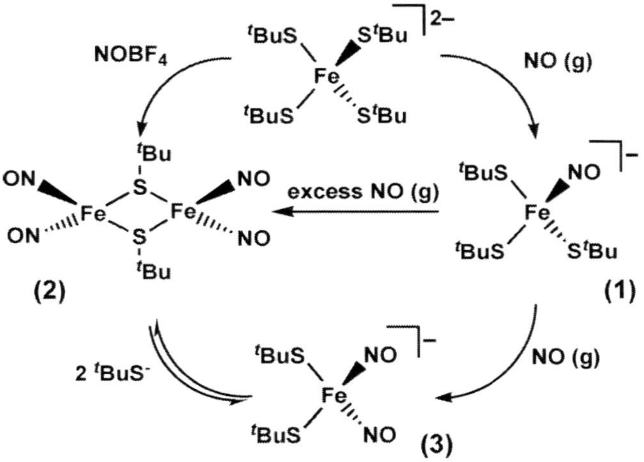

Scheme 3.1. Scheme of synthesis of complex DNIC [Fe(StBu)2(NO)2] [35].

3.2.4. Dinuclear Complexes

Synthesis of mono- and dinuclear nitrosyl complexes, including (PhS)2Fe(NO)2 Fe(H⁺bme-daco)(NO)2 (H+bme-daco - N-protonated bismercapto-ethanediazacyclooctane), N2S2- Fe(NO and {(í-SRS)[Fe2(NO)4]}n (n)1, 2), has been reported [35]. The nitrosyl complexes 1 and 2 can be chemically converted to the DNIC [Fe(StBu)2(NO)2]- (Scheme 3.1).

A series of Roussin's red salt esters [$Fe_2(\mu\text{-}RS)_2(NO)_4$] (R = n-Pr (1), t-Bu (2), 6-methyl-2-pyridyl (3) and 4,6-dimethyl-2-pyrimidyl (4)) were synthesized by the reaction of $Fe(NO)_2(CO)_2$ with thiols or thiolates [36]. Complexes 1–4 were characterized by IR, UV-vis, 1H-NMR, electrochemistry, and single-crystal X-ray diffraction analysis. Molecular structure of a Roussin's red salt esters complex is given in Figure 3.9.

A strong spin exchange (the antiferromagnetic coupling) between the two iron centers is expected because relative short Fe(1)–Fe(1a) distance of ca. 2.70 Å in the complex [36]. As a consequence, Roussin's red salt esters are diamagnetic and EPR silent and demonstrate NMR spectra. Only reduced species, [$Fe_2(\mu\text{-}RS)_2(NO)_4]^-$, exhibits an isotropic signal ESR at g = 1.998–2.004 without hyperfine splitting in the temperature range 180–298 K. At 110 K, this complex displays an axial EPR signal at $g_\perp$ = 2.007 and $g_\parallel$ = 1.916 (Figure 3.10).

According to the author's calculation, as one can see in Figure 3.11, on the singly occupied molecular orbit (SOMO) for the complexes, 60–63% of the electron delocalize on two iron atoms, 25.0% on two sulfur atoms, and only 2% four NOs.

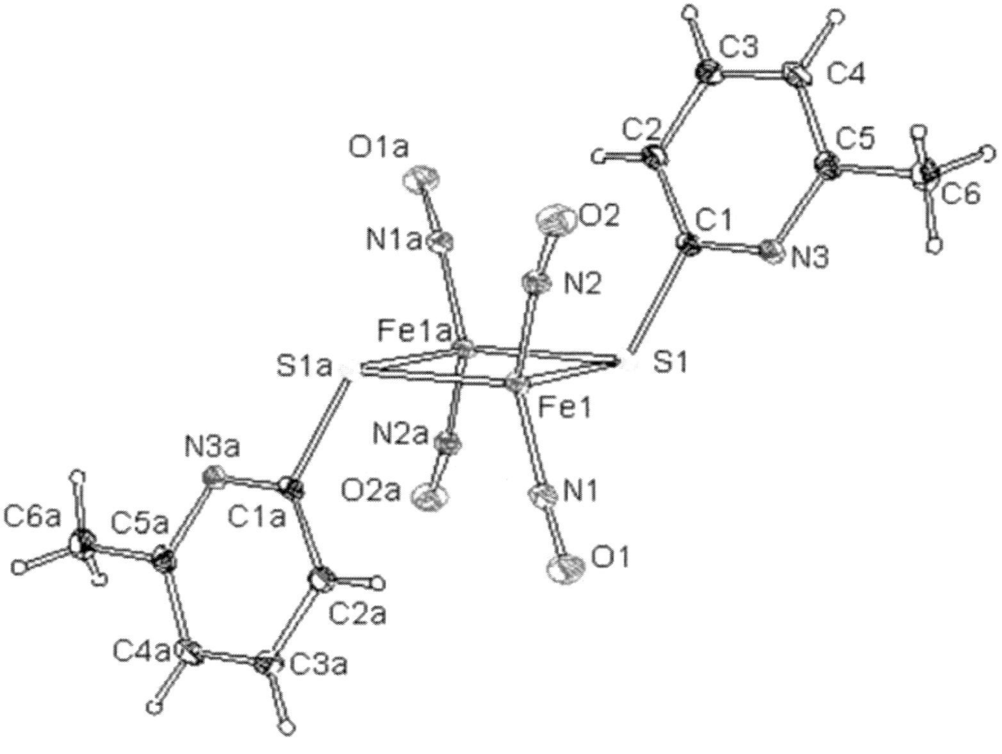

Figure 3.9. Molecular structure of complex 3 with thermal ellipsoids drawn at the 30% probability, symmetry code: a = 1 −x, 1 −y, −z [36]. With permission from the Royal Chemical Society.

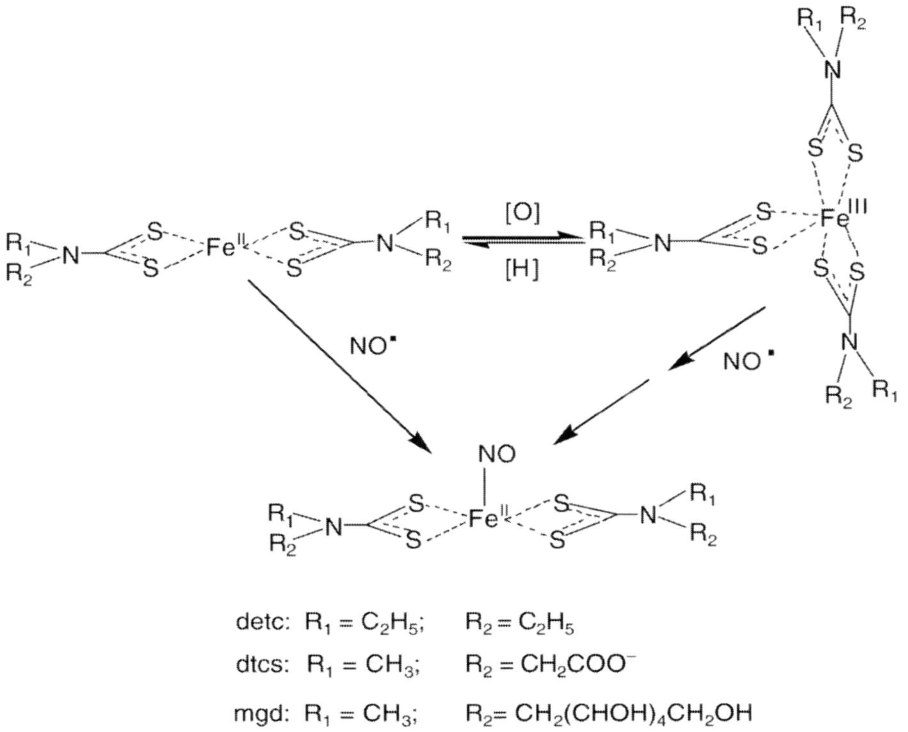

Figure 3.10. (a) EPR spectra of complex 1− at 180 K and (b) complex 1− at 110 K [36]. With permission from the Royal Chemical Society.

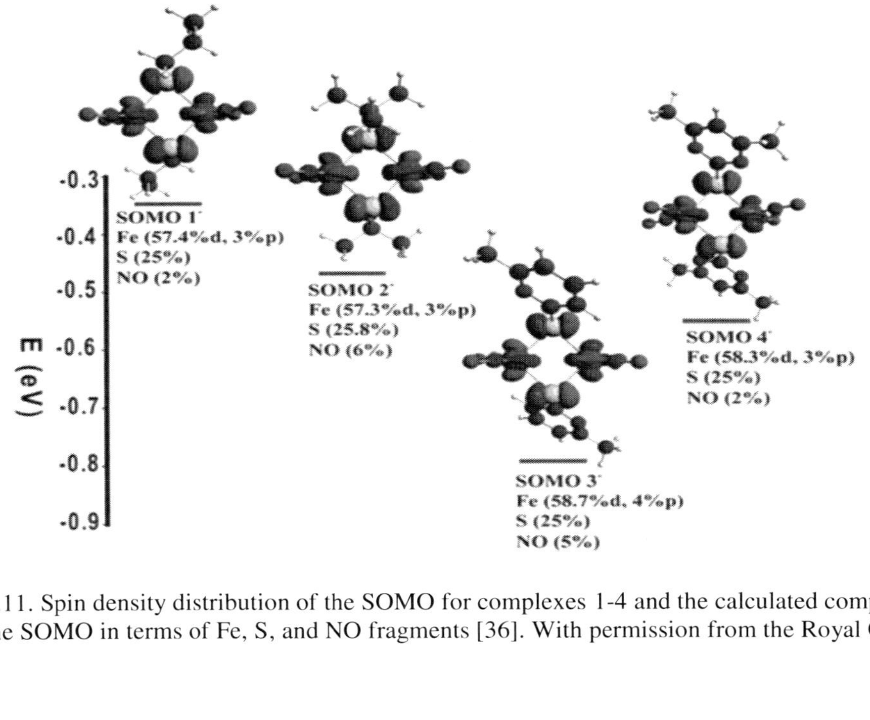

Figure 3.11. Spin density distribution of the SOMO for complexes 1-4 and the calculated composition (%) of the SOMO in terms of Fe, S, and NO fragments [36]. With permission from the Royal Chemical Society.

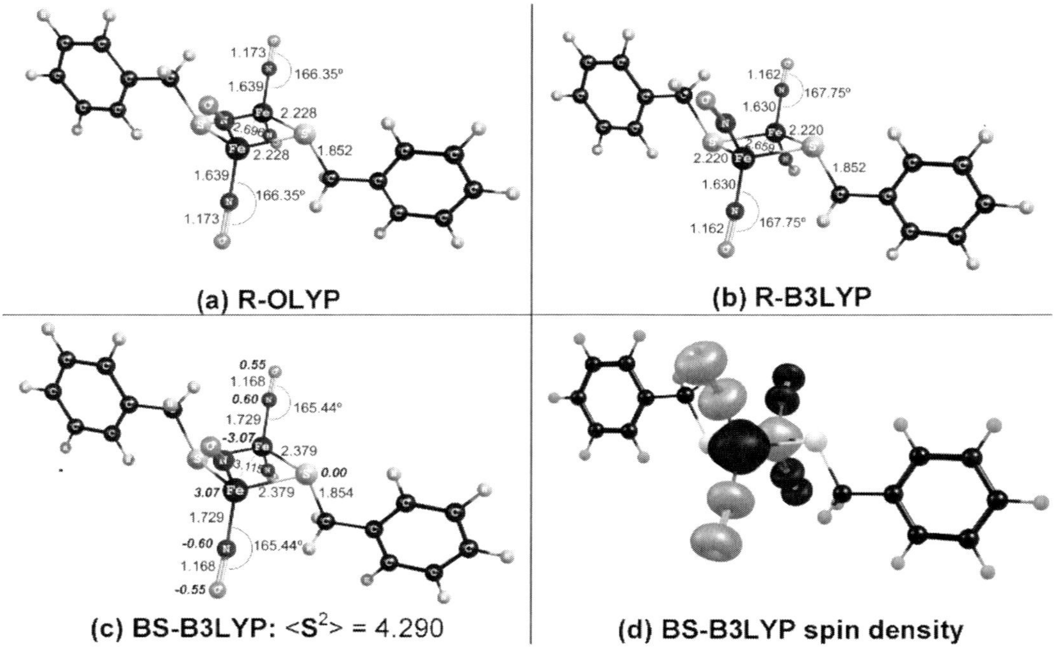

Figure 3.12. Selected DFT/6-311G(d,p) results for Roussin's red salt ester. Distances (Å) and Mulliken spin populations are indicated in normal and bold fonts, respectively. A contour of 0.03 e/Å3 has been used for the spin density plot [37]. Copyrights American Chemical Society 2009.

Three nitrosylated binuclear clusters [Fe2(NO)2(Et-HPTB) (O2CPh)]$^{2+}$ (1), Et-HPTB=N,N,N',N' -tetrakis-(N-ethyl-2-benzimida-zolylmethyl)-2-hydroxy-1,3-diaminopropane), [Fe(NO)2{Fe(NO)(NS3)}-S,S'] (2), and Roussin's red salt anion [Fe2(NO)4(μ-S)2]$^{2-}$ (3) (Figure 3.12) were studied using broken-symmetry density functional theory (DFT, chiefly OLYP/STO-TZP) calculations [37]. In addition, the Noodleman and Yamaguchi formulas were used to evaluate the Heisenberg coupling constants (J), where the J values refer to the following Heisenberg spin Hamiltonian: H=JSA·SB. (JAB = 184 cm^{-1}).

Figure 3.13 illustrates Heisenberg spin ladder for the Roussin's red salt ester complex (1) [37].

The calculation suggests that the nitrosylated iron–sulfur clusters feature some exceptionally high *J* values relative to the non-nitrosylated {2Fe2S} and {4Fe4S} clusters: the Heisenberg *J* for **1** is small ($\approx 10^2$ cm^{-1}), while complexes **2** and **3** exhibit *J* values that are at least an order of magnitude higher ($\approx 10^3$ cm^{-1}).

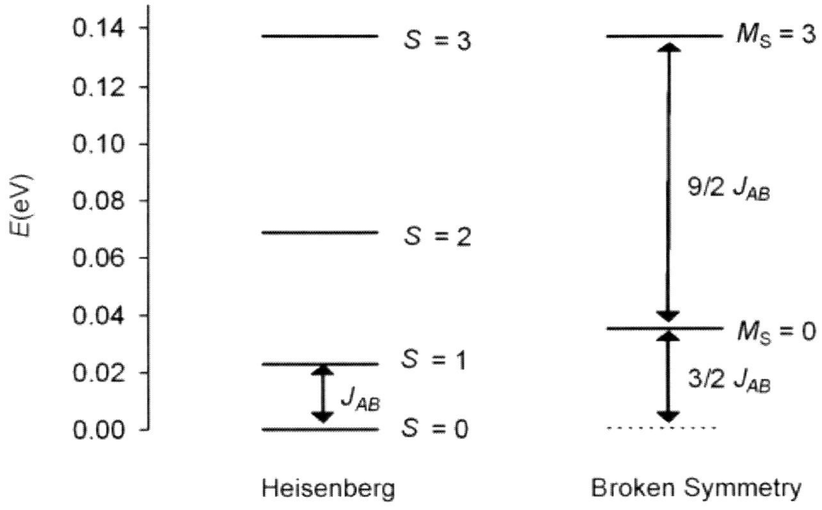

Figure 3.13. Heisenberg spin ladder for (1) showing the relative positions of the pure and broken-symmetry spin states (JAB = 184 cm−1) [37]. Copyrights 2009 American Chemical Society.

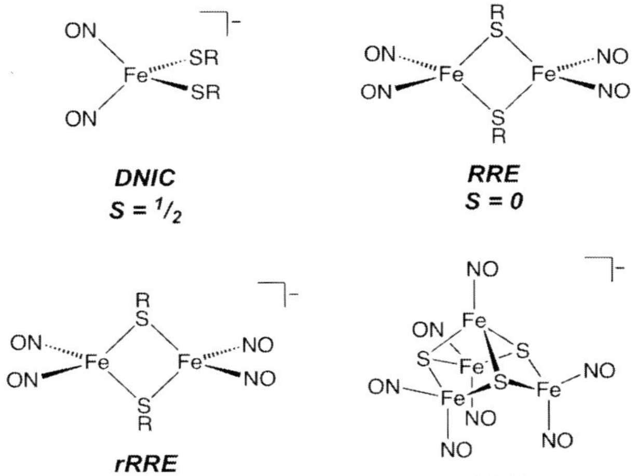

Figure 3.14. Iron initrosyl species of different numbers of NO and spin S [38]. Copyrights 2012 American Chemical Society.

3.2.5. Polynuclear Nitrosyl Complexes

Chemical structures of chosen polynuclear nitrosyl complexes of various number of NO molecules and different spin are presented in Figure 3.14 [38].

Schematic representation of Heisenberg coupling constants in Roussin's black salt anion, [Fe4(NO)7(μ3-S)3]− (2) and of the four broken-symmetry states (MS = 0, 1, 2, 3), are shown in Figures 3.15 and 3.16, respectively [39]. In the Figure, J12 corresponds to the interaction between the apical iron and a basal iron, and J22 refers to that between any two basal iron centers. The basal–basal coupling constant J22 was found to be small (≈ 102 cm−1); the apical–basal coupling constant J12 is some forty times higher (≈ 4000 cm−1).

For complex (2), the calculation indicates Fe spin populations of −1.18 for the apical iron and 0.62 for each basal iron. The theoretical analysis yielded an electronic structure of this complex and corresponding the magnetic orbital pairs (Figure 3.17).

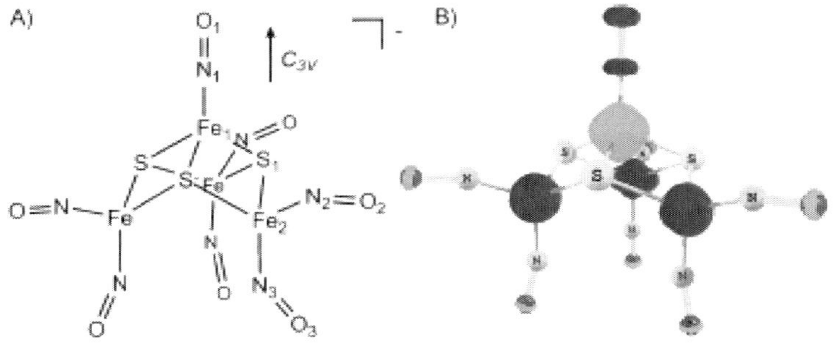

Figure 3.15. Roussin's black salt anion (2). A) Atom numbering and symmetry axis; B) spin density for the broken-symmetry MS = 0 geometry (ADF, OLYP/TZP, COSMO, C3v, contour value 0.02 eÅ−3) [39]. With permission from Wiley-VCH Verlag, 2010.

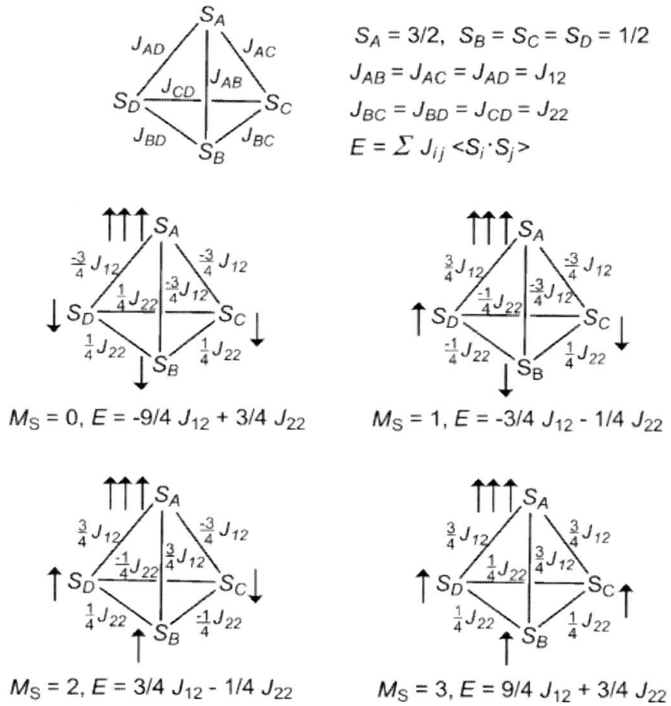

Figure 3.16. Schematic representation of Heisenberg coupling constants in 4 and of the four broken-symmetry states (MS = 0, 1, 2, 3). [39]. With permission from Wiley-VCH Verlag, 2010.

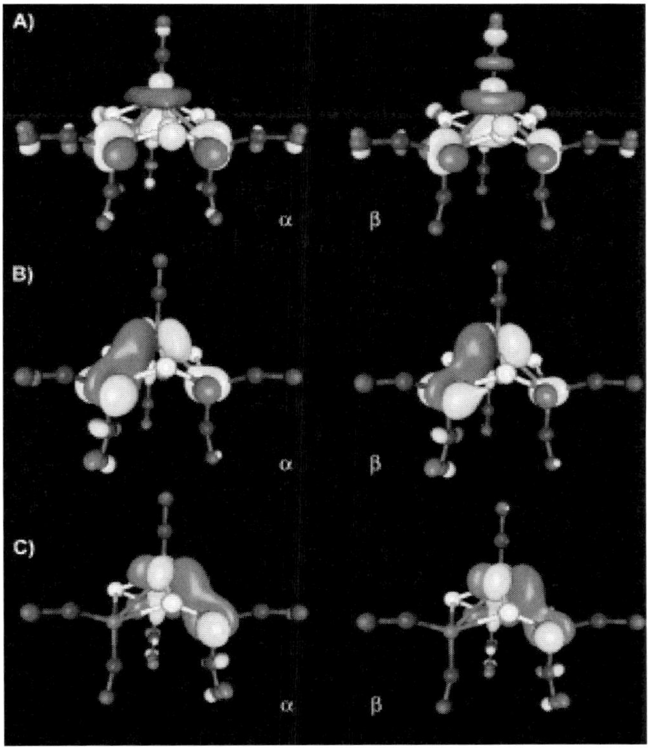

Figure 3.17. Magnetic orbital pairs in complex (2) (OLYP/TZVP, ORCA, contour value of 0.035). Overlap values: A) 0.909 (orbital pair no. 128); B) 0.946 (no. 127); C) 0.946 (no. 126). (See details in [39]. With permission from Wiley-VCH Verlag, 2010).

The following conclusions were formulated: (1) for the apical $S = 3/2$ $\{FeNO\}^7$ unit three unpaired electrons locate in the d_{z^2}, $d_{x^2-y^2}$, and d_{xy} orbitals; (2) the d_{xz} and d_{yz} electrons are involved in π-bonding interactions with NO; (3) coupling of these SOMO electrons to the d_{z^2}-based SOMOs of the basal irons leads to three magnetic orbital pairs with overlap values of 0.909, 0.946, and 0.946; (4) the antiferromagnetic coupling between iron and nitrosyl in the four FeNO units involves four electron pairs for each $\{Fe(NO)_2\}^9$ unit and two pairs for the $\{FeNO\}^7$ unit; (5) finally, for these 14 orbital pairs overlap values are from 0.987 to 0.997.

Examples of several "exotic" polynuclear nitrosyl complexes were presented in [40]:

2
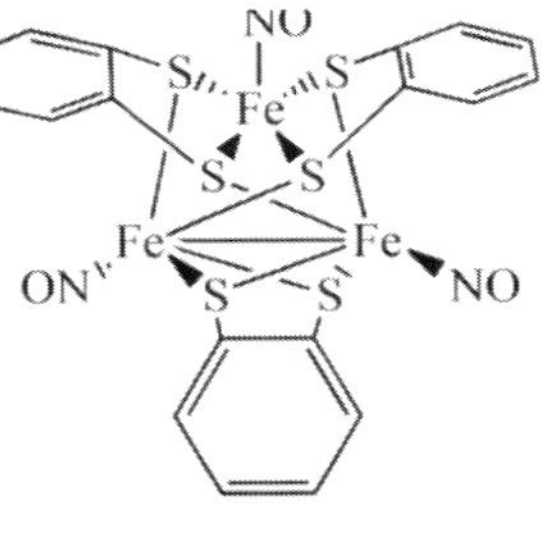

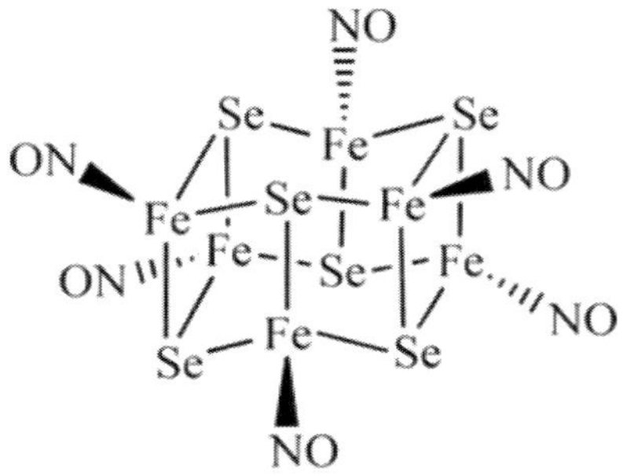

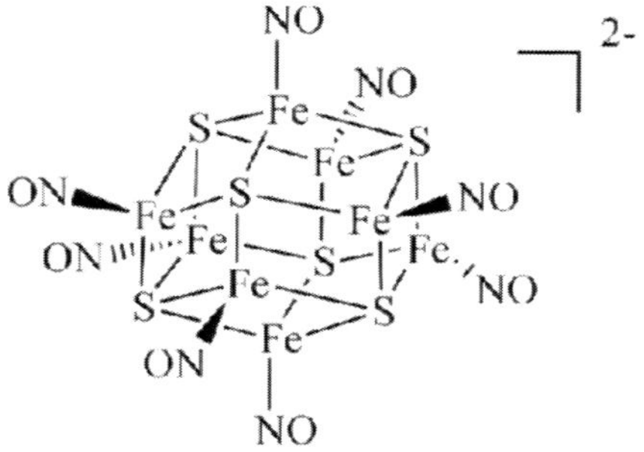

3.2.6. Nitrosyl Iron Complexes with Thiol Ligands, as Potential Precursors of Nitric Oxide

With a goal to develop strategy to establish relationship structure—physicochemical properties and to deliver NO to biological targets—Aldoshin and Sanina research group has synthesized and investigated by an arsenal of physical methods a number of nitrosyl iron complexes with thiol ligands, as potential precursors of NO [41–55]. The structure and physicochemical properties of the obtained compounds were studied by a complex of advance physical methods, such as the X-ray analysis, Mössbauer, absorption, IR, EPR spectroscopy, cyclic voltammograms (CVA), amperometry, and magnetochemistry as well as the chemical kinetics.

3.2.6.1. Mononuclear Complexes

The mononuclear complex $[Fe(SC_2H_3N_3)(SC_2H_2N_3)(NO)_2]\cdot 0.5H_2O$ (I) of the ESR spectrum g-factor "g = 2.03" family with protonated and deprotonated ligands, which provide a neutral structure and are connected *via* the sulfur atom, has been synthesized and characterized by X-ray diffraction, IR, EPR, Mössbauer spectroscopy, and magnetochemistry [41–43]. X-ray structures of the mononuclear sulfur-nitrosyl complex I and similar complexes were synthesized and investigated in [41]. The IR spectra of the ligand in the thione form exhibit characteristic "thioamide" bands at 1570–1395 (band I), 1420– 1260 (band II), 1140–940 (band III), and 800–700 cm–1 (band IV). The bands are mainly caused by N=C stretching vibrations, and N–H deformation vibrations. The C=S stretching vibrations also contribute considerably to the intensity of these bands. The Mössbauer spectrum of complex I has a doublet structure with the following parameters: δFe = 0.188(1) mm s^{-1}, Δ*E*Q = 1.118 (1) mm s^{-1}, and Γ = 0.258(2) mm s^{-1} at 296 K. The EPR spectrum of polycrystals of the complex 1 is typical of an axial anisotropy of *g*-factor ($g_{\perp}$ = 2.04, g|| = 2.02) at *T* = 100–300 K. The effective magnetic moment per iron atom was found as 1.77 μB. This value is near to the theoretical value for spin *S*= 1/2 (μ_{eff} = 1.73 μB).

From precision X-ray analysis of mononuclear dinitrosyl iron complex $[Fe(SC_2H_3N_3)(SC_2H_2N_3)(NO)_2]\cdot0.5H_2O$ at low temperatures, distribution maps of deformation electron density have been obtained [44]. The results indicated that Fe–S bonds can be described as interatomic interactions of a peak–peak type, while Fe–NO bonds as interactions of a peak–hole type. Geometries of cationic and neutral chlorine-containing DNICs in the complex were optimized by quantum chemical calculations, using the ORCA program suite by means of local functional BP86 employing TZVP basis. A new cationic DNIC with thiourea of the formula $[Fe(SC(NH_2)_2)_2(NO)_2]Cl\cdot H_2O$ (I) been prepared and its X-ray structure has been established [45]. The following physical parameters of the complex were reported: the EPR spectrum g-factors are $g_{\perp}$ = 2.04 and g_{II} = 2.02; the values of ^{57}Fe Mössbauer spectrum parameters are quadrupole splitting ΔEQ = 1.167 mm/s, isomeric shift δFe = 0.194 mm/s and the width of the absorption lines Γ = 0.243(1) mm/s at 296 K; the stretching vibrations values in IR spectra are ν_{NO}= 1807, 1744 cm^{-1}. Amount of NO released upon dissolution of complex I in aqueous solutions in anaerobic conditions at pH 7 at T = 25°C (concentration of II is 1.86 10^{-5} M) for 100 seconds was found to be 35 nM.

Various versions of the density functional method were employed for detail description of the electronic and molecular structures of the $Fe(NO)_2$ fragment in the $Fe(SC(NH_2)_2(NO)_2^+$ cation of mononuclear nitrosyl iron complex [46]. The Fe–NO bond energy in complex I of 30–40 kcal/mole was calculated using the selected functional, through the energy of the simplest reaction:

$$[Fe(SC(NH_2)_2)_2(NO)_2]^+ \rightarrow [Fe(SC(NH_2)_2)_2(NO)]^+ + NO$$

In the calculation, it was taken in consideration that for this structure, a through-atom index is strong. The orbital mixing of antibonding orbitals of the Fe–N and N–O bonds was suggested. Results showed that NO exists in the complex as a neutral ligand, the Fe–N bond is covalent polar and has 1.5 order, and, owning to strong p-interaction, the electron density is distributed along Fe–N–O.

A new analog of the active site of mononuclear dinitrosyl [1Fe–2S] proteins, $[C_3N_2H_8SFe(NO)_2Cl][Fe(NO)_2(C_3N_2H_8S)_2]^+Cl^-$ (III) (was synthesized and studied by X-ray diffraction, IR, Mossbauer, EPR spectroscopy, electrochemistry, and quantum chemical calculations [47]. The complex III synthesis was performed by reaction of NO with an aqueous mixture of iron(II) sulfate and N-ethylthiourea in acidic medium. The following Mossbauer spectrum parameters for complex I at 85 K were obtained: isomer shift δFe(1) = 0.308(1) mm s^{-1} and quadrupole splitting ΔEQ(1) = 1.031(3) mm s^{-1}. According to the experimental EPR spectra of III the number of spins per Fe ion is 0.76, that is correspond to the fraction of paramagnetic Fe ions with S = 1/2. Spontaneous generation of NO from complex III in protic media (in water and in 1% v/v water solution of DMSO) in anaerobic solutions at 25 °C and pH 7.0 was detected by the electrochemical method.

3.2.6.2 Binuclear Complexes

A series of neutral paramagnetic nitrosyl binuclear iron complexes with bridging azaheterocyclic ligands of "l-SCN"-type

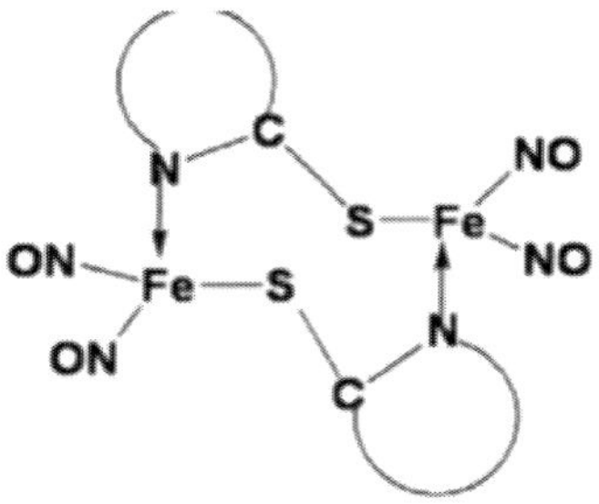

were synthetized [48]. The complexes reduction potential values were measured by cyclic voltammetry method. In addition, quantum-chemical modeling of the processes accompanying the reduction of the species was performed employing the Gaussian 09 program, version D by means of local functional BP86 and 6-311++G//6-31G basis. The experimental and theoretical results allowed to prediction of the stability of the forming anions and suggest mechanism of transferring the neutral complexes of this structural type into ionic forms

Geometrical and electronic structures of neutral paramagnetic binuclear nitrosyl iron complexes with azaheterocyclic thiolyls [$Fe_2(\mu\text{-}SR)_2)(NO)_4$], bridging ligands: aminomercaptotriazolyl, R = $C_2N_3H(NH_2)$ (**1**), mercaptoimidazolyl, $RC_3N_2H_3$ (**2**), methylmercaptoimidazolyl, R = $C_3N_2H_2CH_3$ (**3**), and dihydromercaptoimidazolyl, R= $C_3N_2H_5$ (**4**) have been calculated by the methods of density functional, B3LYP, and PBE [49]. The calculation found that coordination of bridging ligands corresponds to S–C–N type. The complexes' effective magnetic moment was estimated at about 2.5 Bohr magneton. The interaction of Fe atoms spins is involved in antiferromagnetic coupling with corresponding spin exchange. The electronic configuration of the $Fe(NO)_2$ unit with one unpaired electron forms due to binding of spin 3/2 of Fe^+d^7 center with oppositely oriented spins 1/2 of two NO groups was theoretically predicted. Theoretical structures of the complexes agree with the complexes' experimental X-ray structures. The calculated IR spectrum was obtained by Lorentz convolution with half-width 2 cm^{-1}. Experimental data on correlation of isomeric shift of the Mossbauer spectra of dinuclear nitrosyl complexes with Fe-S distance, obtained by X-ray analysis were summarized.

In work [50], the structure and physicochemical properties of tetranitrosyl iron complex with benzimidazole-2-thiolyl, [Fe2[SC7 H5N2]2[NO]4] · 2C3H6O (1) (Figure 3.18) obtained in the reaction of thiosulfate tetranitrosyl iron complex with 2-mercaptoimidazole has been studied by the methods of X-ray, Mossbauer, IR-, mass spectroscopy, and SQUID-magnetometry.

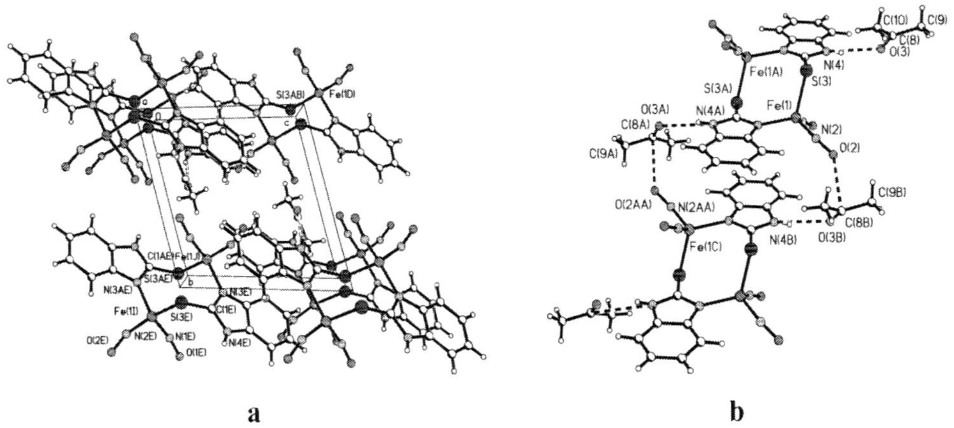

Figure 3.18. (a) Projection of the crystalline packing on plane ac displaying the molecules layers in the crystals of 1. (b) Layer fragment. Dotted lines show the van der Waals contacts and hydrogen bonds [50]. With permission from the Royal Chemistry Society.

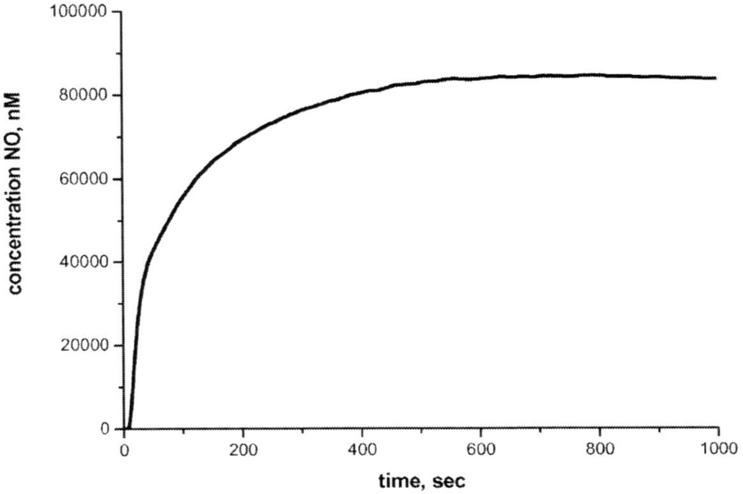

Figure 3.19. Concentration of NO generated upon decomposition of sulfur–nitrosyl iron complexes 1 (0.4\10-5 mol L-1) in 1% DMSO water solutions at 25°C vs. time [50]. With permission from the Royal Chemistry Society.

Measurement of temperature dependence of magnetic susceptibility of the dimers at constant magnetic field 1 kOe yielded the values of exchange integral J = -33.1 cm^{-1}. After dissolving in a protonic solvent, the complex spontaneously generates NO (Figure 3.19).

Magnetic properties of a series of binuclear iron nitrosyl complexes, $Fe_2(SR)_2(NO)_4$ were investigated using the magnetic susceptibility ESR techniques [51]. The temperature dependences of magnetic susceptibility was measured at T = 2–300K in static magnetic field H = 1000 Oe exploring a MPMS 5XL Quantum Design SQUID magnetometer. The temperature decrease from 300K for the $Fe_2(SR)_2(NO)_4$ complexes led to the growth of magnetic susceptibility χ = M/H, which followed the Curie law for Fe ions weakly interacting in dimers. The experiments indicated that the antiferromagnetic interaction in dimers appears when the temperature decreased down to 60–100 K. The absolute values of exchange integral *J* magnetic centers in the dimer from 30 cm^{-1} to 60 cm^{-1}, which depend on the Fe-Fe distance, were determined from the experimental curves using the Bleaney–Bowers expression. $Fe_2(SR)_2(NO)_4$. Dependence *J* on distance between iron ions was discussed. Finally, contribution of direct and indirect exchange interaction to binuclear iron complexes was evaluated.

The magnetic moment (M) of four salts of a new family of DNICs with thiocarbamide: monocationic $[Fe(SC(NH_2)_2)_2$ $(NO)_2]Cl$ H_2O (1), dicationic $[Fe(SC(NH_2)_2)_2(NO)_2]$ $2SO_4H_2O$ (2), and $[Fe(SC(NH_2)_2)_2(NO)_2]_2[Fe_2(S_2O_3)_2(NO)_4]$ (3), and a co-crystal of cationic and neutral dinitrosyl iron complexes with N-ethylthiocarbamide $[C_3N_2H_8SFe(NO)_2^{-}$ $Cl][Fe(NO)_2(C_3N_2H_8S)_2]Cl_{-}$ (4) was measured using at a Quantum Design MPMS 5XL SQUID magnetometer [52]. The experiments were carried out at T = 2–300 K in static magnetic field B = 1 kOe. The magnetic field dependence of the magnetic moment was also measured at B = 0–5 T at 2 K. Strong intermolecular antiferromagnetic interactions was proved by demonstration of a strong decrease of the magnetic moment at low temperatures. It was shown that the room temperature value of the effective magnetic moment for complex 4 comprising two S = 1/2—cations is 2.8 B that exceeds the value 2.486 <u>B</u> expected for two non-interacting spins 1/2. The authors concluded that such unusual magnetic behavior can be a result of the presence of unquenched orbital angular momenta on the NO-groups, which, in turn, arise from not fully split $^2\Pi$-term of the NO - group and its Zeeman mixing with the excited $^2\Sigma$-terms.

The structure and physical and kinetics properties of novel nitrosyl iron complexes $[Fe(SC(NH_2)_2)_2(NO)_2]_2SO_4H_2O$(I) and $[Fe(SC(NH_2)_2)_2$ $(NO)_2]_2[Fe_2(S_2O_3)_2(NO)_4]$(II) (Figure 3.20) were studied employing X-ray analysis, Mossbauer, IR, EPR spectroscopy, and amperometry [53].

For complex I, the absorption bands of the IR spectra lie in the frequency region of 1808–1734 cm^{-1}, while complex II is characterized by shift to a more positive region (1822–1732 cm^{-1}). The values of the Mossbauer spectrum parameters at 293 K were found to be as isomeric shift δFe = 0.184mm s^{-1}, quadruple splitting ΔE_Q = 1.165 mm s^{-1}, and

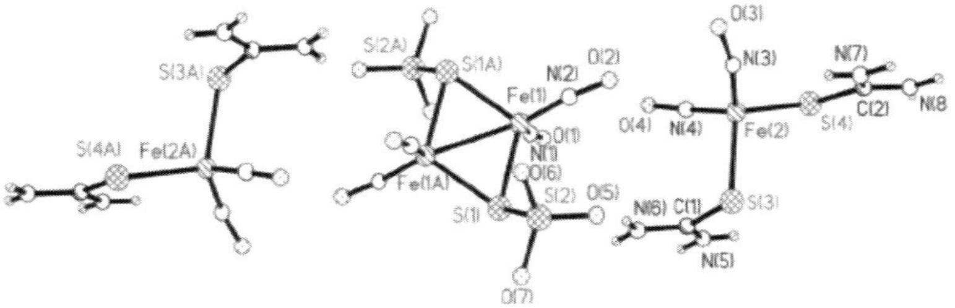

Figure 3.20. The general view of complex II. Selected bond distances (Å) and angles (deg): Fe(2)–N(3) 1.688(2), Fe(2)–N(4) 1.689(2), Fe(2)–S(4) 2.3130(5), Fe(2)–S(3) 2.3353(6), N(4)–O(4) 1.163(2), N(3)–O(3) 1.175(2), O(4)–N(4)– Fe(2) 170.1(2), O(3)–N(3)–Fe(2) 162.5(2), N(3)–Fe(2)–N(4) 114.18(9), S(4)– Fe(2)–S(3) 109.42(2). [53]. With permission from the Royal Society of Chemistry and the Centre National de la Recherche Scientifique.

width of absorption lines $\Gamma = 0.246$ mm s^{-1} for I. The Mossbauer spectrum of complex II indicated a double symmetric quadrupole doublet with parameters $\delta Fe = 0.219$, and $\Delta E_Q =$ 1.08), $\Gamma = 0.250(1)$ mms-1 for the cationic part of the complex, and $\delta Fe = 0.115(1)$, ΔE_Q = 1.33, $\Gamma = 0.230$ mms^{-1} for the anionic part. In the powder state, Lorentz form of the complex I ESR spectra indicates exchange interaction between the iron atoms. In the solution, EPR spectrum of the complex I is the superposition of at least three paramagnetic centers with g-factors 2.0385, 2.032, and 2.015. ESR spectrum of complex II showed the g-factors g 2.039 and g 2.032. Both complexes spontaneously generate NO in aqueous anaerobic solutions at pH 7 with the NO of amount about 32 nM for period of 100 s.

The results of promising applications of the described complexes in biological systems are presented in [41, 47, 55–60].

In work [54] it was shown that complex $[Fe_2(\mu_2\text{-}SC_4H_3N_2)_2(NO)_4]$ $S0_4H_2O]$ (I) reacts with ferrocytochrome -(cyt c Fe^{2+}) to form cyt c Fe^{2+}NO (II) with the rate constant 8.3 $M^{-1}s^{-1}$ at pH 3. The reaction was monitored by optical absorption technique using the spectra bands. The rate constant of NO release from the complex I was estimated as 2.7 10^{-3} s^{-1}. The results suggested that the complex II can serve as a NO depo in cells. Binuclear dinitrosyl-iron complex with glutathione (DNICglu), binuclear tetranitrosyl-iron complex with thiosulfate (TNICthio), binuclear tetranitrosyl-iron complex with aminotriazole (TNICatria), and mononuclear dinitrosyl-iron complex with triazole (DNICtria) activated expression of the soxS and sfiA genes in *Escherichia coli* [56]. The iron chelating agent o-phenanthroline completely inhibited the gene expression induced by all compounds studied. It was suggested that the genetic signal transduction is caused by the complexes themselves, which activate transcriptional proteins by transfer onto them of nitrosyl-iron groups $[Fe^+(NO^+)_2]$.

NO donors (di-and trinitrosyl iron complexes with synthetic ligands) induced expression of the *E. coli sfiA* gene belonging to the SOS regulon and exerted a mutagenic effect on *Salmonella typhimurium* TA1535 [57]. These effects were fully or significantly inhibited by the iron(II)-chelating agent o-phenanthrolin, depending on the mono-or binuclear structure of the ligands [57]. The cytotoxic activity of fifteen nitrosyl iron complexes, as NO donors, was tested on lines of human tumor cells of different histogenesis [41]. Experiments indicated that the preparation has a selective cytotoxic activity relative to lines of tumor cells of various histogenesis. For the four structural types of the most cytotoxic compounds, high anticancer activity in experimental models of tumors in mice *in vivo* was revealed.

The findings of examining the antihypertensive effect of the synthetic analogue of the endogenous NO donors suggested that the dinitrosyl iron complex is highly effective in treating in patients with grades 2–3 hypertension and uncomplicated hypertensive crisis [58]. The antihypertensive effect of the drug persists for eight hours after its injection. The drug is tolerated by patients and causes an insignificant number of side effects. To established the relationship between the expression of O6-methylguanine-DNA methyltransferase and cell sensitivity to CysAm, and apoptosis-inducing capacity human tumor cells, antiproliferative activity of binuclear tetranitrosyl [Fe-S] complex with cysteamine $[Fe_2\ (S(CH)_2\ NH3)_2\ (NO)_4]\ 2.5H_2O$ human was examined [59]. It was found that CysAm induced apoptosis via activation of caspases 3 and 7.

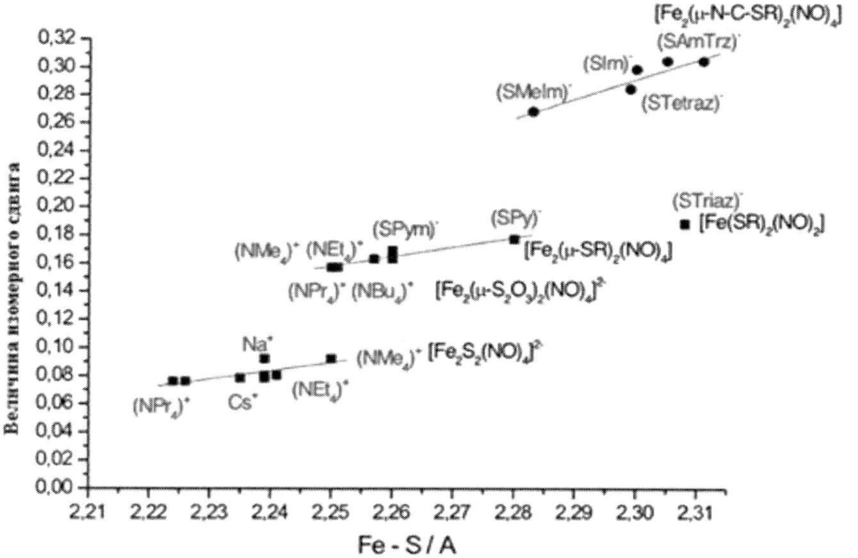

Figure 3.21. Dinuclear nitrosyl complexes: isomeric shift of the Mossbauer spectra versus Fe-S distance. Triaz is triazol (Private communication from S.M. Aldoshin and N.A. Sanina).

Kinetics of release of NO from complex $Fe_2(\mu_2\text{-}SC_4H_3N_2)_2(NO)_4]$ (C_4H3N_2S-pyrimilin-2thiophtolate) in absence and presence Hb was investigated [60]. In the absence of Hb, the kinetics showed lag-periods that testified to the complex mechanism of reaction presumably under the scheme:

$[Fe_2(\mu_2\text{-}SC_4H_3N_2)_2(NO)_4] + 2H_2O \rightarrow [Fe(SR)(H_2O)(NO)_2]$.

$2[Fe(SR)(H_2O)(NO)_2] \rightarrow [Fe(SR)(H_2O)(NO)_2(\mu_2\text{-}SR)Fe(H_20)(NO)] + NO$

In the presence of Hb, the NO molecule rapidly and irreversibly binds to Hb to form HbNO. In addition, a reversible covalent binding NO to the protein functional moieties (most probably thiol groups) can occur. Thus, hemoglobin can provide complete and sustained absorption of NO.

Experimental data for isomeric shift of the Mossbauer spectra versus Fe-S distance is presented in Figure 3.21

REFERENCES

[1] Ford PC, Fernandez BO, MD. Mechanisms of Reductive Nitrosylation in Iron and Copper Models Relevant to Biological Systems. *Chem. Rev.* (2005) 105, 2439-2456.

[2] Abhik Ghosh. Metalloporphyrin-NO Bonding: Building Bridges with Organometallic Chemistry. *Acc. Chem. Res.* (2005) 38 943-954.

[3] Hoshimo M, Ozawa K, Seki H, Ford PC. Photochemistry of nitric oxide adducts of water-soluble iron(III) porphyrin and ferrihemoproteins studied by nanosecond laser photolysis. *J. Am. Chem. Soc.* (1993) 115, 9568-9575.

[4] Kharitonov VG, Bonaventura J, Sharma VS. (1996) in *Methods in Nitric Oxide Research* (Feelisch, M., and Stamler, J. S., eds) 1st Ed., pp. 39-45, John Wiley & Sons, Chichester.

[4] Kharitonov VG, Sharma, V Magde, D, Koesling D. Kinetics of Nitric Oxide Dissociation from Five- and Six-Coordinate Nitrosyl Hemes and Heme Proteins, Including Soluble Guanylate Cyclase. *Biochemistry* (1997) 36, 6814-6818.

[5] Fernandez BO, Lorkovic IM, Ford PC. Mechanisms of Ferriheme Reduction by Nitric Oxide: Nitrite and General Base Catalysis. *Inorg. Chem.* (2004) 43, 5393.

[6] Kurtikyan TS, Gulyan GM, Martirosyan GG, Lim MD, Ford PC. Reactions of Nitrogen Oxides with Heme Models. Spectral and Kinetic Study of Nitric Oxide Reactions with Solid and Solute FeIII(TPP) (NO3). *J. Amer. Chem. Soc.* (2005) 127 6216-6224.

[7] Laverman LE, Wanat A, Oszajca J, Stochel G, Ford PC, van Eldik R. Mechanistic Studies on the Reversible Binding of Nitric Oxide to Metmyoglobin. *J. Am. Chem. Soc.* (2001) 123, 28.

[8] Scheele JS, Eric Bruner E, Kharitonov VG, Pavel Martasek P, Linda J. Roman RJ, Bettie Sue Siler Masters, Vijay S. Sharma VS, and Douglas Magde D. Kinetics of NO Ligation with Nitric-oxide Synthase by Flash Photolysis and Stopped-flow. *Spectrophotometry.* (1999) 274, 13105-13110.

[9] Brunori M. Nitric oxide moves myoglobin centre stage. *Trends Biochem. Sci.* (2001) 26, 209.

[10] Andersen JF, Balfour C, Shokhireva TKh, Champagne DE, Walker FA, Montfort WR. Kinetics and Equilibria in Ligand Binding by Nitrophorins 1-4: Evidence for Stabilization of a Nitric Oxide-Ferriheme Complex through a Ligand-Induced Conformational Trap. *Biochemistry* (2000) 39, 10118-10131.

[11] Yoo, Byung-Kuk, Kruglik, Sergei G, Lamarre, Isabelle, Martin, Jean-Louis, Negrerie, Michel. Absorption Band III Kinetics Probe the Picosecond Heme Iron Motion Triggered by Nitric Oxide Binding to Hemoglobin and Myoglobin. *J Phys. Chem.* B (2012) 116, 4106-4114.

[12] Enemark JH. Feltham RD. Principles of structure, bonding, and reactivity for metal nitrosyl complexes. *Coord. Chem. Rev.* (1974) (13), 339-406.

[13] Nalbandyan RM, Vanin AF, Blumenfeld LA. EPR sinals of a new type in yeast cells, In: Abstracts of the Meeting "*Free Radicals Processes in Biolog ical Systems,*" Moscow, 1964, p. 18.

[14] Vanin AF, Nalbandyan RM. Free radical states with unpaired elec tron localization on sulfur atom in yeast cells, *Biofiz. Rus.* (1966) 11, 16778-17179.

[15] Vanin AF, Liu XP, Samouilov A, Stukan RA, Zweier JL. Redox properties of iron-dithiocarbamates and their nitrosyl derivatives: Implications for their use as traps of nitric oxide in biological systems. *Biochim. Biophys. Acta* (2000) 1474, 365-377.

[16] Vanin, A.F Huisman A, Van Faassen E.E. Iron dithiocarbamate as spin trap for nitric oxide detection: pitfalls and successes. *Methods Enzymol.* (2002) 359, 27- 42.

[17] Lu Ch, Koppenol WH. Redox cycling of iron complexes of N N(dithiocarboxy)sarcosine and N-methyl-d-glucamine dithio-carbamate. *Free Radical Biology & Medicin*e (2005) 39, 1581 - 1590.

[18] Vanin AF, Poltorakov, Mikoyan VD, Kubrina LN, van Faassen E. Why iron-dithiocarbamate ensure detection of nitricoxide in cells and tissues. *Nitric Oxide, Biol. Chem.* (2006) 15, 295-311.

[19] Fujii S, Yoshimura T, Kamada H. Nitric oxide trapping efficiencies of water-soluble iron(III) complexes with dithiocarbamate derivatives. *Chem. Lett.* (1996)785- 786.

[20] Mikoyan VD, Burgova EN, Borodulin RR, Vanin AV. The binuclear form of dinitrosyl iron complexes with thiol-containing ligands in animal Tissues. *Nitric Oxide* (2017) 62, 1-10.

[21] Vanin AF, Kiladze SV, Kubrina LN. On factors determining the formation of dinitrosyl non-heme iron complexes in animal organs in vivo, *Biofiz. Rus.* 22 (1977) 850-855.

[22] Vanin, AF. Dinitrosyl iron complexes with thiol-containing ligands as a "working form" of endogenous nitric oxide *Nitric Oxide Biol. Chem.* (2016) 54, 156-229.

[23] Vanin AF. Dinitrosyl iron complexes with thiolate ligands: Physicochemistry, biochemistry and physiology. *Nitric Oxide Biol. Chem.* (2009) 21 1-13.

[24] Vanin AF. Dinitrosyl iron complexes with thiol-containing ligands as a base for new-generation drugs (Review). *Open Conf. Proc. J* (2013) 4, 31-37.

[25] Vanin AF, Poltorakov AP, Mikoyan VD, Kubrina LN, Burbaev DSh. Polynuclear water-soluble dinitrosyl iron complexes with cysteine or glutathione ligands: Electron paramagnetic resonance and optical studies. *Nitric Oxide Biol. Chem.* (2010) 23 1236-1249.

[26] Vanin AF, Burbaev DSh, Electronic and spatial structures of water-soluble dinitrosyl iron complexes with thiol-containing ligands underlying their activity to act as nitric oxide and nitrosonium ion donors. *Biophys. J.* (2011), 878236.

[27] Hickok JR, Sahni S, Shen H, Arvindt A, Antoniou C, Fung LW, Thomas DD. Dinitrosyl iron complexes are the most abundant nitric oxide- 14 derived cellular adducts: Biological parameters of assembly and disappearance. *Free Rad. Biol. Med.* (2011) 51, 1558e1566.

[28] Lok HC, Sahni S, Richardson V, Kalinowski DS, Z. Kovactvic Z, Lane DJR, Richardson DR. Glutathione S-transferase and MRP1 form an integrated system involved in the storage and transport of dinitrosyl-dithiolate iron complexes in cells, *Free Rad. Biol. Med.* (2014) 75, 14e29.

[29] Vanin, AF. Dinitrosyl iron complexes with thiol-containing ligands as a base for developing drugs with diverse therapeutic activities: Physicochemical and biological substantiation. *Biophysics* (Moscow, Russian Federation) (2017), 62(4), 509-531.

[30] Vanin, AF. Dinitrosyl iron complexes with thiol-containing ligands as a "working form" of endogenous nitric oxide. *Nitric Oxide Nitric Oxide Biol. Chem.* (2016) 54, 15-29.

[31] Vanin AF, Poltorakov AP, Mikoyan VD, Kubrina LN, DSh, Polynuclear water soluble dinitrosyl iron complexes with cysteine o glutathione: electron para magnetic resonance and optical studies, Nitric Oxide, *Biol. Chem.* (2010) 23, 136-149.

[32] Wu SC, Lu CY, Chen YL, Lo FC, Wang TY, Chen YJ, Yuan SS, Liaw WF, Wang YM. Water-Soluble Dinitrosyl Iron Complex (DNIC): a Nitric Oxide Vehicle Triggering Cancer Cell Death via Apoptosis. *Inorg. Chem.* (2016) 55, 9383−9392.

[33] Chiang CY, Miller ML, Reibenspies JH, Darensbourg MY.., Reibenspies Joseph H, and Marcetta Y. Darensbourg Bismercaptoethanediazacyclooctane as a N2S2 Chelating Agent and Cys-X-Cys Mimic for Fe(NO) and Fe(NO)2. *J Am. Chem. Soc.* (2004) 126, 10867-10874.

[34] Rauchfuss TB, Weatherill TD. Roussin's Red Salt revisited: reactivity of Fe2(.mu.-E)2(NO)4- (E = S, Se, Te) and related compounds. *Inorg. Chem.* (1982) 21, 827-830.

[35] Harrop TC, Song D, Lippard SJ. Interaction of Nitric Oxide with Tetrathiolato Iron(II) Complexes: Relevance to the Reaction Pathways of Iron Nitrosyls in Sulfur-Rich Biological Coordination Environment. *J. Am. Chem. Soc.* (2006) 128, 3528-3529.

[36] Wang R, Camacho-Fernandez MA, Xu W, Zhang J, Li L. Neutral and reduced Roussin's red salt ester [Fe2(μ-RS)2(NO)4] (R = n-Pr, t-Bu, 6-methyl-2-pyridyl and 4,6-dimethyl-2-pyrimidyl): Synthesis, X-ray crystal structures, spectroscopic, electrochemical and density functional theoretical investigations. *Dalton Trans.* (2009) 777-786.

[37] Hopmann KH, Conradi J. Abhik GhoshBroken-Symmetry DFT Spin Densities of Iron Nitrosyls, Including Roussin's Red and Black Salts: Striking Differences between Pure and Hybrid Functionals. *J. Phys. Chem.* B, (2009) 113, 10540-10547.

[38] Tinberg CE, Tonzetich ZJ, Wang H, Loi H. Do LH, Yoda Y, Cramer SP, Stepard SJ. Nitric Oxide with a Biological Rieske Center. *J Am Chem Soc.* (2010) 132, 18168-18176.

[39] Hopmann KH, Noodleman L, Ghosh A. Spin Coupling in Roussin's Red and Black Salts. *Chemistry.* (2010); 16, 10397-10408.

[40] Li L, Li L. Recent Advances in Multinuclear Metal Nitrosyl Complexes. *Coord. Chem. Rev.* (2016) 306, 678-700.

[41] Aldoshin SM, Sanina NA, Davydov MI, Chazov EI. A new class of nitric oxide donors. *Herald of the Russian Academy of Sciences* (2016) 86,158-163.

[42] Aldoshin SM, Sanina NA. Functional nitrosyl complexes—A new class of nitric oxide donors for the treatment of socially significant diseases. In Grigor'ev AI, Vladimirov Yu A. (eds). *Fundamental science to medicine: Biophysical medical technologies.* Moscow, MAKS press, 2015, pp 72-102 (In Russian).

[43] Sanina NA, Rakova OA, Aldoshin SM, Shilov GV. Structure of the neutral mononuclear dinitrosyl iron complex with 1,2,4-triazole-3-thione [Fe(SC2H3N3)(SC2H2N3)(NO)2]·0.5H2O. *Mendeleev Commun.* (2004) 14, 7-8].

[44] Aldoshin SM, Lyssenko KA, Antipin MYu, Sanina NA, Gritsenko VV. Precision X-ray study of mononuclear dinitrosyl iron complex [Fe(SC2H3N3)(SC2H2N3)(NO)2]·0.5H2O at low temperatures. *J. Mol. Structure* (2008) 87, 309-315.

[45] Sanina NA, Aldoshin SM, Shmatko NYu, Korchagin DN, Shilov GV, Ovanesyan NS, Kulikov AV. Mesomeric tautomerism of ligand is a novel pathway for synthesis of cationic dinitrosyl iron complexes: X-ray structure and properties of nitrosyl complex with thiourea. *Inorganic Chemistry Communications* (2014) 49, 44-4.

[46] Emel'yanova NS, Shmatko Nyu, Sanina NA, Aldoshin SM. Quantum-chemical study of the Fe(NO)2 fragment in the cation of mononuclear nitrosyl iron complex [Fe(SC(NH2)2)2(NO)2]Cl_H2O. *Computational and Theoretical Chemistry* (2015) 1060, 1-9.

[47] Sanina NA, Shmatko NYu, Korchagin DV, Shilov VD, Terent'ev AA, Stupina TS, Balakina AA, Komleva NA, Ovanesyan NS, Alexander V. Kulikov AV, Aldoshin SM. A new member of the cationic dinitrosyl iron complexes family incorporating

N-ethylthiourea is effective against human HeLa and MCF-7 tumor cell lines. *J. Coord. Chem.* (2016) 69, 812-825.

[48] Sanina NA, Kniazkina EV, Manzhos RA Emel'yanova NS, Krivenko AG, Aldoshin SM. Redox reactions of binuclear tetranitrosyl iron complexes with bridgingN-C-S ligands. *Inorganica Chimica Acta* (2016) 449, 61-68.

[49] Shestakov AF, Shul'ga YuM, Emel'yanova NS, Sanina NA, Rudneva TN, Sergei M. Aldoshin SM, Ikorskii VN, Ovcharenko VI. Experimental and theoretical study of the arrangement, electronic structure and properties of neutral paramagnetic binuclear nitrosyl iron complexes with azaheterocyclic thyolyls having 'S-C-N type' coordination of bridging ligands. *Inorg Chim. Acta* (2009) 362, 2499-2504.

[50] Sanina N, Roudneva T, Shilov G, Morgunov R, Ovanesyan N, Aldoshin S. Structure and properties of binuclear nitrosyl iron complex with benzimidazole-2-thiolyl. *Dalton Trans.,* 2009 1703-1706.

[51] Aldoshin SM, Morgunov RB, Sanina NA, Kirman MV. Contribution of direct and indirect exchange interaction to binuclear iron complexes. *Materials Chemistry and Physics* (2009) 116, 589-592.

[52] Aldoshin SM, Morgunov RB, Andrei V. Palii AVNatal'ya Yu. Shmatko NYu, NatalSanina NA. Study of Magnetic Behavior of Salts of Cationic Dinitrosyl Iron Complexes with Thiocarbamide. and its Derivatives. *Appl. Magn. Reson.* (2015) 46, 1383-1393.

[53] Sanina NA, Aldoshin SM, Shmatko NYu, Denis V. Korchagin DN, Shilov GV, Knyazkina EV, Ovanesyan NS, Kulikov AV. Nitrosyl iron complexes with enhanced NO donating ability: synthesis, structure and properties of a new type of salt with the DNIC cations [Fe(SC(NH2)2)2(NO)2]+. *New J.Chem.* (2015) 39, 1022.

[54] Sanina NA., Syrtsova LA, Psikha BL, Shkondina NI, Rudneva TN, Kotel'nikov AI, Aldoshin SM. *Reaction of ferrocytochrome and deoxyhemoglobin with nitrosyl complex* [Fe2(μ2-SC4H3N2)2 (NO)4]S04H2O. Izvestiya Akademii Nauk, Seriay Khimicheskaya. (2010) N9 12 2148 -2159.

[55] Vasilieva SV, Moshkovskaya EYu, Sanina NA, Aldoshin SM, Vanin A F. Genetic signal transduction by nitrosyl-iron complexes in Escherichia coli. *Biochemistry* (Moscow, Russian Federation) (Translation of Biokhimiya (Moscow, Russian Federation) (2004) 69, 883-889.

[57] Zhukova OS, Sanin NA, Fetisova KV, Gerasimova GK. "Cytotoxic effect of nitrosyl iron complexes on human tumor cells in vitro," *Ross. Bioter. Zh.* (2006) 5, 14.

[58] Gosteev AYu, Zorin AV, Rodnenkov OV, Dragnev AG, Chazov EI. Effects of the synthetic analog of endogenous nitrogen oxide (II) donators: A dinitrosyl iron complexes preparation in hypertension patientswith intact hypertensic crises, *Ter. Arkh.,* (2014) 9, 49.

[59] Zhukova OS, Smirnova ZS, Chikileva IO, Kiselevskii MV. Antiproliferative Activity of a New Nitrosyl Iron Complex with Cysteamine in Human Tumor Cells In Vitro. *Bulletin of Experimental Biology and Medicine* (2017) 162, 583-588.

[60] Syrtsova LA, Sanina NA, Shestakov AF, Shkondina NI, Rudneva TN, Emel'yanova NC, Kotel'nikov AI, Aldoshin SM. Influence of hemoglobin on NO-donor activity of Fe2(μ2-SC4H3N2)2(NO)4] (C4H3N2S-pyrimilin-2thiophtolate). *Izvestiya Akademii Nauk, Seriay Khimicheskaya.* (2010) N9 12 2148-2159.

Chapter 4

NITRIC OXIDE ANALYSIS

ANNOTATION

Nitric oxide (NO) is essential in a significant number of vital normal and pathological physiological processes. Nowadays a whole arsenal of chemical, physical, and biological methods spanning the NO picomolar-to-micromolar concentration range in physiological milieus has been developed and widely used. Nevertheless, real-time monitoring of NO dynamics under physiological conditions of nano and less molar concentrations is still a challenging analytical problem. This has been attributed to the labile nature of the NO molecule with half-life of the order of seconds, which both rapidly diffuses through the medium and readily reacts with many scavenger targets.

4.1. INTRODUCTION

The scope of importance of NO spans from normal numerous biological and physiological processes to pathophysiology of the cardiovascular, nervous pulmonary, hypertension, chronic obstructive lung diseases, asthma, atherosclerosis, wound inflammation, etc. [1]. NO content in human exhalation and liquids can be used for diagnostic purposes. NO is a powerful signaling molecule capable of modulating cytokine production in the immune response and in wound healing.

In physiological conditions, scavenger targets for include oxygen, superoxide, amino and mercapto compounds (e.g., cysteine residues, glutathione), hemeproteins, (e.g., hemoglobin, myoglobin), free radicals, and so forth. Although a variety of methods for NO detection have been proposed (Figure 4.1) [1–11], they are not generally amenable to real-time *in situ* measurements of endogenous NO in nano and picomolar concentrations.

Data on NO detection techniques with associated limits of detection and detection ranges were summarized in [3].

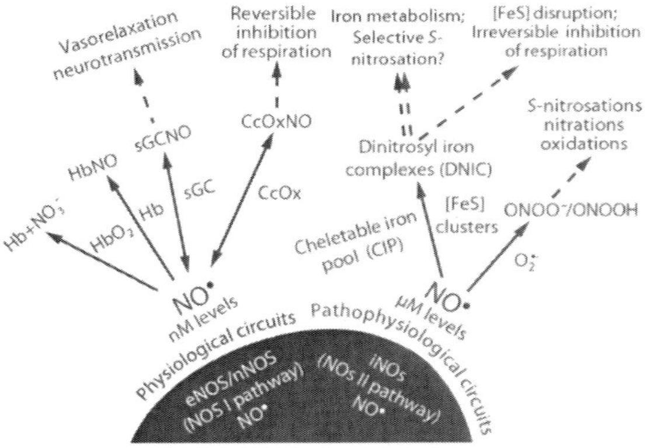

Figure 4.1. Commonly used methods of nitric oxide analysis and action [5]. Copyrights 2012 American Chemical Society.

4.2. Indirect Assay of Nitric Oxide—Nitrite Analysis

4.2.1. Optical Spectroscopy

Because of its simplicity, availability, and the intense color of an adduct, optical spectroscopy for a long time served as main practical tools for nitrite and NO analysis in chemistry and biology [2–19].

The Griess diazotization reaction by Peter Griess, on which the Griess reagent relies was the first assay for nitrite, can be used for an indirect assay of NO [2–10]. Nowadays, Sigma-Aldrich sells N-(1-Naphthyl)ethylenediamine dihydrochloride

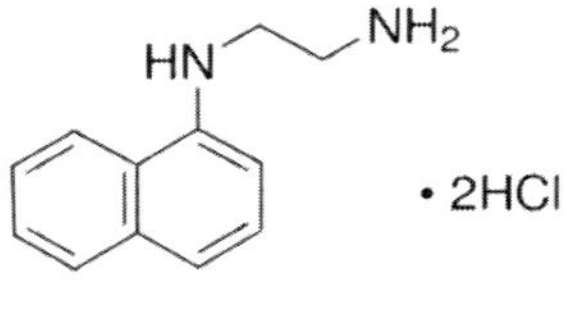

as the Griess reagent for determination of sulfonamide and nitrite. A number of new modified Griess reagents and corresponding reaction assay kits for biological samples were suggested [9, 10]. A major limitation of Griess Reaction Assay is relatively low, as compared with other methods, is sensitivity to measure nitrite concentration no less than ~0.5 μM level. In addition, the assay is not highly specific and can be liable to interference by environment compounds, such as ascorbate, thiols, phosphate, NG-nitro-L-arginine, and heparin.

Nitrite concentration can be measured by a number of additional methods, such as colorimetric measurements [6], chemiluminescence assay [7], fluorescent assay using fluorescence indicator, 2,3-diamino-naphthalene and cadmium reduction of nitrate $NO_3^-$to nitrite NO_2^- [8], the enzymatic conversion of nitrate into nitrite including the reaction modification, Griess reaction coupled to high-performance liquid chromatography including the fluorometric method, which involves pre-column derivatization of nitrite with 2,3-diaminonaphthalene [9], gas chromatography-mass spectrometry employing pentafluorobenzyl derivatives [10], and electrochemical detection [11].

Since oxyhemoglobin has a high affinity for NO and upon binding is converted quantitatively to methemoglobin (MetHb), spectrophotometric determination of NO using hemoglobin is employed for analysis of NO concentrations between 300 nM and 30 μM [12]. In frame of this sensitivity, selective methods for detection of NO using cytochrome *c*–doped xerogel [13], heme-nitric oxide or oxygen-binding domain (H-NOX) [14], Cu(II) cyanine complex[15], and the hemoglobin-trapping technique [16] and others [17–19] were developed.

4.3. Nitric Oxide Fluorescence Analysis

The fluorescence technique has a sensitivity from mM up to a single molecule and is widely available for routine analysis and precision investigations, and therefore, gains many advantages over other physical methods used in biology and medicine [20–48].

4.3.1. Organic Fluorophores

Classical fluorophores, such as anthracene, coumarin, BODIPAY, fluorescein, rhodamine, and cyanine are routinely employed in biology for analysis of small molecules [21]. In the last decades, the essential progress was achieved in design of novel fluorescent methods of analysis and real-time monitoring of NO [20–35]. A number new non-cytotoxic water-soluble and membrane-permeable probes, including two-photon fluorophores excitable at low-energy wavelengths, were synthesized and investigated employing advanced spectroscopic techniques. One of the general basis of methods for NO analyses is the conversion of a fluorescent silent probe to a compound of high-fluorescent quantum yield. This objective can be achieved in various ways. One of the methods of such a conversion is to expand the aromatic structure of a probe with markedly increased fluorescence intensity of the NO product. For example, the molecular probe o-phenylenediamine- Phe-Phe-OH (1) can react with NO with turning on very weak fluorescence emission at 440 nm to high intensity fluorescence at 367 (Figure 4.2) [29]. In phosphate-buffered saline buffer, this probe was applied for detecting NO within the range of 6 nM–12 μM.

The stable, water-soluble, nonfluorescent FA-OMe can sense NO and form the intensely fluorescent product dA-FA-OMe via reductive deamination of the aromatic primary amine (Figure 4.2) [29]. The turn-on fluorescent signals were performed by suppression of quenching the photoinduced electron transfer.

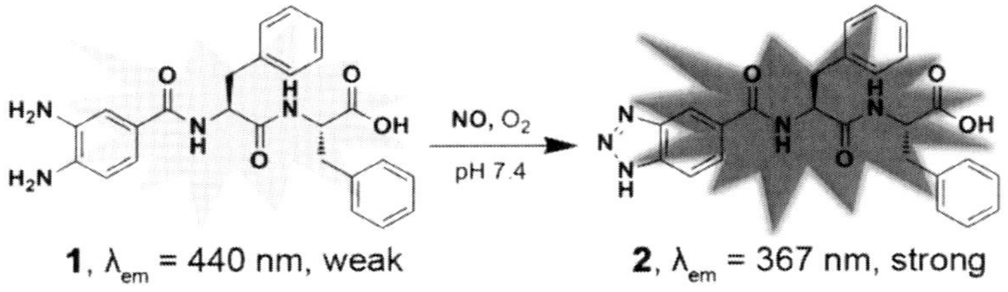

Figure 4.2. Proposed Mechanism of FA-OMe with NO under aerobic conditions to form dA-FA-Ome [29]. Copyrights 2012 American Chemical Society.

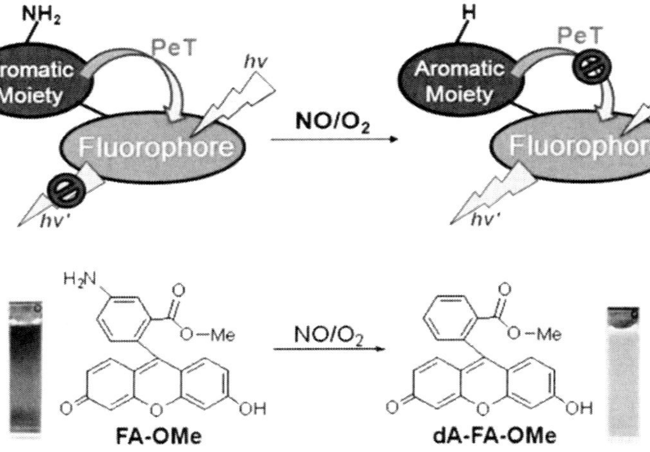

Figure 4.3. Proposed mechanism by which FA-OMe reacts with NO under aerobic conditions to form dA-FA-OM [29]. Copyrights 2012 American Chemical Society.

A novel sensing mechanism for NO detection displays a rapid and linear response to NO with a red-shifted 1500-fold turn-on signal from a dark background and was introduced in [28]. High selectivity was observed against other reactive oxygen/nitrogen species, pH, and various substances that interfere with existing probes. Another mechanism to rise the fluorescent yield of a probe is removal of a quenching group. For example, reductive deamination of the aromatic primary amine of nonfluorescent FA-OMe by NO forms the intensely fluorescent product dA-FA-Ome (Figure 4.3) [29]. In this study, the NO detection limit was shown as 44 nM. According to suggested mechanism based on the density functional theory calculations, in FA-Ome photoinduced electron transfer is suppressed. The suppression is taken off in fluorescent product dA-FA-Ome.

Monitoring NO in subcellular compartments by a hybrid probe based on rhodamine spirolactam and SNAP-tag was described in [30]. By connection of O6-benzylguanine (BG) to an "o-phenylenediamine- locked" rhodamine spirolactam responsive to nitric oxide (NO), a novel substrate (TMR-NO-BG) of genetically encoded SNAP-tag has been constructed (Figures 4.4 and 4.5). Fluorescence detection of TMRNO-SNAP (0.5 μM) at 564 nm was used for the evaluation of specificity.

By connection of O6-benzylguanine (BG) to an "o-phenylenediamine- locked" rhodamine spirolactam responsive to NO, a novel substrate (TMR-NO-BG) of genetically encoded SNAP-tag has been constructed (Figures 4.4 and 4.5). Fluorescence detection of TMRNO-SNAP (0.5 μM) at 564 nm was used for the evaluation of specificity.

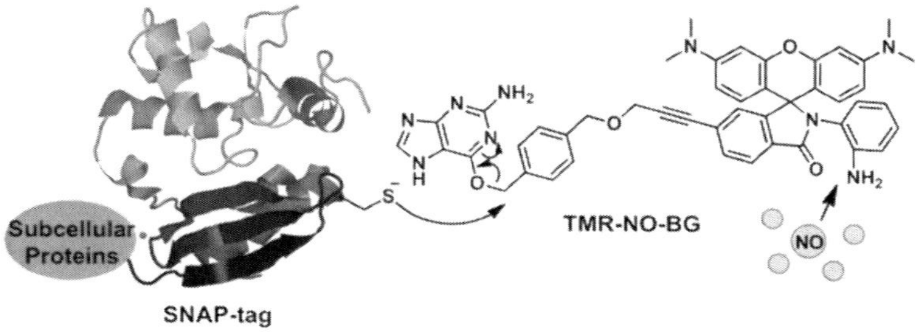

Figure 4.4. A novel substrate (TMR-NO-BG) of SNAP-tag [30]. Copyrights 2012 American Chemical Society.

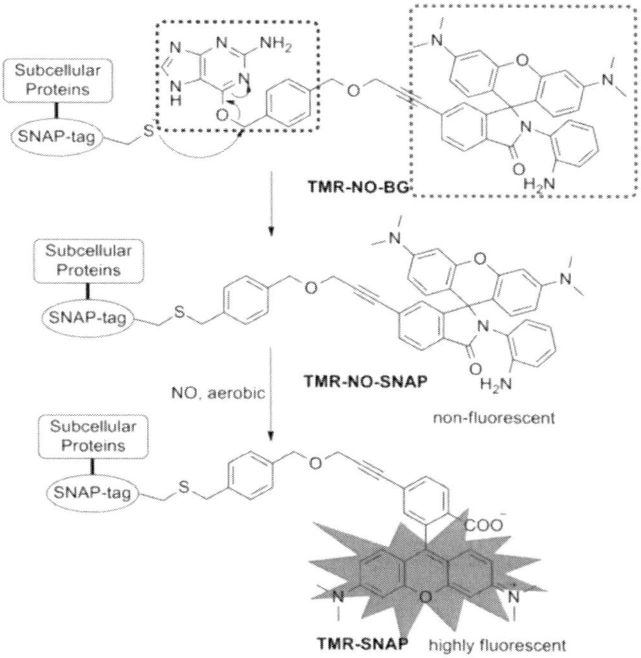

Figure 4.5. Mechanism of protein-targetable NO sensing [30]. Copyrights 2012 American Chemical Society.

Monitoring NO in subcellular compartments by hybrid probe based on rhodamine spirolactam and SNAP-tag was described in [31].

Deamination of the probes, based on pyrylium salts, o-on pyrylium salts, and o-hydroxyamine in which the o-hydroxyamine was used as the NO-reactive moiety, leads to a highly fluorescent product [31]. The probe utilizes emission of fluorescence following the internal charge transfer between dimethyloamino and cyano groups upon reaction with NO. Selective and sensitive fluorescence detection of NO based on synthesized rhodamine B selenolactone (RBSe)

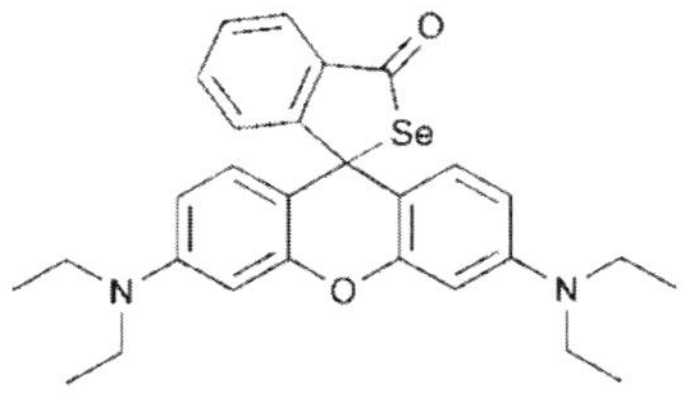

was reported [32]. The fluorescence of RBSe is switched when NO reacts with the Se atom. Two-photon microscopy with two near-infrared photons offers a number of advantages over one-photon microscopy, including increased depth of penetration (>500 μm), localized excitation, high spatial resolutions, low photo damage, and low cellular auto-fluorescence [33]. Utilizing two-photon microscopy made it possible to detect NO in live cells and live tissues at a depth of 90–180 μm. In experiments described in [33], a two-photon fluorescent probe, for example QNO (Figure 4.6), adopting a quinoline derivative as the fluorophore and an o-phenylenediamine moiety as the receptor for NO linked with glycinamide was designed. A photoinduced electron transfer mechanism provides twelvefold fluorescence enhancement toward NO.

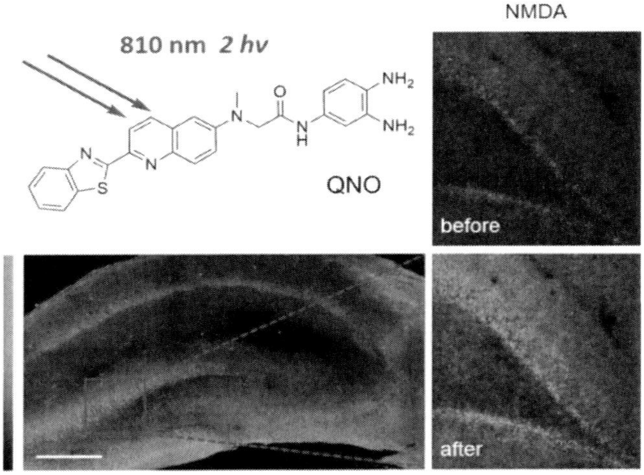

Figure 4.6. Fluorescent probe QNO based on o-diamine functionality and schematic illustration of two-photon excitation of the probe [33]. Copyrights 2014 American Chemical Society.

Density functional theory in combination with the polarizable continuum model showed that both fluorescent intensity and two-photon absorption are enhanced when fluorophores QNO

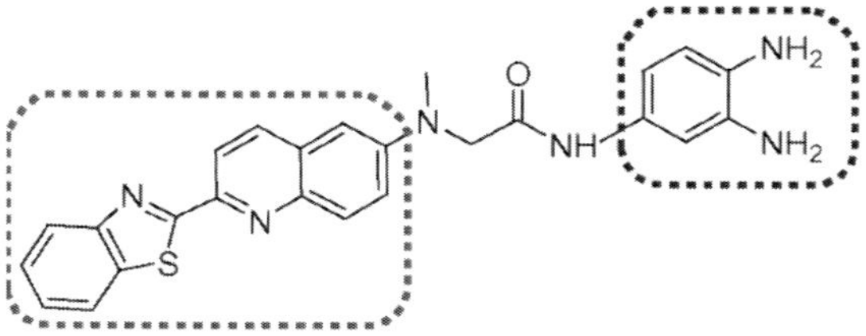

and LNO

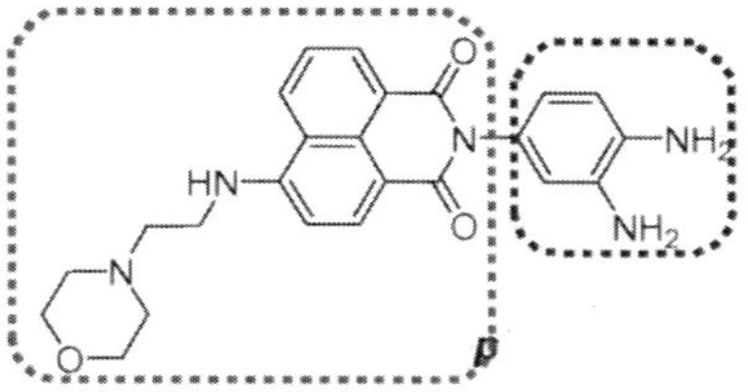

react with NO [34].

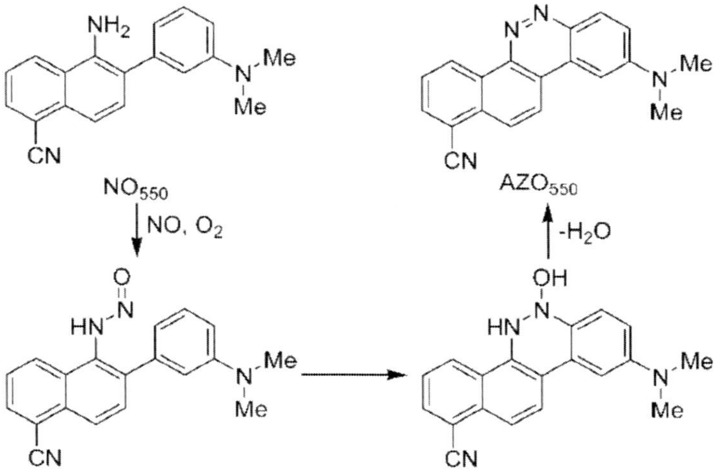

1 2

Figure 4.7. Probe NO_{550} and the proposed stepwise route to AZO_{550} [35]. Copyrights 2014 American Chemical Society.

For probe ONO-T, the following characteristics were reported: for the one-photon absorption λ^{1}_{max} = 408 nm and emission λ^{Fl}_{max} = 535 nm while for two-photon absorption λ^{2}_{max} = 810 nm.

A novel sensing mechanism for NO detection with a compound (NO_{550}), which after reaction with NO displays a rapid and linear response to NO with a red-shifted 1500-fold turn-on signal from a dark background, was established (Figure 4.7) [35].

4.3.2. Metalloorganics

Due to practically unlimited variety of structures and properties, organometal-based fluorescent probes provide for new possibilities in the field [36–41]. Several mechanisms of metal-transition NO-detecting probes are presented in Figures 4.8–4.11 [36]. In paramagnetic transition of metal complexes, the fluorescent emission is quenched by paramagnetic metal ions (Figure 4.8a). Reduction of metal (say CuII) by binding nitric oxide can restore the emission (Figure 4.8b). Both displacing the fluorophore and reducing the paramagnetic metal center by reductive nitrosylation of NO can also occur (Figure 4.8c).

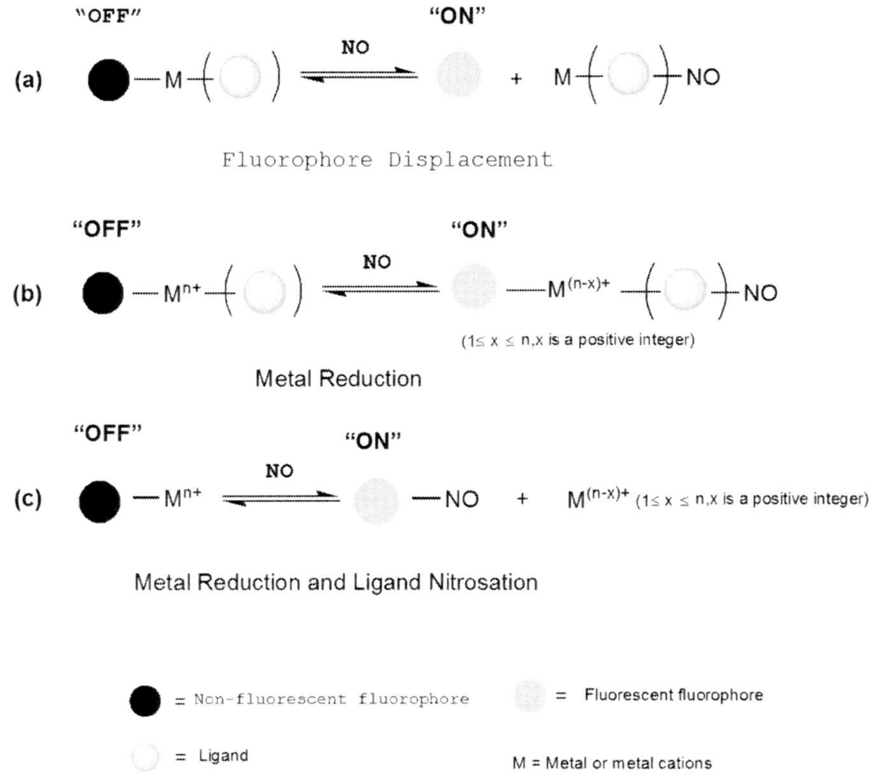

Figure 4.8. Mechanisms of metal-transition NO detecting probes [36]. With permission from the Royal Society of Chemistry.

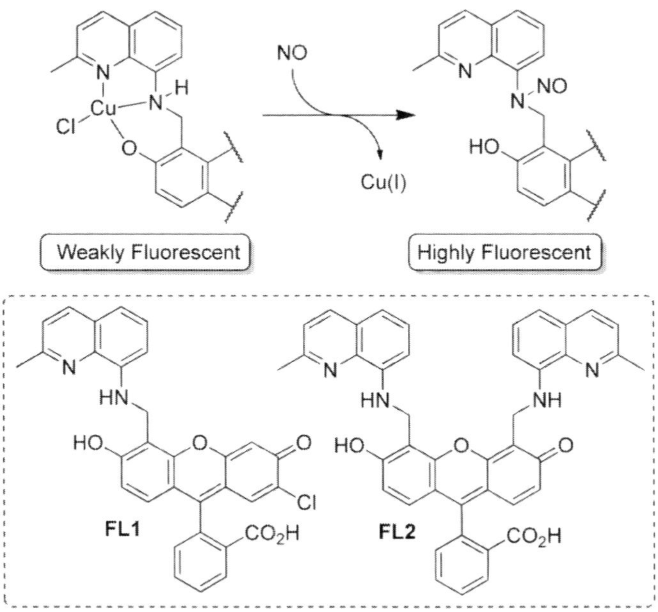

Figure 4.9. Schematic showing the mechanism of fluorescence response for Cu(II)-based NO probes, such as FL1 and FL2 [38]. Copyrights 2014 American Chemical Society.

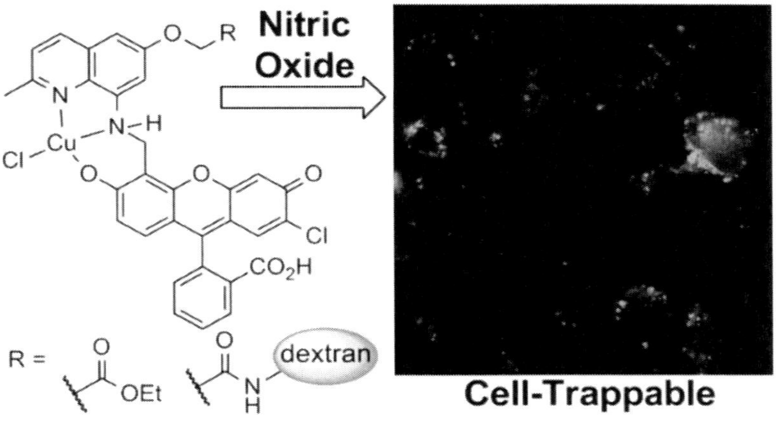

Figure 4.10. Structure of cell trappable complex CuFL1E [40]. Copyrights 2012 American Chemical Society.

Fluorescent turn-on probes for NO based on seminaphtho-fluorescein Cu(II) complexes were prepared and spectroscopically characterized [38]. Emission maxima between 550 and 625 nm is achieved at excitations between 535 and 575 nm. Upon treatment with NO under anaerobic conditions, the Cu(II) ion is reduced to Cu(I) with concomitant nitrosation of the secondary nitrogen to form the fluorescent nitrosamine product to yield a 20–45-fold increase (Figure 4.9). The probes are found to be highly selective for NO over NO_3^-, NO_2^-, HNO, $ONOO^-$, NO_2, OCl^-, and H_2O_2.

In the frame of the strategy developed in [36, 39], binding NO to fluorescent silent probes containing Co(II), Fe(II), Ru(II), and Rh(II) causes fluorophore fragment displacement reaction accompanied by strong increase of fluorescence intensity. As an example, the reaction of copper(II) complex CuRBT, having a ring-closed rhodamine-containing tripodal ligand with NO, features a 700-fold fluorescent enhancement toward NO from a dark background [39]. In this method, the detection limit of NO in aqueous solution is about 1 nM.

Two new nitrogen fluorescent probes for NO highly selective over other reactive oxygen were constructed [40]. These structures consisting of chitosan diazeniumdiolates as NO donors and a CuFL1 as a fluorophore are based on either incorporation of hydrolyzable esters or conjugation to aminodextran polymers (Figure 4.10).

A new CuFL (2-{2-chloro-6-hydroxy-5-[(2-methyl-quinolin-8-ylamino)-methyl]-3-oxo-3*H*-xanthen-9-yl}-benzoic acid)–CS (chitosan) NS diazeniumdiolates system consisting of NO donors and highly-sensitive NO probes was reported [41]. Under physiological conditions, the CuFL–CS NS diazeniumdiolates can release NO. A double quantum dot nanocomposite, as a NO ratiometric fluorescent probe, CdSe/SiO_2-CdTe nano composite was constituted [46]. NO was introduced into the nanocomposite system, followed by detecting fluorescence emission spectrum of the system. The linear relationship between ratiometric fluorescent intensity of CdSe/SiO_2-CdTe and NO concentration was shown.

4.3.3. Methods Based on Utilization of Super Molecules

4.3.3.1. Dual Fluorescent-Nitroxide Probes

An approach to detection of NO based on the phenomenon of the intramolecular fluorescent quenching of the fluorophore fragment by the nitroxide in a dual fluorescent-nitroxide supermolecula was developed [26, 42, 43]. Specifically, pyrene-nitronyl (PN)

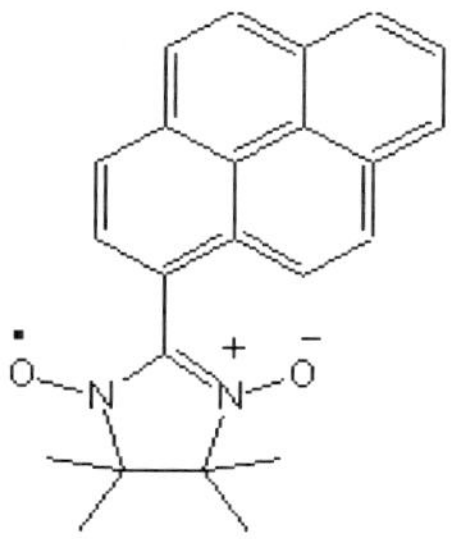

reacts with NO to yield a pyrene-imino nitroxide radical (PI) (Figure 1.5). Conversion of PN to PI results in a change of the electron paramagnetic resonance spectrum from a five-line pattern (two equivalent N nuclei) into a seven-line pattern (two nonequivalent N nuclei) accompanied by an eightyfold increase in the fluorescent intensity). The experiments indicate that the fluorescent measurements enable detection of nanomolar concentrations of NO compared to a sensitivity threshold of only several micromolar concentration for the EPR technique (Figures 1.6 and 4.11).

4.3.3.2. Fluorescence Inductive–Resonance Method of Analysis (FIRMA1)

The principle of a fluorescent NO analysis based on the phenomena of inductive-resonance energy transfer from a donor to an acceptor is to construct a donor acceptor pair, in which the donor is a fluorophore (a stilbene probe tethered to myoglobin) in exited singlet state and the acceptor is an NO trap (hemin) (Figure 4.12) [24, 25, 44, 45]. The proposed methods meet the following strict requirements: (1) the hemin NO trap provides high-speed formation of the corresponding nitronyl complex and very high stability and (2) modern fluorescence techniques are able to detect very weak signals, down to single molecule.

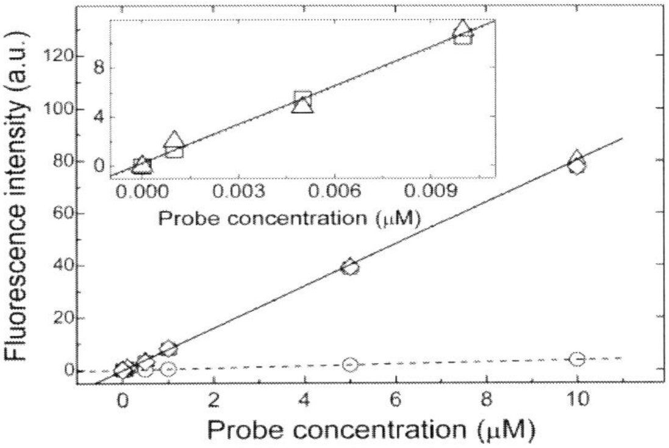

Figure 4.11. Integrated fluorescence measured before (circles) and after a 5-h incubation of PN with excess N-acetyl-Snitrosopenicillamine (10 μM; squares) and a 10-min incubation with NO (10 μM; diamonds). The fluorescence of the pyrene–imino (triangles) is measured for comparison [43]. With permission from Elsevier.

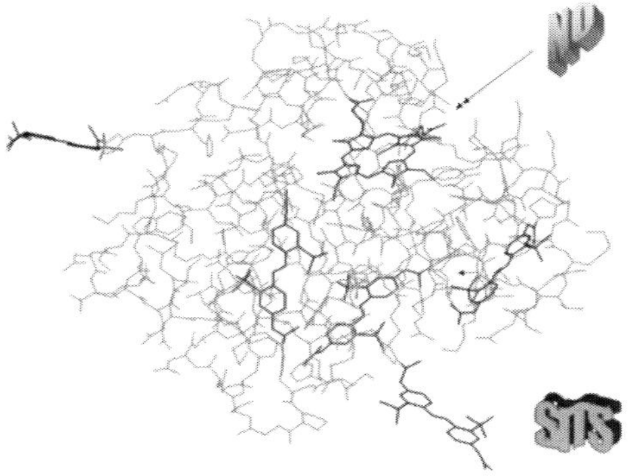

Figure 4.12. Molecule of myoglobin with attached fluorescence label SITS (Mb(Fe^{2+})-SITS) [24].

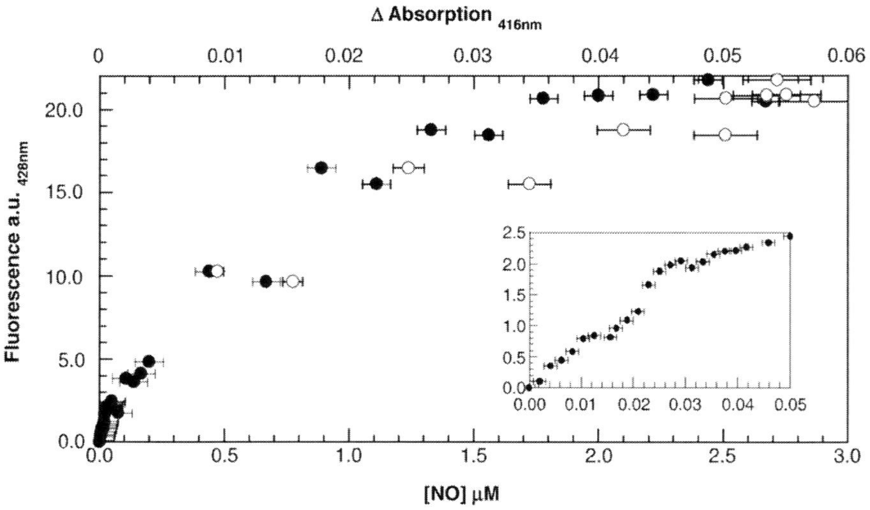

Figure 4.13. Calibration curve drawn as fluorescent intensity versus absorption (upper x-axis), which is proportional to NO concentration (lower x-axis). The graph inset is an extended version of the low concentration part of the calibration curve [24]. With permission from Springer Nature.

The obvious consequence of the theory of inductive-resonance energy transfer is that trapping NO by the hemin Fe^{2+} will lead to a change in absorption spectrum and, in consequence of this, to a change of the overlap integral and the intensity of fluorescence. The latter makes possible the detection of small quantities of NO by the fluorescence technique of very high sensitivity. This method was tested using myoglobin covalently modified by stilbene label 4-acetamido-4′-isothiocyanatostilbene- 2,2′-disulfonic acid (SITS) (Figures 4.12 and 4.13). The rapid trapping of NO by myoglobin results in the change of the absorbance of myoglobin, which causes the change of both the absorbance–fluorescence overlap integral value and the fluorescence intensity of myoglobin–SITS.

The change in emission intensity of the stilbene fragment, versus an increasing concentration of NO precursors, clearly demonstrated the spectral sensitivity required to monitor the formation of a heme–NO complex in a concentration range of 10 nM – 2 μM (Figure 4.14).

4.4. CHEMILUMINESCENCE

Nitric oxide can be detected in the gas phase by the chemiluminescence of its reaction with ozone [47–50]:

$$NO + O_3 \rightarrow NO_2 + O_2 + h\nu$$

This method is the most sensitive chemical assay currently available for the detection of NO up to 50 attomole, $5x10^{-17}$ mole. Nevertheless, competition reactions of NO with other species in a system under investigation is expected.

The following detailed mechanism of these reactions was proposed: [48]

$$NO+O_3= NO^*_2(^2B_1)+O_2 \quad (1a)$$

$$NO+O_3= NO_2(^2A_1)+O_2 \quad (1b)$$

$$NO^*_2= NO_2+hv \quad (2)$$

$$M+NO_2= NO_2+M \quad (3)$$

The rate constants of the primary steps are $k_{1a} = (7.6 \pm 1.5)\times 10^{11}$ exp $(-4180 \pm 300/RT)$ and $k_{1b}=(4.3 \pm 1.0)\times 10^{11}$ exp $(-2330 \pm 150/RT)$ cm^3 $mole^{-1}$ s^{-1}.

The activated NO_2-luminesces broadband visible to infrared light as it reverts to a lower energy state, which can readily be detected by a luminometer (for example, a chemiluminescence detector for the analysis of NO designed in [51]. In the detector, conversion of NO to NO_2, employing the oxidant chromium trioxide, followed by detection of chemiluminescence in the reaction of NO_2 with an alkaline luminol/H_2O_2 solution was used. The presence of H_2O_2 is found to enhance the sensitivity of NO_2 detection by a factor of ~20.

Another series of chemiluminescent methods used for quantitation of NO generation is based on reaction catalyzed by enzyme luciferase [53]:

Luciferin + O_2 + ATP → Luciferase Oxyluciferin + CO_2 + AMP + PPi + light
The reaction occurs in two steps:

luciferin + ATP → luciferyl adenylate + PP_i
luciferyl adenylate + O_2 → oxyluciferin + AMP + light

Luciferyl adenylate singlet-excited singlet state emits light upon relaxation to its ground state. For example, yellow-green light (λ_{max} = 565 nm) is produced by fireflies for reaction with the luciferin.

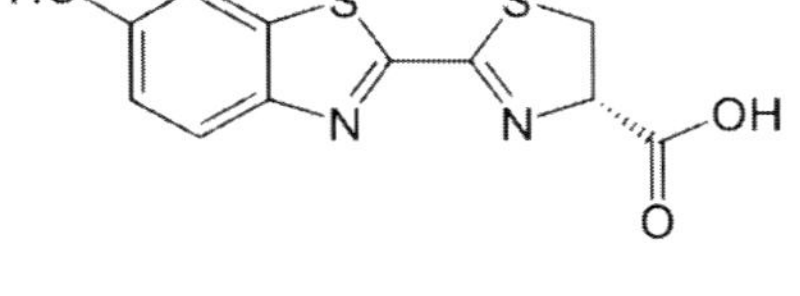

The method of NO assay developed in [53] is based on the reaction catalyzed by glyceraldehyde 3-phosphate dehydrogenase (GAPDH), whose product is used as a substrate for phosphoglycerate kinase, generating adenosine 5′-triphosphate (ATP), which is an essential cofactor for the firefly luciferase bioluminescent reaction. The range from 10 to 100 nM of NO, with limits of detection and quantitation of 4 and 15 nM, respectively, can be detected.A highly sensitive chemiluminescence approach to direct NO detection in aqueous solutions using a natural NO target, soluble guanylyl cyclase (sGC), which catalyzes the conversion of guanosine triphosphate to guanosine 3′,5′-cyclic monophosphate, and inorganic pyrophosphate was introduced in [54]. Converting inorganic pyrophosphate into ATP catalyzed by ATP sulfurylase is accomplished by light emission from the ATP-dependent luciferin–luciferase reaction. Detected by an LB9505 luminometer (Berthold Analytical Instruments, Nashua, NH, USA).

4.5. Electrochemistry, Cyclic Voltammetry, and Amperometry

4.5.1. General

Electrochemistry studies chemical reactions that use or make electricity: electrochemical techniques, cyclic voltammetry and amperometry, together with fluorescence, are the most sensitive and commonly employed analytical method for monitoring nitric oxide. Real-time monitoring and the ability for high-spatial resolution, wide variation of electrodes materials, and its modification to enhance selectivity and sensitivity provide specific advantages [3, 55–64]. Nevertheless, it is necessary to take in consideration that the relatively high-working potential required to oxidize NO (+0.7 to 0.9 V versus Ag/AgCl), for different types of electrodes may lead to interference from NO_2^-, ascorbic acid, uric acid, dopamine, CO, etc. presenting in high concentrations [61].

4.5.2. Electrochemical Nitric Oxide Sensors

Most electrochemical NO sensors are based on the oxidation of NO to NO_2 though electrocatalytic reduction of NO also has been reported. Starting from pioneering work of Shibuki [62] in which a Teflon©-coated platinum was used as a working electrode, a number of selective membranes have since been employed to fabricate electrochemical sensors. Among them are cellulose acetate, collodion/ polystyrene, polycarbazole, Nafion, polyeugeno, polydimethylsiloxane, phenylenediamine [3], polysiloxane cross-linked LB films [63], the xerogel composed of methyltrimethoxysilane and (aminoethylaminomethyl) phenethyltrimethoxysilane, and Nafion [64]. Other approaches for NO electrocatalytical measurement include modifying electrodes with metallorganics, such as hemin, hemin proteins, nickel planar macrocyclic complex [65], Co(II) phthalocyanine [66], films with central ions of Fe, Co, Cu, and Mn [67], and others.

With specific molecular functionalization, semiconductor-based field-effect transistors (FETs) can be conFigured as NO sensors with reduced dimension for real-time detection of NO [68–74]. The following functionalized composites used for NO analysis have been reported: Graphene Gold Nanoparticle Composite [68], carbon fiber [69], nonorganic-semiconductor GaAs [70], ZnO–PPy nanocomposite modified Pt electrode [72], and cytochrome c modified-conducting polymer microelectrode oxide [73].

A composite film containing horseradish peroxidase and kieselguhr, after the deposition on a pyrolytic graphite electrode, exposed high catalytic activity toward the reduction of NO [74]. The peak current related to NO was linearly proportional to its concentration from 2.0×10^{-7} to 2.0×10^{-5} M. The electrochemical reduction of NO was catalyzed, employing an electrode modified with hemoglobin–DNA films [75]. Experimental results revealed that the peak current related to NO is linearly proportional to its concentrations starting from 2.9 μM. In work [76], myoglobin of the horse heart was incorporated on multi-walled carbon nanotubes (MWNTs) and immobilized at a glassy carbon (GC) electrode surface. Electrochemical methods were employed to characterize the microsensor electrochemical behavior and the protein activity in the reduction of NO. The results also indicated that MWNTs can promote the direct electron transfer between Mb and the electrode. Experimental results reveal that the peak current related to NO is linearly proportional to its concentration in the range of 2.0×10^{-7}–4.0×10^{-5} M, and the detection limit is 8.0×10^{-8} M.

4.5.3. Cyclic Voltammetry

Nitric oxide sensors based on electrocatalytic platforms such as ruthenium (colloids, nanoparticles, and nanotubes) and carbon (pastes and nanotubes), acting as catalytic sites for NO oxidation, was fabricated and investigated by cyclic voltammetry and amperometry both in the solution phase and gas phase [77]. Functionalized graphene sheets (FGSs) monolayer electrodes were fabricated employing different heat treatments (Figure 4.14) [78]. Electrocatalytic properties of FGSs in NO sensing were determined by cyclic voltammetry, which exhibits an NO oxidation peak potential of 794 mV (vs 1 M Ag/AgCl). Porous FGS electrodes indicated a stronger apparent electrocatalytic effect as compared to platinized electrode.

A hemin-modified electrode for cyclic voltammetry was designed and utilized to study its electrocatalytic reaction with NOx [79]. An electrochemical NOx sensor was fabricated based on the incorporation of hemin on a ZnO–PPy nanocomposite modified Pt electrode and cyclic voltammetry (CV) were employed to characterize the NOx sensor. The electrocatalytic response of the sensor was proportional to the NO_xconcentration in the range of 0.8 to 2000 μM ($r^2 = 0.9974$) with a sensitivity of 0.04 μA μM^{-1} cm^{-2} and detection limit of 0.8 μM for the hemin–ZnO–PPy–Pt electrode. The experiments indicated that the presence of ZnO–PPy nanocomposite facilitates the direct electron transfer from hemin to the Pt electrode. The authors concluded that ZnO–PPy nanocomposite combined the advantages of polymers as p-type semiconductors and high electron mobility of inorganic n-type semiconductors. Highly sensitive voltammetric biosensor for NO based on its high affinity with Hb was designed [80]. It was shown that the catalytic reduction of oxygen occurs via direct exchange of electrons of film-entrapped Hb with the electrode. In addition, NO induced a cathodic potential shift of the catalytic reduction peak of oxygen which was proportional to the logarithm of NO concentration ranging from 4.0×10^{-11} to 5.0×10^{-6} mol/L. The Nafion/lead-ruthenate pyrochlore chemically modified electrode was fabricated and employed for investigation of the dual sensing activity toward NO_2- oxidation and NO reduction using CV, ac-impedance spectroscopy and flow injection analysis [81]. The calibration curve was linear in the range of 100 nM-100 microM and 800 nM-63.3 μM, and the detection limit was 4.8 nM and 15.6 nM for NO_2- and NO, respectively.

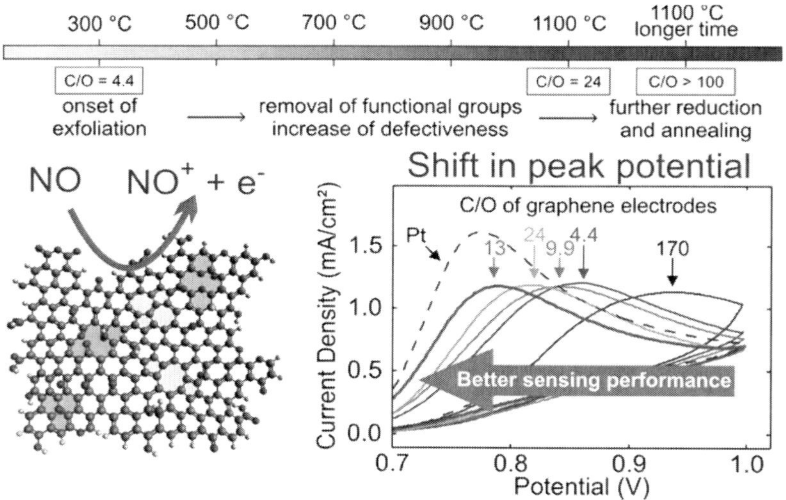

Figure 4.14. Functionalized graphene sheets (FGSs) monolayer electrodes: (top) temperature employing different heat treatments during their fabrication; (left) schematic view of FGS, cyclic voltammetry with the FGS monolayer electrodes [78].

Direct electrochemistry and electrocatalysis of nano-sized gold nanoparticle–graphene composite electrode modified with Hb was investigated by the CV method [82]. The separation of anodic and cathodic peak potentials in a phosphate buffer solution (pH 7.0) was 81 mV, indicating a fast electron transfer reaction. The experiments showed the linear response range of NO concentration of 0.72-7.92 μM. Based on the direct electron transfer of cyt *c*, determination of NO with the electrode, modified with the cyt *c*-bonded a functionalized-conducting polymer (poly-TTCA poly-TTCA and Nafion film), was achieved using CV and chronoamperometry [83]. Cyclic voltammograms showed a reduction peak at −0.7 V and corresponding calibration plot for the NO concentration range of 2.4–55.0 μM and the detection limit 13 nM. A biomimetic NO microsensor (carbon fiber microelectrode) was designed using the electrodeposition of Chit (chitosan, a linear ß-1,4-polysaccharid), crosslinking with MWCNTs and hemin by 1-[3- (Dimethylamino)propyl]-3-ethylcarbodiimide [84]. Nitric oxide analysis was performed by CV and square wave voltammetry techniques. Square wave voltammetry revealed a NO reduction peak at -0.762 V vs. Ag/AgCl that increased linearly with NO concentration between 0.25 and 1 μM.

4.5.4. Amperometry

Amperometry in chemistry is detection based on an electric current or changes in an electric current [85-88]. Below, efficiency of amperometry in the NO analysis is illustrated by several examples. An amperometric NO microsensor based on polycondensation of the xerogel composed of methyltrimethoxysilane and (aminoethylaminomethyl) phenethyltrimethoxysilane and Nafion was constructed. The sensor exhibits a sensitivity to NO gas of 0.17 ± 0.02 pA/nM (from 25 to 800 nM, detection limit of 25 nM (S/N = 3), and a response time of 9 seconds [86]. Higher selectivity over nitrite, ascorbic acid, uric acid, and acetaminophen has been also reported. An amperometric microsensor for sensing NO was prepared based on a dual recessed electrode possessing Pt microdisk modified with electro-platinization and the following coating with fluorinated xerogel [87]. The high selective and sensitive microsensor was used for amperometric sensing of NO (106 ± 28 pA μM^{-1}, n = 10, at +0.85 V applied vs Ag/AgCl). The sensor showed good selectivity over common biological interferents.

An improved planar amperometric NO sensor with enhanced selectivity over carbon monoxide, using formation of an oxide film on the inner platinum working electrode was reported [89]. The sensor exhibits the selectivity coefficients (log $K_{NO,j}$) at various pH values range from −0.08 at pH 2.0 to −2.06 at pH 11.7, with average NO sensitivities of 1.24 nA/μM and a limit of detection of < 1 nM. A new type of hybrid, organic-semiconductor, electronic sensor based on measurement electrical resistivity changes as a result of NO binding to a layer of hemin molecules was reported [90]. The sensor provides a fast and simple direct detecting NO at concentrations down to 1 μM in a physiological aqueous (pH=7.4) solution at room temperature.

Remarkable advances of electrochemical techniques were in full extend demonstrated in [91]. A sensor for real-time monitoring of NO based on hemin-functionalized graphene field-effect transistors with single atom thickness, and the highest carrier mobility a subnanomolar sensitivity was designed. A graphene FET device was prepared by the following procedure: (1) mechanical exfoliation onto a silicon substrate with a 300-nm thermal oxide layer, (2) fabrication titanium-gold thin film source, (3) drain and solution gate electrodes, using e-beam lithography, (4) vacuum metal deposition, and (5) a lift-off process. Immobilization of hemin was performed via the non-covalent functionalization through p–p stacking interaction. Ultraviolet–visible absorption and Raman spectroscopies and atomic force microscope were used to characterize the immobilization of hemin molecules on graphene. A typical graphene device has an active area of 0.5x0.5x0.5x2 mm^2, sizes of source and drain electrodes are 2x5 mm^2 (excluding the electrical leads), and the size of a gate electrode is 50x50 mm^2. The microfluidic polydimethylsiloxane channel has a width of 1 mm, height of 0.5mm and length of 1 cm [91]. High sensitivity of the porous haemin-functionalized ZnO sensor to NO concentration was illustrated [91]. Response of the sensors was exposed to several cycles of NO concentrations (30 ppm and 10 ppm). Real-time detection of NO released from living cells (20 nM Bradykinin) was also demonstrated.

A NO gas sensor, which was functionalized by hemin with the reaction between hydroxyl group (–OH) on the ZnO surface and carboxyl group (–COOH) of the probe, was designed [85]. The sensor makes it possible to detect NO concentration from 1 ppm to 30 ppm for a few seconds. In order to increase the surface area, the ZnO film was made porous by a template technique where films with three-dimensional periodic submicrometre spherical voids were arranged in a close-packed structure. In the porous sample, both the sensitivity and response speed are essentially improved. This method has high selectivity. A reaction mechanism of the electron transfer between the hemin Fe^{3+} molecule and ZnO followed by trapping of NO was suggested.

A novel hemin-functionalized InAs planar resistors or molecularly controlled resistors (MOCSERs), microsensor for NO in gas phase, was prepared and characterized using X-ray photoelectron spectroscopy, atomic force microscopy, and spectroscopic ellipsometry [93]. The covalent attachment of hemin to both In and As sites and the strong electronic coupling of the InAs surface electrons with surface-attached hemin molecules provide high sensor sensitivity (Figure 4.15). Figure 4.16 presents the hybrid InAs-hemin sensor response, which was found to be proportional to the NO concentration. The response of the device to NO spikes at 300 K in N_2 over a 2-100 ppb exposure range after background signal drift and contact resistance removal was shown. Changes in resistance as a function of various gases→ species are illustrated in Figure 4.17.

The suggested hemin-NO chemical interactions was given as a sequence of the reactions:

$$\text{Hemin } Fe^{3+} + e \longrightarrow \text{Hemin } Fe^{2+}$$
$$\text{Hemin } Fe^{2+} + NO \leftrightarrow \text{HeminFe}^{2+}\ NO$$
$$\text{Hemin } Fe^{2+}\ NO + NO \leftrightarrow \text{Hemin } Fe^{2+}\ N_2O_2$$

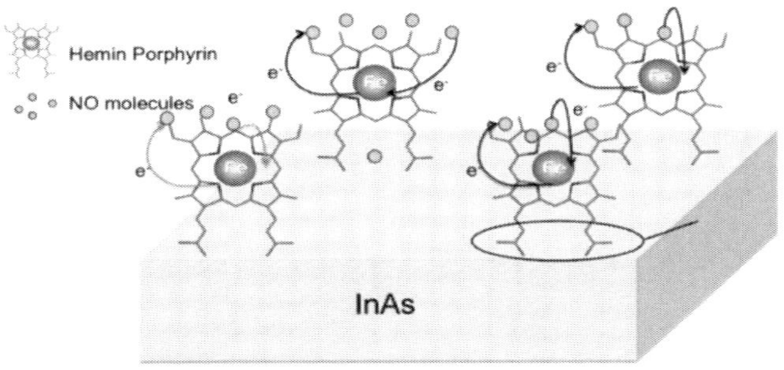

Figure 4.15. Schematic illustration of the hybrid InAs hemin sensor [93]. Copyrights 2012 American Chemical Society.

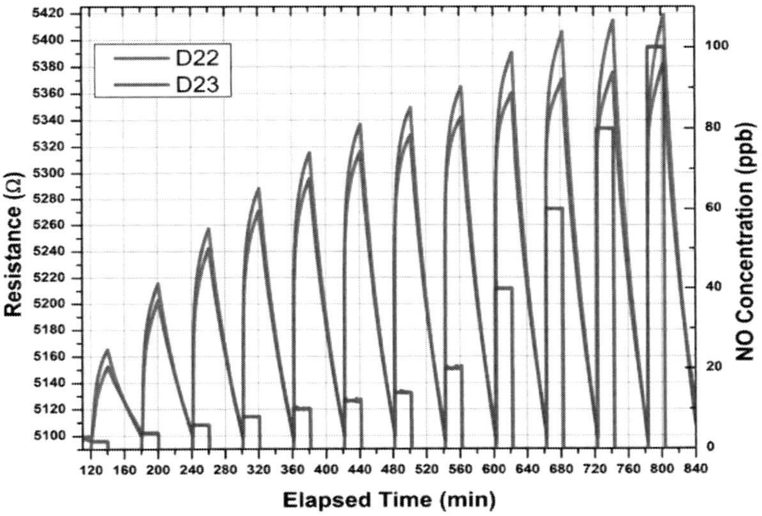

Figure 4.16. MOCSER response (two devices) to NO as a function of time (see details in [93]. Copyrights 2012 American Chemical Society.

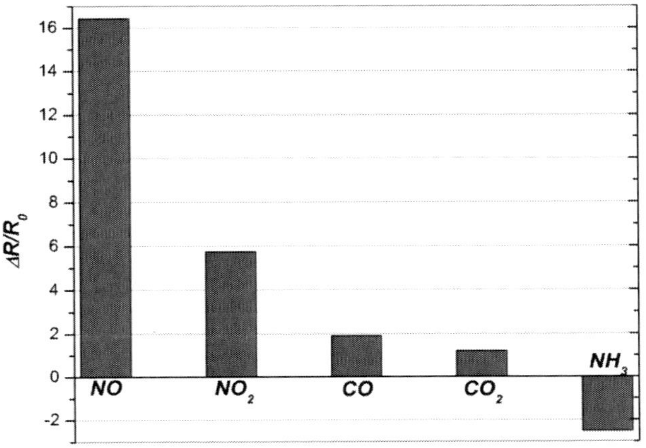

Figure 4.17. Change in resistance as a function of various gases→ species at 2.5 ppm concentration [93]. Copyrights 2012 American Chemical Society.

In the latter reaction, the formation of a N-N bond between two NO molecules presumably occurs by the electrophilic attack of free NO on one of the bound NO molecules, leading to the oxidation of Fe^{2+}.

4.6. Other Nitric Oxide-Detection Techniques

The spectroscopic and electrochemical methods described above are the most commonly employed techniques for measuring NO. Other less frequently employed approaches that have been reported include mass spectrometry [94], X-ray photoelectron spectroscopy [95], quantum cascade infrared laser spectroscopy [96], mechatronics sensoring [97], and the quartz crystal microbalance technique [98]. Space constraints do not permit us to describe these techniques in detail. The fast, specific, and non-invasive online detection of NO using a combination of membrane inlet mass spectrometry and restriction capillary inlet mass spectrometry was developed [94]. This approach allows discriminate nitrogen isotopes and simultaneously measures NO and O_2 from the same sample. X-ray photoelectron spectroscopy (XPS) is based on irradiating a material with a beam of X-rays while measuring the kinetic energy and number of electrons that escape from the top 0 to 10 nm of the material being analyzed at ultra-high vacuum. [https://en.wikipedia.org/wiki/X-ray photoelectron_spectroscopy]. A highly sensitive technique is reported to detect NO using XPS of a silicon-bound iron heme-like complex [95]. In this work, hematin was covalently attached to a self-assembled monolayer of an hydroxyalkylphosphonic acid grown on the native oxide surface of silicon (Figure 4.18). After reaction with NO, a new, distinct peak is observed in the N1s spectrum, at approximately 5.5 eV higher binding energy, which is attributed to the heme-bound NO. On the basis of measurements of surface loading of hematin species using quartz crystal microgravimmetry and XPS, detection of $\leq$ 50 picomoles of NO in the sampled region can be accomplished.

Quantum cascade lasers (QCLs) are semiconductor lasers that emit in the mid- to far-infrared portion of the electromagnetic spectrum from multiple quantum well heterostructures [https://en.wikipedia.org/wiki/ Quantum_cascade_. In such structures, electrons undergo tunnel intersubband transitions and photons are emitted. To detect atmospheric NO at spectral lines at 1900.07 cm^{-1}, a QCL operating in both continuous-wave (cw) mode (-9°C) and pulsed mode (+45°C) has been used [96]. A quartz-enhanced photoacoustic sensor (QEPAS) for NO detection using a mid-infrared fiber-coupled QCL near 5.2 μm was reported [92]. For light delivery to the quartz tuning fork, a tiny piezoelectric element converting the acoustic-wave-induced mechanical vibration to the gas-absorption associated electrical signal, the photoacoustic radiation was coupled into an InF_3 fiber. The photoacoustic absorption line with a stronger absorption coefficient located at 1900.08 cm^{-1} was selected. The measurement of QEPAS 2f signals amplitude for different NO concentrations showed the linearity of the sensor from 30 ppm to 500 ppm.

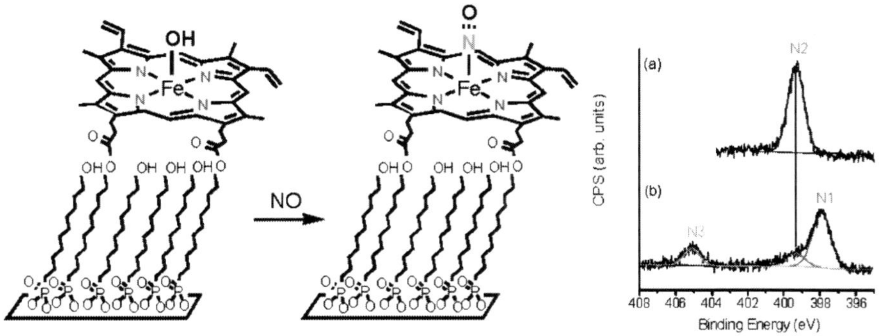

Figure 4.18. Schematic of the surface modification strategy [95]. Copyrights 2007 American Chemical Society.

A QCM measures a mass variation per unit area, down to a level of below 1 μg/$cm^2$by measuring the change in frequency of a quartz crystal resonator. In the acoustic resonator, frequency measurements can be made to high precision. Paper [98] presents the properties of indium–tin oxide (ITO)-covered QCMs as a NO gas sensors. The sensor featured ITO thin films of ~100 nm as the receptor to sense the gas NO. A quartz crystal with frequency of 10 MHz was found to be sensitive to the gas addition or removal to compound growth/decay in NO concentration scale 53-1800 ppm.

A coherent anti-Stokes Raman scattering (CARS) setup has been developed to detect contamination of atmospheric nitrogen by NO [99]. A frequency-doubled Nd:YAG laser (l = 532 nm, bandwidth 0.05 cm^{-1}) serves as pump for a dye laser (bandwidth, 0.03 cm^{-1}), which is tunable between 585 and 615 nm. Nitric oxide CARS spectra have been measured with high spectral resolution in a temperature range from 300 to 800 K. The detection limit of NO is on the order of 0.25% in nitrogen under atmospheric pressure.

Of particular interest is the analysis of NO at single cell levels. In review [100], for three categories of analytical techniques enabling NO detection at single cell levels, namely, fluorescence microscopy, capillary electrophoresis with laser-induced fluorescence detection, and electrochemistry, the basic principles, performance, applications, and Figures of merits and limitations are presented.

4.7. Nitric Oxide Spin Trapping

Spin trapping is designed for an investigation of processes with participation of short living particles and reactive free radicals bearing unpaired electrons [101–108]. The technique is based on the reaction of a molecule (*spin trap*) with formation and sufficiently more persistent paramagnetic species, called the *spin adduct*, most commonly nitroxide or complex of paramagnetic metal. The spin adduct can be detected by EPR spectroscopy. From the EPR spectrum of the nitroxide or paramagnetic complex, it is usually possible to identify the trapped species. The examples of various types of spin traps, namely, iron dithiocarbamate complexes, specific organic compounds, and heme-proteins are given below.

The dithiocarbamate-iron spin traps

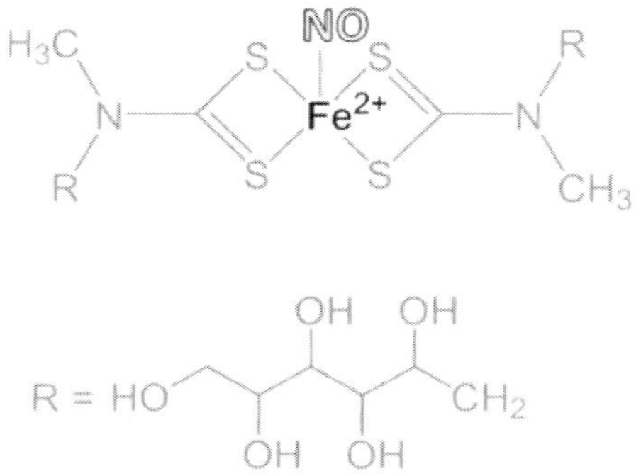

were employed in combination with a liposome-encapsulating technique and ESR spectroscopy The liposome membrane forms a physical barrier between the spin-trap, catalytic enzymes (i.e., nitricoxidesynthase and nitrate reductase) and most substrates, while permitting the diffusion of NO. 5-Hydroxy-2,2,6,6-tetramethyl-4-(2-methylprop-1-en-yl)cyclohex-4-ene-1,3-dione behaves as an efficient trap for both NO and NO_2 radicals in the presence of oxygen, yielding EPR observable nitroxide and alkoxynitroxide [104].

3,5-dibromo-4-nitrosobenzene sulfonate (DBNBS) has been used in combination with EPR spectrometry to trap NO [105]. In the presence of oxygen, the reaction between DBNBS and NO yields a radical product, which gives rise to an EPR signal consisting of three lines with an = 0.96 mT. The negative ion fast atom bombardment–mass spectrometry data suggested that the radical product is the monosodium electrostatic complex with the dianion, bis(2,6-dibromo-4-sulfophenyl) nitroxyl. The reaction mechanism of trapping of NO by DBNBS 3,5- was also investigated in [106]. Nitronyl nitroxide reacts with NO, yielding an imino nitroxide and nitrogen dioxide [107] (Figures 1.5 and 1.6)

It was reported that nitroalkanes (RCH_2NO_2) undergo deprotonation and rearrange to an aci anion ($RHC{=}\ NO_2^-$), which may function as a spin trap in alkaline solutions [108]. Suitability of aci anions of a series of nitroalkanes (CH_3NO_2, $CH_3CH_2NO_2$, $CH_3(CH_2)_2NO_2$, and $CH_3(CH_2)_3NO_2$) to spin-trap nitric oxide (*NO) was established by ESR spectroscopy. The structure of the adducts was verified using ^{15}N-labeled *NO. The adduct ESR spectra show triplet due to splitting on ^{14}N ($I = 1$) in 14NO/aci nitro adducts, which was replaced by a doublet due to ^{15}N ($I = 1/2$) in ^{15}NO/aci nitro adducts.

4.8. NOVEL HIGH-SENSITIVE AND HIGH-SELECTIVE FLUORESCENCE METHODS FOR REAL-TIME MONITORING OF NITRIC OXIDE IN BIOLOGICAL SYSTEMS IN THE PICOMOLELESS RANGE

4.8.1. Physical Background

Although a variety of methods for NO detection have been proposed, they are not generally amenable to real-time measurement of endogenous NO *in situ*. The principle of proposed analysis of nitroxide is design systems of donor acceptor pairs in which the donor is a fluorophore in the excited singlet state and the acceptor of light energy is hemin (system 1), or excited fluorophore is the electron acceptor and hemin is the electron donor (system 2). The proposed approach is based on well-known independent physical phenomena, namely, singlet-singlet inductive-resonance energy transfer from an exited donor to acceptor (in the system 1), or on effect of spin state of acceptor on electron transfer from donor and intersystem crossing (in the system 2). The proposed methods of NO analysis obey to the following strict requirements: (1) NO trap needs to provide high-speed formation of the corresponding nitronyl complex and high stability in terms of experience in the presence of oxygen.Hemin Fe^{2+} is fully consistent with is this requirement in certain conditions; and (2) for the detection of the formation of this complex, a highly sensitive method should be used. Modern fluorescence techniques allow to detect very weak signals, down to single molecule.

Hemin-Fe^{2+} having an intense Soret peak in the blue wavelength region of the visible spectrum ranging around 400 nm can serve as a light energy acceptor or electron donor. As a fluorescent donor it is suggested to utilize fluorophores with corresponding characteristics (stilbene derivatives, for example, with absorption range 300-770 nm) and size-tunable CdSe quantum dots [109]. These particles having diameters of 2.1 nm, 2.5 nm, 2.9 nm, 4.7 nm, and 7.5 possess unique optical properties, namely, to absorb and to emit light in a wide range of lengths of waves. The above-mentioned physical phenomena and optical characteristics of the potential donor and acceptor compounds form the basis of two fluorescent methods namely: fluorescence inductive-resonance method of analysis (FIRMA2) and fluorescence spin exchange method of analysis (FSEMA)

4.8.2. Fluorescence Inductive-Resonance Method of Analysis (FIRMA2)

FIRMA 2 is a development of the approach of analysis of NO (FIRMA1) described in section 4.3.3.2. and references [24, 44]. The suggested method was also based on phenomena of the singlet-singlet energy transfer by inductive resonance mechanism, due to the dipole-dipole interaction between the excited fluorescent donor (D*) and nonexcited acceptor (A), according to the Forster theory rate constant of the energy transfer [110].

$$k_r = 9\,\kappa^2\, ln10\ 128\pi^5 n^4\, N_A\tau_D\ \Phi e\ r^6 \int \epsilon A(\lambda) I^F{}_D(\lambda)\lambda^4 d\lambda \qquad (4.1)$$

where τ_D is the mean lifetime of D*, N_A is Avogadro's constant, r is the distance between the donor and the acceptor, κ is the relative orientation of the donor emission dipole moment and the acceptor absorption, and integral $\int$ is the spectral overlap of the donor emission spectrum and the acceptor absorption spectrum. Förster distance (r_0) of this pair of donor and acceptor is the distance at which the energy transfer efficiency is 50%

The proposed method is based on the obvious consequence of the theory, that trapping NO by the hemin Fe^{2+} will lead to a change in absorption spectrum and, in consequence of this, change of the overlap integral $\int$ and, therefore, the intensity of fluorescence. The latter makes possible the detection of small quantities of NO by the fluorescence technique of very high sensitivity. A pair of hemin (absorption in the Soret region) and a corresponding CdSe quantum dot (Figure 4.21) obeys the requirement of efficient overlap of the donor emission spectrum and the acceptor absorption spectrum.

Figures 4.19 and 4.20 illustrate schematically two donor -acceptor (D-A) models, having a stilbene derivative as a donor and hemin as an acceptor, with direct D-A energy transfer and transfer via a light harvesting antenna, correspondingly. The latter is an analog of the photosynthetic light-harvesting antenna (Figure 4.21).

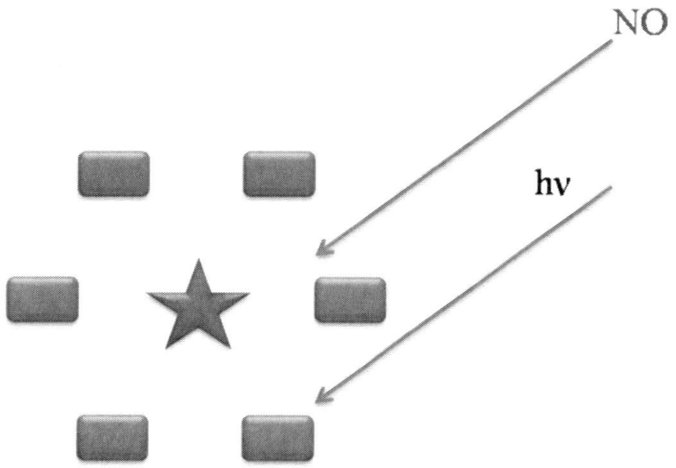

Figure 4.19. Direct energy transfer from a fluorescent donor (rectangle) to the hemin acceptor (star) (Model 1).

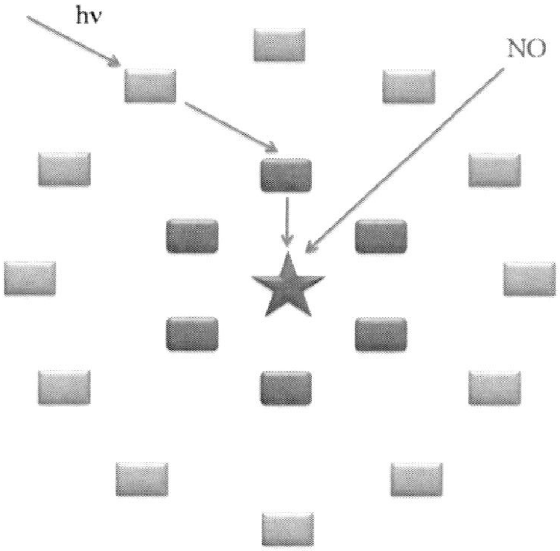

Figure 4.20. Energy transfer via inductive resonance exciton cascade (light-harvesting antenna) (Model 2). Rectangles are fluorescent donor; star is hemin.

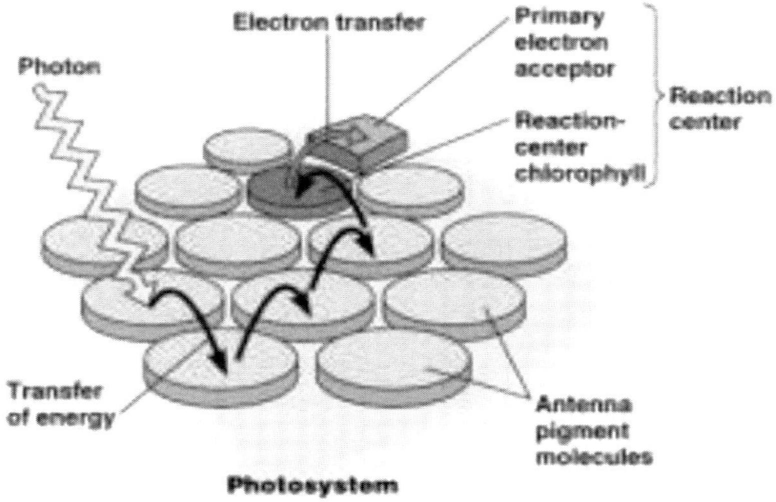

Figure 4.21. Photosynthetic inductive resonance exciton cascade (antenna).

The harvesting light system is expected to increase sensitivity of detection of NO trapped by the hemin acceptor. In a supermolecule (Figure 4.22), donor (excited CdSe quantum dots) connects with acceptor (hemin) pair connects with a bridge (Model 3). The length of the bridge should approximately correspond to the Forster distance (R_0). A pair of hemin (Soret peak) and a corresponding CdSe quantum dot [109] fits s the requirement of overlap of the donor emission spectrum and the acceptor absorption spectrum.

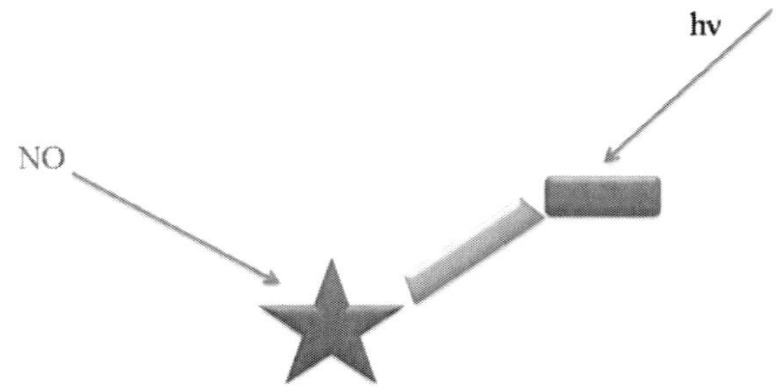

Figure 4.22. Donor (excited CdSe quantum dots, rectangle) acceptor (hemin, star) pair connected with a bridge (trapezium) (Model 3).

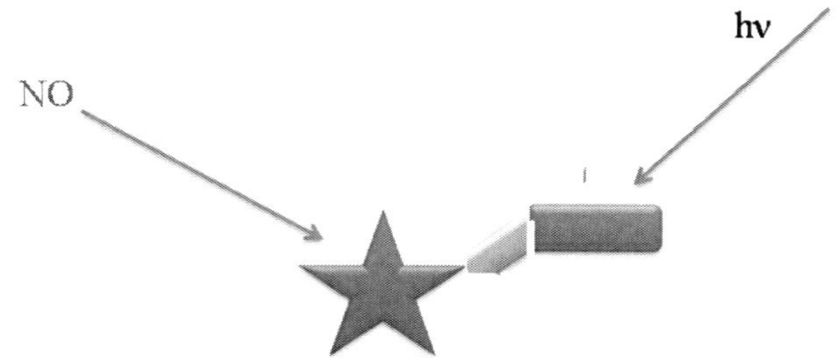

Figure 4.23. Electron donor (hemin Fe2^{+}, star), electron acceptor (an exited fluorophore (rectangle) connected via a bridge optimum for exchange interaction (Model 4).

4.8.3. Fluorescence Spin Exchange Method of Analysis

The method is based on measuring the difference in efficiency of spin exchange (electron transfer, intersystem crossing) in supermolecules, consisting of NO trap (hemin Fe^{2+}) and a fluorophore, by detecting fluorescence in absent and present NO. Binding of NO by the trap fragment should influence on efficiency of the spin exchange between excited fluorophore and the nitryl complex and, as a consequence, on the fluorescence intensity. In the frame of this conception, supermolecules composed with acceptor (an excited fluorophore) and donor (hemin Fe2+) segments tethered via a bridge can be designed (Model 4).

For non-adiabatic process, the rate constant of electron transfer (ET) in a donor acceptor pair is given by equation 4.2: [111]

$$k_{ET} = \frac{2\pi V^2}{h\sqrt{4\pi\lambda k_B T}} \exp\left[-\frac{(\lambda + \Delta G_0)^2}{4\lambda k_B T} \right] \tag{4.2}$$

where ΔG_0 is the standard Gibbs energy, λ is the reorganization energy, and V is the resonance integral approximately exponentially dependent on the distance between donor and acceptor. A change in these parameters after the NO binding should lead to a change in the ET rate and intensity of fluorescence. A similar effect is expected in the case of the intersystem crossing mechanism. Intersystem crossing (ISC) is a process in which a singlet state nonradioactively transforms into a triplet state. According to theory [49], the intersystem crossing efficiency is also dependent on the distance between the donor and acceptor.

The ET rate can be exponentially dependent on the distance between donor and acceptor. An example is dependence of maximum rate constant of ET on the edge-edge distance between the donor and acceptor center in photosynthetic reaction centers (RCs) of bacteria and plants (Figure 4.24) [112]. At a distance of about 12-14 Å, the ET rate is comparable with the rate of spontaneous emission of fluorophore acceptor (typically, $1/\tau_D = 10^8 - 10^9\ s^{-1}$). Thus, this distance would be optimum for detection effect of change of fluorescence after the NO trapping by hemin Fe^{2+}.

In the framework of FSEMA, to prevent a competing effect of the inductive resonance process, which is very strong at distances of 12–14 Å, is indispensably to choose a fluorophore with a minimum overlap of its emission spectrum and the acceptor absorption spectrum. The menu of the size-tunable fluorescence spectra of CdSe quantum dots [109] offers the way for the selection of a corresponding pair in a supermolecule.

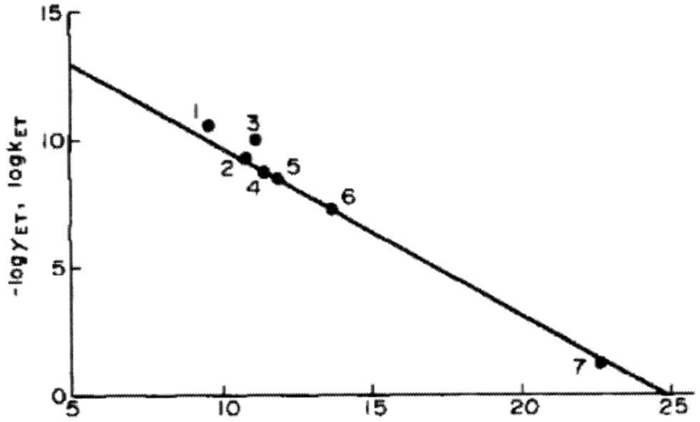

Figure 4.24. Dependence of maximum rate constant of ET on the edge-edge distance between the donor and acceptor centers in photosynthetic RCs of bacteria and plants (see details in [112]).

Suggested methods can be employed for research of processes of NO flux and adsorption in cells, tissues, organs, animals, biological liquids and the human body, in the future. The methods can make basis for design of portable simple and cheap biosensor for NO for wide use in research laboratories and medical practice.

REFERENCES

[1] Louise J, Ignarro LJ. *Nitric Oxide.* 2nd Edition. Elsevier Inc. 2010.

[2] Griess P. Bemerkungen zu der Abhandlung der HH. Weselky und Benedikt Ueber einige Azoverbindungen. *Ber. Deutsch Chem. Ges.* (1879) 12, 426-428.

[3] Hetrick EM, Schoenfisch MH. Analytical Chemistry of Nitric Oxide. *Annu. Rev. Anal. Chem.* (2009) 2, 409-2433.

[4] Jie Sun XZ, Broderick M, Fein H. Measurement of Nitric Oxide Production in Biological Systems by Using Griess Reaction Assay. *Sensors* (2003) 3, 276-284.

[5] Toledo JC Jr, Augusto O. Connecting the Chemical and Biological Properties of Nitric Oxide. *Chem. Res. Toxicol.* (2012) 25, 975-989.

[6] Moshage H, Kok B, Huizenga JR, Jansen PL. Nitrite and nitrate determination in plasma: a critical evaluation. *Clin. Chem.* (1995) 41, 892-896.

[7] Yang F, Troncy E, Francoeur M, Vinet B, Vinay P, Czaika G, Blaise G. Effect of reducing agents and temperatures on conversion of nitrite and nitrate to nitric oxide and detection of NO by chemiluminescence. *Clin. Chem.* (1997) 43, 657-662.

[8] Casey TE, Hilderman RH. Modification of the cadmium reduction assay for detection of nitrite production using fluorescence indicator 2,3-diaminonaphthalene. *Nitric Oxide* (2000) 4, 67-74.

[9] Jobgen WS, Jobgen SC, Li H, Meininger CJ, Wu G. Analysis of nitrite and nitrate in biological samples using high-performance liquid chromatography. *J Chromatogr B Analyt Technol Biomed Life Sci* (2007) 15;851, 71-82.

[10] Tsikas D. Analysis of nitrite and nitrate in biological fluids by assays based on the Griess reaction: Appraisal of the Griess reaction in the L-arginine/nitric oxide area of research. *J. Chromatogr. B Analyt Technol Biomed Life Sci.* (2007) 851, 51-70.

[11] Zhang X.; Broderick M. Electrochemical NO sensors and their applications in biomedical research. *Biomedical Significance of Nitric Oxide.* International Scientific literature, Inc., 2003.

[12] Zhang Y, Samson FE, Nelson SR, Pazdernik TL. Nitric oxide detection with intracerebral microdialysis: Important considerations in the application of the hemoglobin-trapping technique. *J. Neurosci. Methods* (1996) 68, 165-173.

[13] Aylott JW, Richardson DJ, Russell DA. Optical biosensing of gaseous nitric oxide using spin-coated sol-gel thin films. *Chem. Mater.* (1997) 9, 2261-2263.

[14] Boon EM, Marletta MA. Sensitive and selective detection of nitric oxide using an H-NOX domain. *J. Am. Chem. Soc.* (2006) 128,10022-10023.

[15] Dacres H, Narayanaswamy R. A new optical sensing reaction for nitric oxide. *Sens. Actuators B* (2003) 90, 222-229.

[16] Zhang Y, Samson FE, Nelson SR, Pazdernik TL. Nitric oxide detection with intracerebral microdialysis: Important considerations in the application of the hemoglobin-trapping technique. *J. Neurosci. Methods* (1996) 68,165-173.

[17] Aylott JW, Richardson DJ, Russell DA. Optical biosensing of gaseous nitric oxide using spin-coated sol-gel thin films. *Chem. Mater.* (1997) 9, 2261-2263.

[18] Boon EM, Marletta MA. Sensitive and selective detection of nitric oxide using an H-NOX domain. *J. Am. Chem. Soc.* 2006 128:10022-10023.

[19] Dacres H, Narayanaswamy R. A new optical sensing reaction for nitric oxide. *Sens. Actuators B* (2003) 90, 222-229.

[20] Schwendemann J, Sehringer B, Noethling C, Zahradnik HP, Schaefer WR. Nitric oxide detection by DAF (diaminofluorescein) fluorescence in human myometrial tissue. *Gynecol Endocrinol.* (2008) 24:306-311.

[21] Li H, Wan A.Fluorescent probes for real-time measurement of nitric oxide in living cells. *Analyst* (2015) 140, 7129.

[22] Chan J, Chang CJ. Reaction-based small-molecule fluorescent probes for chemoselective bioimaging. *Nature Chemistry* (2012) 4, 973-984.

[23] McQuade LE, Lippard SJ, Fluorescent probes to investigate nitric oxide and other reactive nitrogen species in biology (truncated form: fluorescent probes of reactive nitrogen species. *Curr. Opin. Chem. Biol.* (2010) 14, 43-49.

[24] Liu Xiaomei, Liu Shuang, Liang Gaolin. Fluorescence turn-on for the highly selective detection of nitric oxide in vitro and in living cells. *Analyst* (Cambridge, United Kingdom) (2016), 141, 2600-2605.

[25] Likhtenshtein GI. Novel fluorescent methods for biotechnological and biomedical sensoring: assessing antioxidants, reactive radicals, NO dynamics, immunoassay, and biomembranes fluidity. *Applied biochemistry and biotechnology* (2009) 152, 135-155.

[26] Likhtenshtein GI. Stilbene Molecular Probes as Potential Materials for Bioengineering: Real-Time Analysis of Antioxidants, and Nitric Oxide, Immunoassay in Solution and Biomembrane Fluidity. In Jin Yun, Dehual Zeng (Eds) *Materials Science and Chemical Engineering.* Trans Tech Publications, Singapore, pp. 718-723, (2013).

[27] Likhtenshtein GI. Dual phluorophore-nitroxides as a tool for analysis of antioxidants, nitric oxide and superoxide in biological liquids and tissues, *International Conference "Recent Advantages in Sensoring, Signals and materials,"* November 3-5, 2010, Faro, Portugal, pp 162-168, (2010).

[28] Ma S, Fang DC, Ning B, Li M, He L, Gong B. The rational design of a highly sensitive and selective fluorogenic probe for detecting nitric oxide. *Chem. Commun.* (2014) 50, 6475-6478.

[29] Shiue TW, Chen YH, Wu CM, Singh G, Chen HY, Hung CH, Liaw WF, Wang, YM. Nitric Oxide Turn-on Fluorescent Probe Based on Deamination of Aromatic Primary Monoamines. *Inorg. Chem.* (2012) 51, 5400-5408.

[30] Wang C, Song X, Han Z, Li X, Xu Y, Xiao YXiao, Monitoring Nitric Oxide in Subcellular Compartments by Hybrid Probe Based on Rhodamine Spirolactam and SNAP-tag. *ACS Chemical Biology* (2016) 11, 2033-2040.

[31] Beltrán A, Burguete MI, Abánades DR, Pérez-Sala D, Luis SV, Galindo F. Turn-on fluorescent probes for nitric oxide sensing based on the *ortho*-hydroxyamino structure showing no interference with dehydroascorbic acid. *Chem Commun* (2014) 50, 3579-3581.

[32] Sun C, Shi W, Song Y, Chen W, Ma H. An unprecedented strategy for selective and sensitive fluorescence detection of nitric oxide based on its reaction with a selenide. *Chem.Commun.* (2011) 47, 8638-8640.

[33] Dong CH, Heo S, Chen S, Kim HM, Liu Z. Fluorescent probes based on o-diamine functionality 27, *Anal. Chem.* (2014) 86, 308-311.

[34] Zhang Yu-Jin, Yang Wen-Jing, Wang Chuan-Kui. Effect of fluorophore on sensing NO for newly synthesized PET-based two-photon fluorescent probes. *Chemical Physics* (2016), 468, 37-43.

[35] Yang Y, Seidlits SK, Adams MM, Lynch VM, Schmidt CE, Anslyn EV, Shear JB. A Highly Selective Low-Background Fluorescent Imaging Agent for Nitric Oxide. *J. Am. Chem. Soc.* (2010) 132, 13114-13116.

[36] Li H, Wan A. Fluorescent probes for real-time measurement of nitric oxide in living cells. *Analyst* (2015) 140, 7129-7141.

[37] Hilderbrand SA, Lim MH, Lippard SJ. Metal-Based Turn-On Fluorescent Probes for Sensing Nitric Oxide. *Acc. Chem. Res.* (2007) 40, 41-51.

[38] Pluth Michael D, Chan Maria R, McQuade Lindsey E, Lippard Stephen J. Seminaphthofluorescein-Based Fluorescent Probes for Imaging Nitric Oxide in Live Cells. *Inorg. Chem* (2011) 50, 9385-9392.

[39] Hu X, Wang J, Zhu X, Dong D, Zhang X, Wu S, Duan C. A copper(II) rhodamine complex with a tripodal ligand as a highly selective fluorescence imaging agent for nitric oxide. *Chem. Commun.* (2011) 47, 11507-11509.

[40] Pluth MD, McQuade LE, Lippard SJ. Cell-Trappable Fluorescent Probes for Nitric Oxide Visualization in Living Cells. *Org. Lett.* (2010) 12, 2318-2321.

[41] Tan L, Wan A, Li H. Fluorescent chitosan complex nanosphere diazeniumdiolates as donors and sensitive real-time probes of nitric oxide. *Analyst* (2013) 138, 879-886.

[42] Likhtenshtein GI. Electron Spin in Chemistry and Biology: Fundamentals, Methods, Reactions Mechanisms, Magnetic Phenomena, *Structure Investigation.* Springer, 2016.

[43] Lozinsky EM, Martina LV, Shames AI, Uzlaner N, Masarwa A, Likhtenshtein GI, Meyerstein D., Martin VV, Priel Z. Detection of NO from pig trachea by a fluorescence method, *Analytical Biochemistry* (2004) 326, 139-145.

[44] Chen O, Uzlaner, N, Priel Z, Likhtenshtein GI. Novel fluorescence method for real-time monitoring of nitric oxide dynamics in nanoscale concentration. *Journal of Biochemical and Biophysical Methods* (2008) 70, 1006-1013.

[45] Likhtenshtein GI. Stilbenes: Application in Chemistry, *Life Science and Material Science.* WILEY-VCH, Weinhem, 2009.

[46] Sun, Jie; Li, Na; Wang, Xiaojing; Wang, Bing; Sun, Jingyong; Wu, Zhongyu; Liu, Teng; Li, Huijuan; Li, Ling; Qin, Guizhi; Double quantum dot nanocomposite nitric oxide ratiometric fluorescent probe, and its preparation. *Faming Zhuanl Shenqing* (2016), CN 105885849 A 20160824.

[47] Clough PN, Thrush BA. Mechanism of chemiluminescent reaction between nitric oxide and ozone Trans. *Faraday Soc.* (1967) 63, 915-925.

[48] Dunham AJ; Barkley RM, Sievers RE. Aqueous nitrite ion determination by selective reduction and gas phase nitric oxide chemiluminescence. *Anal. Chem.* (1995) 67, 220-224.

[49] Barbosa RM, Lourenco CF, Santos RM, Pomerleau F, Huettl P, Gerhardt GA, Laranjinha GA. In vivo real-time measurement of nitric oxide in anesthetizedrat brain. *Methods Enzymol.* (2008) 441:351-367.

[50] *Berthold Analytical Instruments,* Nashua, NH, USA.

[51] Robinson JK, Bollinger MJ, Birks JW. Luminol/H_2O_2 chemiluminescence detector for the analysis of nitric oxide in exhaled breath. *Anal. Chem.* (1999) 71, 5131-5136.

[52] Corey MJ. *Coupled Bioluminescent Assays: Methods, Evaluations, and Applications.* John Wiley & Sons, 2008.

[53] Marques SM, Esteves da Silva JCG. A nitric oxide quantitative assay by a glyceraldehyde 3-phosphate dehydrogenase/phosphoglycerate kinase/firefly luciferase optimized coupled bioluminescent assay. *Anal Methods* (2014) 6, 3741-3750.

[54] Woldman YY, Sun J, Zweier JL, Khramtsov VV. Direct chemiluminescence detection of nitric oxide in aqueous solutions using the natural nitric oxide target soluble guanylyl cyclase. *Free Radical Biology & Medicine* (2009) 47, 1339-1345.

[55] Jiang S, Cheng R, Wang X, Xue T, Liu Y, Nel A, Huang Y, Duan X. Real-time electrical detection of nitric oxide in biological systems with sub-nanomolar sensitivity. *Nat Commun.* (2013) 4, 2225.

[56] Dedigama A, Angelo M, Torrione P, Kim TH, Wolter S, Lampert W, Atewologun A, Edirisoorya M, Collins L Thomas F, Kuech TF, Losurdo M, Bruno G, Brown A. Hemin-Functionalized in As-Based High Sensitivity Room Temperature NO Gas Sensors. *J. Phys. Chem.* C (2012) 116, 826-833.

[57] Koh WCA, Rahman MA, Choe ES, Lee DK, Shim YB. A cytochrome c modified-conducting polymer microelectrode for monitoring in vivo changes in nitric oxide. *Bioelectronics* (2008) 23, 1374-1381.

[58] Bedioui F, Villeneuve N. Electrochemical nitric oxide sensors for biological samples: Principle, selected examples and applications. *Electroanalysis* (2003) 15, 5-18.

[59] Davies IR, Zhang XJ. Nitric oxide selective electrodes. *Methods Enzymol.* (2008) 436, 63-95.

[60] Fan C, Li G, Zhu J, Zhu D. A reagentless nitric oxide biosensor based on hemoglobin-DNA films. *Anal. Chim. Acta* (2000) 423, 95-100.

[61] Shin JH, Weinman S, Schoenfisch MH. Sol-gel derived amperometric nitric oxide microsensor. *Anal. Chem.* (2005) 77, 3494-3501.

[62] Shibuki K. An electrochemical microprobe for detecting nitric oxide release in brain tissue. *Neurosci. Res.* (1990) 9 - 69-76.

[63] Kato D, Sakata M, Hirayama C, Hirata Y, Mizutani F, Kunitake M. [1] Selective permeation of nitric oxide through two dimensional polysiloxane cross-linked LB films. *Chem. Lett.* (2002) 31, 1190-1191.

[64] Shin JH, Weinman S, Schoenfisch MH. Sol-gel derived amperometric nitric oxide microsensor. *Anal. Chem.* (2005) 77, 3494-3501.

[65] Pereira-Rodrigues N, Albin V, Koudelka-Hep M, Auger V, Pailleret A, Fethi Bedioui F. Nickel tetrasulfonated phthalocyanine based platinum microelectrode array for nitric oxide oxidation. *Electrochem. Comm.* (2002) 4, 922-927.

[66] Oni J, Diab N, Reiter S, SchuhmannW. Metallophthalocyanine-modified glassy carbon electrodes: Effects on film formation conditions on electrocatalytic activity towards the oxidation of nitric oxide. *Sens. Actuators* B (2005) 105, 208- 213.

[67] Mao L, Yamamoto K, Zhou W, Jin L. Electrochemical nitric oxide sensors based on electropolymerized film of M(salen) with central ions of Fe, Co, Cu, and Mn. *Electroanalysis* (2000) 12, 72-77.

[68] Xu Miao-Qing, Wu Jian-Feng, Zhao Guang-Chao. Direct Electrochemistry of Hemoglobin at a Graphene Gold Nanoparticle Composite Film for Nitric Oxide Biosensing. *Sensors* (Basel) (2013) 13, 7492-7504.

[69] Hrbac J, Gregor C, Machova M, Kralova J, Bystron T, Cíz M, Lojek A. Nitric oxide sensor based on carbon fiber covered with nickel porphyrin layer deposited using optimized electropolymerization procedure. *Bioelectrochemistry* (2007) 71, 46-53.

[70] Wu DG, Cahen D, Graf P, Naaman R, Nitzan A, Shvarts D. Direct Detection of Low-Concentration NO in Physiological Solutions by a New GaAs-Based Sensor. *Chem.—Eur. J.* (2001) 7, 1743-1749.

[71] Rei Vilar M, El-Beghdadi J, Debontridder F, Naaman R, Arbel A, Ferraria AM, Botelho Do Rego AM. Mater Development of nitric oxide sensor for asthma attack prevention. *Sci. Eng.* C (2006) 26, 253.

[72] Prakash S, Rajesh S, Singh SR, Karunakaran C, Vasu V. Electrochemical incorporation of hemin in a ZnO-PPy nano-composite on a Pt electrode as NOx sensor. *Analyst* (2012) 137, 5874-5880.

[73] Koh WCA, Rahman MA, Choe ES, Lee DK, Shim B. A cytochrome c modified-conducting polymer microelectrode for monitoring in vivo changes in nitric oxide. *Biosensors and Bioelectronics* (2008) 23, 1374-1138.

[74] Shang LB, Liu XJ, Fan CH, Li GX. A Nitric Oxide Biosensor Based on Horseradish Peroxidase/Kieselguhr Co-Modified Pyrolytic Graphite Electrode. *Ann Chim* (2004) 94, 457-462.

[75] Fan C, Li G, Zhu J, Zhu D. A reagentless nitric oxide biosensor based on hemoglobin-DNA films, *Anal. Chim. Acta* (2000) 423, 95- 100.

[76] Zhang L, Zhao GC, Wei XW, Yang, ZS. A Nitric Oxide Biosensor Based on Myoglobin Adsorbed on Multi-Walled Carbon Nanotubes. *Electroanalysis* (2005) 17, 630-634.

[77] Peiris W, Pubudu M. New Generation of Electrochemical Sensors for Nitric Oxide; Ruthenium/Carbon-Based Nanostructures and Colloids as Electrocatalytic Platforms *ETD Archive* (2009) 234.

[78] Liu YM, Punckt C, Pope MA, Gelperin A, Aksay IA. Electrochemical Sensing of Nitric Oxide with Functionalized Graphene Electrodes. *ACS Appl. Mater. Interfaces* 5, 12624-12630.

[79] Prakash S, Rajesh S, Singh SR, Karunakaran C, Vasu V. Subash Prakash, Seenivasan Rajesh, Sarkkarai Raja Singh, Chandran Karunakaran, Veerapandy Vasuc. Electrochemical incorporation of hemin in a ZnO-PPy nanocomposite on a Pt electrode as NOx sensor. *Analyst* (2012) 137, 5874-5880.

[80] Fan CH, Liu XJ, Pang JT, Li GX, Scheer, H. Highly sensitive voltammetric biosensor for nitric oxide based on its high affinity with hemoglobin. *Analytica Chimic Acta* (2004) 523, 225-228.

[81] Zen J, Kumar AS, Wang H. A dual electrochemical sensor for nitrite and nitric oxide. *Analyst* (2000) 125, 2169-2172.

[82] Xu MQ, Wu JF, Zhao GC. Direct Electrochemistry of Hemoglobin at a Graphene Gold Nanoparticle Composite Film for Nitric Oxide Biosensing. *Sensors (Basel)* (2013) 13, 7492-504.

[83] Koh WCA, Rahman MA, Choe ES, Lee DK, Shim YB. A cytochrome c modified-conducting polymer microelectrode for monitoring in vivo changes in nitric oxide. *Bioelectronics* (2008) 23, 1374-1381.

[84] Santos RM, Rodrigues MS, Laranjinh J, Barbosa RM. Biomimetic sensor based on hemin/carbonnanotubes/chitosan modified microelectrode for nitric oxide measurement in the brain. *Biosensors and Bioelectronics* (2013) 44, 152-159.

[85] Chien-Cheng Liu, Jan-Hao Li, Cheng-Chung Chang, Yu-Chiang Chao, Hsin-Fei Meng, Sheng-Fu Horng, Cheng-Hsiung Hung, Tzu-Ching Meng. Electrochemical incorporation of hemin in a ZnO-PPy nanocomposite on a Pt electrode as NOx sensor. *J. Phys. D: Appl. Phys.* (2009) 42, 155105 (4pp).

[86] Shin JH, Weinman S, Schoenfisch MH. Sol-gel derived amperometric nitric oxide microsensor. *Anal. Chem.* (2005) 77, 3494-3501.

[87] Jungmi Moon, Yejin Ha, Misun Kim, Jeongeun Sim, Youngmi Lee, Minah Suh. Dual Electrochemical Microsensor for Real-Time Simultaneous Monitoring of Nitric Oxide and Potassium Ion Changes in a Rat Brain during Spontaneous Neocortical Epileptic Seizure. *Anal. Chem.* (2016) 88, 8942-8948.

[88] Liu, Zhonggang; Nemec-Bakk, Ashley; Khaper, Neelam; Chen, Aicheng. Sensitive Electrochemical Detection of Nitric Oxide Release from Cardiac and Cancer Cells via a Hierarchical Nanoporous Gold Microelectrode. *Analytical Chemistry* (Washington, DC, United States) (2017), 89), 8036-8043.

[89] Jensen GC, Zheng Z, Meyerhoff ME. Amperometric Nitric Oxide Sensors with Enhanced Selectivity Over Carbon Monoxide via Platinum Oxide Formation Under Alkaline Conditions. *Anal. Chem.*, *85*, 10057-10061.

[90] Wu DG, Cahen D, Graf P, Naaman, R, Nitzan A, Shvarts D. Direct Detection of Low-Concentration NO in Physiological Solutions by a New GaAs-Based Sensor. *Chem.—Eur. J.* (2001) 7, 1743-1749.

[91] Jiang S, Cheng R, Wang X, Xue T, Liu Y, Nel A, Huang Y, Duan X. Real-time electrical detection of nitric oxide in biological systems with sub-nanomolar sensitivity. *Nature Communications* (2013) 4, 2225.

[92] Ren, Chao Shi, Zhen Wang, Chenyu Yao. QEPAS nitric oxide sensor based on a mid-infrared fiber-coupled quantum cascade laser. *Optical Fiber Sensors Conference* (OFS), 2017 25th 24-28 2017, Jeju, South Korea, Publisher: IEEE.

[93] Dedigama A, Angelo M, Torrione P, Tong-Ho Kim, Wolter V, Lampert W, Atewologun A, Edirisoorya M, Collins L, Kuech TF, Losurdo M, Bruno G, April Brown A. Hemin-Functionalized InAs-Based High Sensitivity Room Temperature NO Gas Sensors. *J. Phys. Chem.* C (2012) 116, 826-833.

[94] Conrath U, Amoroso G, Kohle H, Sultemeyer DF. Non-invasive online detection of nitric oxide from plants and some other organisms by mass spectrometry. *Plant J.* (2004) 38, 1015-1022.

[95] Dubey M, Bernasek SL, Schwartz J. Highly sensitive nitric oxide detection using X-ray photoelectron spectroscopy. *J. Am. Chem. Soc.* (2007) 129, 6980-6981.

[96] McManus JB, Nelson DD, Herndon SC, Shorter JH, Zahniser MS, Blaser S, Hvozdara L, Muller A, Giovannini M, Faist J. Comparison of cw and pulsed operation with a TE-cooled quantum cascade infrared laser for detection of nitric oxide at 1900 cm^{-1}. *Appl. Phys.* B (2005) 85, 235-241.

[97] Di Franco C, Elia A, Spagnolo V, Scamarcio G, Lugarà PM, Ieva E, Cioffi N, Luisa Torsi L, Bruno G, Losurdo M, Garcia MA, Wolter SD, Brown A, Ricco M. Optical and Electronic NO_x Sensors for Applications in Mechatronics. *Sensors* (2009) 9, 3337-3356.

[98] Zhang J, Hu J, Zhu ZQ, Gong H, O'Shea SJ. Quartz crystal microbalance coated with sol-gelderived indium-tin oxide thin films as gas sensor for NO detection. *Colloids Surf.* A (2004) 236, 23-30.

[99] Privett BJ, Shin JH, Schoenfisch MH. Electrochemical nitric oxide sensors for physiological measurements. *Chem. Soc. Rev.* (2010) 39, 1925-1935.

[100] Ye X, Rubakhin SS, Sweedler JV. Detection of nitric oxide in single cells. *Analyst* (2008) Apr;133, 423-433.

[101] Janzen EG, Blackburn BJ. Detection and identification of short-lived free radicals by electron spin resonance trapping techniques (spin trapping). Photolysis of organolead, -tin, and -mercury compounds. *J. Am. Chem. Soc.* (1969) 91, 4481-4490.

[102] Davies CA, Winyard PG. Assay for S-nitrosothiol compounds using EPR spectrometry. *PCT Int. Appl.* (2002), WO 2002016934 A1 20020228.

[103] Hirsh Donald J, Schieler Brittany M, Fomchenko Katherine M, Jordan Ethan T, Bidle Kay D. A liposome-encapsulated spin trap for the detection of nitric oxide. *Free Radical Biology & Medicine* (2016), 96, 199-210.

[104] Lauricella R, Triquigneaux M, Andre-Barres Ch, Charles L, Tuccio B. 5-Hydroxy-2,2,6,6-tetramethyl-4-(2-methylprop-1-en-yl)cyclohex-4-ene-1,3-dione, a novel cheletropic trap for nitric oxide EPR detection. *Chemical Communications* (Cambridge, United Kingdom) (2010) 46, 3675-677.

[105] Davies CA, Nielsen BR, Timmin G, Hamilton L, Brooker A, Guo R, Symons MCR, Winyard PG. Characterization of the Radical Product Formed from the Reaction of Nitric Oxide with the Spin Trap 3,5-Dibromo-4- Nitrosobenzene Sulfonate. *Nitric Oxide: Biology and Chemistry* (2001). 5, 116-127.

[106] Venpin WKPF, Kennedy EM, Mackie JC, Dlugogorski BZ. Mechanistic Study of Trapping of NO by 3,5-Dibromo-4-Nitrosobenzene Sulfonate. *Industrial & Engineering Chemistry Research* (2012), 51, 14325-14336.

[107] Hogg N. Detection of Nitric Oxide by Electron Paramagnetic Resonance Spectroscopy. *Free Radic Biol Med* (2010) 49, 122-129.

[108] Reszka KJ, Bilski P, Chignell CF. Spin trapping of nitric oxide by aci anions of nitroalkanes. *Nitric Oxide* (2004)10, 53-59.

[109] Smith AM, Nie Sh. Chemical analysis and cellular imaging with quantum dots. *Analyst* (2004) 129, 672-677.

[110] Förster T. Energiewanderung und Fluoreszenz. [Energy migration and fluorescence.] *Naturwissenschaften* (1946) 33, 166-175.

[111] Marcus RA. At the Birth of Modern Semiclassical Theory. *Mol Phys* (2012) 110, 513-516.

[112] Likhtenshtein GI. (1996) Role of orbital and dynamic factors in electron transfer in reaction centers of photosynthetic systems. *J. Photochem. Photobiol. A: Chem.* 96, 79-92.

Chapter 5

NITRIC OXIDE SYNTHASE

ANNOTATION

Nitric oxide produced by NO synthase (NOS) participates in diverse biological and physiological processes, such as vasodilation, neurotransmission, the innate immune response, and other vital processes. The crystallography has paved a hardcore for detailed research of mechanism of the NO synthase, regulation, and purposeful design of inhibitor and activators. The multidomain structure, multistage mechanism, decisive role of electrostatic, and conformation effects make up the uniqueness of the NOS. To study the kinetics and thermodynamics of the process, an arsenal of physicochemical methods was employed.

5.1. INTRODUCTION

Nitric oxide synthases (EC 1.14.13.39) are a family of enzymes catalyzing the production of NO from L-arginine [1–15]:

2 L-arginine + 3 NADPH + 1 H^+ + 4 O_2 2 citrulline + 2 nitric oxide + 4 H_2O + 3 $NADP^+$

NOS can also catalyze superoxide anion production, as a side process. The following mechanism of the reaction has been commonly accepted [1–4]. Electrons from nicotinamide adenine dinucleotide phosphate (NADPH) flows through the reductase

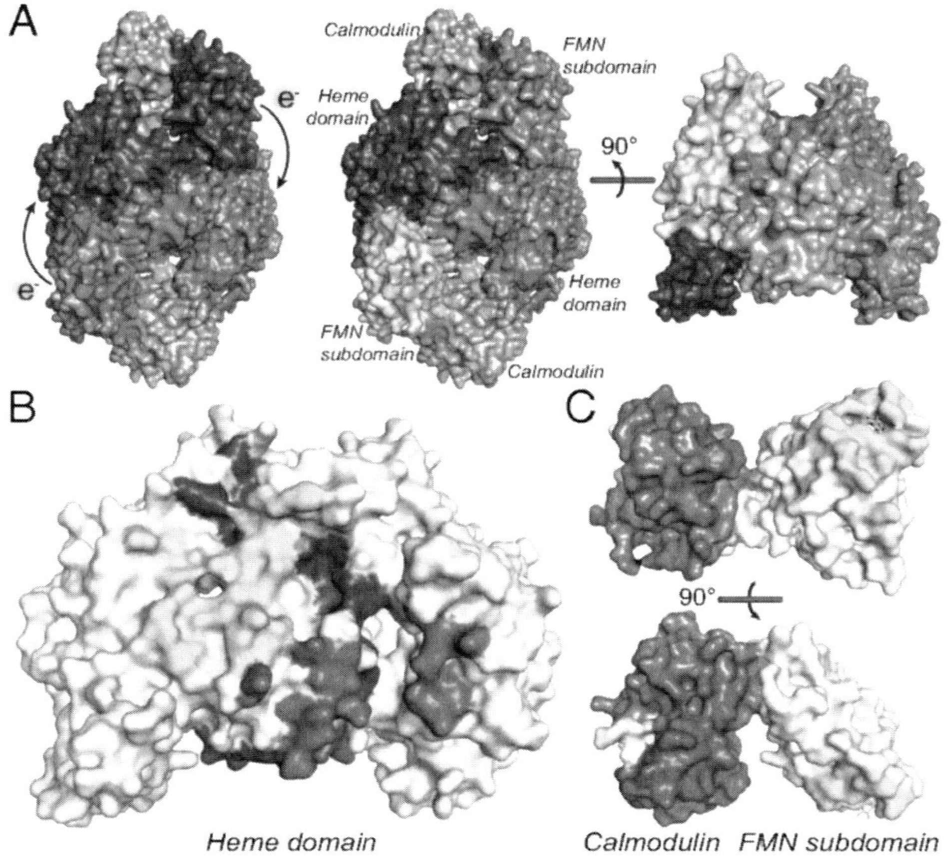

Figure 5.1. Models of the iNOS heme domain complexed with the iNOS FMN subdomain and calmodulin. (A) Representative model (iNOS-3) of the iNOS heme and FMN subdomains bound to calmodulin generated using Rosetta. (Left) iNOS monomers are colored blue and the calmodulins are colored gray. (Center and Right) FMN subdomains are highlighted in brown. (B) iNOS heme domain interaction surfaces with the FMN subdomain and calmodulin as well as the heme domain dimer interface on a surface representation of the iNOS heme domain (PDB ID: 1DWV) (6). (C) iNOS FMN subdomain and calmodulin (blue) surfaces that interact with the iNOS heme domain in the Rosetta models. For B and C, residues were colored if they were present within the interface in at least two of the three iNOS Rosetta models or for calmodulin in the iNOS-3 model as analyzed using the KFC2 server [15].

domain via flavin adenine dinucleotide (FAD) and flavin mononucleotide (FMN) redox carriers to the oxygenase domain. There they interact with the heme iron and tetrahydrobiopterin (BH_4) at the active site to catalyze the reaction of oxygen with L-arginine, generating citrulline and NO as products. Proceeding this process requires the presence of bound Ca^{2+}/Calmodulin (CaM) (Figure 5.1) [4].

The heme with adjacent tetrahydrobiopterin (BH_4)

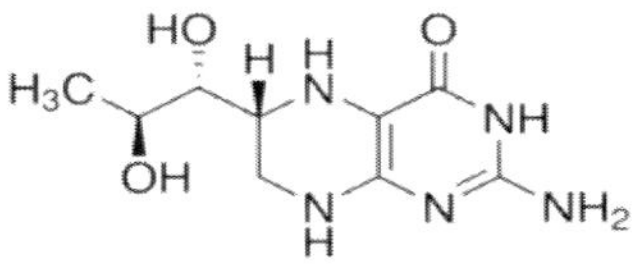

as an electron donor and dioxygen, as oxidizing agent, are key structures essential for NO production.

NOS generates NO from L-arginine, dioxygen, and NADPH in two sequential reactions with *Nω*- hydroxy-L-arginine (NHA) as intermediate [3]:

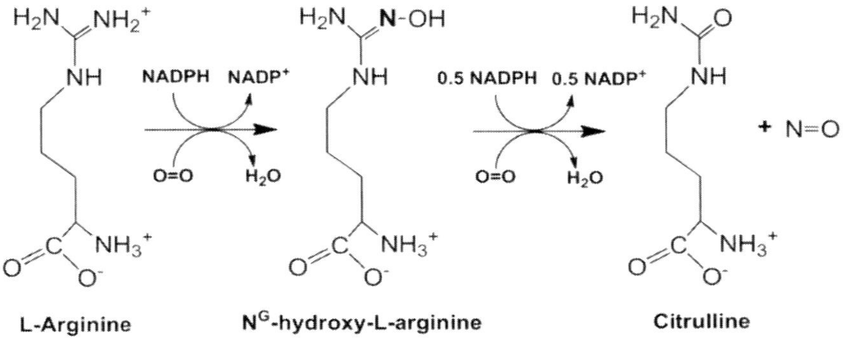

There are three main isoforms of the enzyme, named neuronal NOS (nNOS), inducible NOS (iNOS), and endothelial NOS (eNOS), which differ in their dependence on Ca^{2+}, their expression, and activities. The molecular weight of nNOS, was found to be 200 kD [5]. nNOS contains an additional N-terminal domain, the PDZ domain of 80-90 amino-acids.

5.2. Nitric Oxide Synthase Architecture

Figures 5.2–5.5 schematically show the enzymes oxygenase and reductase domains and other important components of NOS structure and activity [15].

The enzyme functions as a dimer consisting of two identical monomers (Figures 5.1, 5.2 and 5.3) [6]. The association in the presence of haem forms a dimer with NADPH oxidase activity. Low levels of BH_4 lead to the formation of a stable dimer, which catalyzes a simultaneous production of NO and $O_2^{\cdot-}$, while at high-saturated BH_4 levels the enzyme acts as an NO synthase. Each monomer of NOS consists of the oxygenase or heme domain and the reductase domain, which is composed of FMN- and FAD- containing subdomains [6]. The oxygenase domain forms the heme active site and is the site of dimerization for holo-NOS. In the input state, NADPH binds and reduces FAD, which in turn reduces FMN.

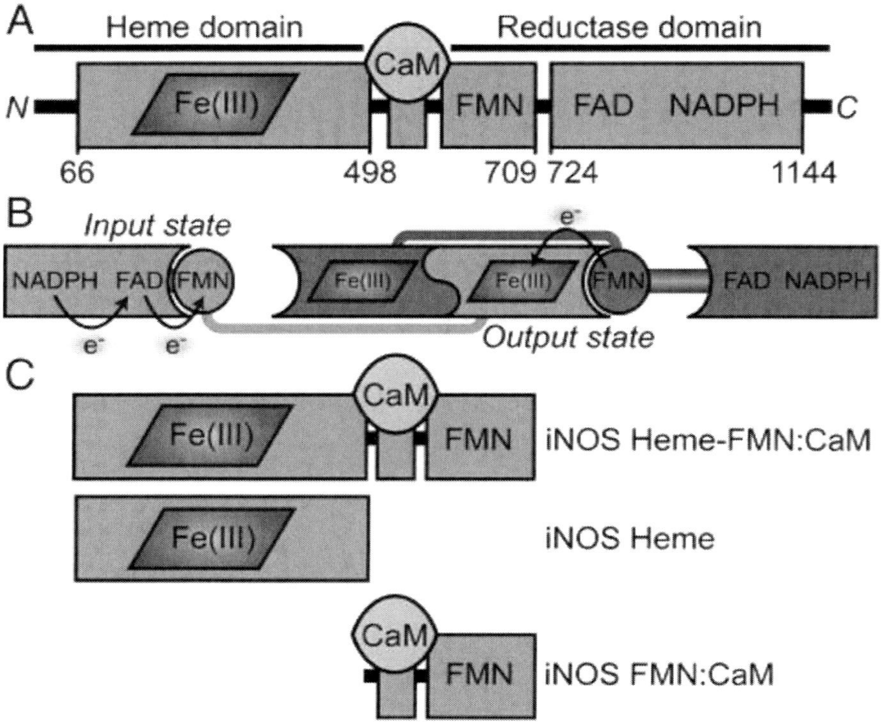

Figure 5.2. Mammalian nitric oxidase synthases. (A) Mammalian nitric oxide synthase domain organization. (B) Nitric oxide synthase electron transfer pathway. In the input state, the FMN subdomain interacts with the FAD subdomain allowing electron transfer to occur between the flavin cofactors. In the output state, the FMN subdomain interacts with the heme domain allowing electron transfer between the FMN and heme. (C) iNOS constructs used in this study. iNOS heme-FMN:CaM consists of residues 66–709 bound to calmodulin. The hyphen in heme-FMN:CaM indicates the native covalent linkage of the heme domain and FMN subdomain, and the colon indicates the noncovalent interaction of calmodulin with iNOS heme-FMN. iNOS heme consists of residues 66–498. iNOS FMN:CaM consists of residues 499–709 bound to calmodulin. All residue numbering refers to residues in murine iNOS [15]. nNOS-bound CaM dynamics and reaction chemistry are kinetically coupled. The graph shows the transient reaction chemistry of nNOS (relative absorbance, red) and dynamics of nNOS-bound CaM (relative FRET efficiency, grey) recorded when mixing excess NADPH with nNOS (bound to CaM), under pseudo first-order conditions. The schematic presented here shows the temporal resolved dynamics of nNOS-bound CaM when nNOS is reduced with NADPH. [15].

Calmodulin (CaM) binding on the a-helical linker between the FMN and oxygenase domain induces a conformational switch toward the output state to transfer electrons from FMN to the heme group. The dynamic structure of the enzyme and optimal flexibility of domains, elucidated by molecular dynamics calculation, provided a precise unique fitting for chemical steps of the processes [6].

The above-mentioned unique features pose NOS in a row along with such sophisticated molecular machines as ribosome and nitrogenase and give rise to and mechanistic features, which are responsible for the physiological and pathophysiological roles of the enzyme (chapters 8–13).

5.3. Nitric Oxide Synthase and Its Fragment's Crystal Structure

Determination of the crystal structure of NOS using the modern X-ray technology in combination with modern physical computational methods appeared to be a remarkable achievement [7–27]. The high-resolution crystallography of the oxygenase, reductase, and calmodulin modules of NOS isoenzymes, coupled with computational approaches, has provided a reliable structural basis for understanding the molecular mechanisms underlying electron transfer, dioxygen activation, and the enzyme activity regulation. All of these steps are crucial to NO production.

Figures 5.3 and 5.4 elucidate a position of each monomer consisting of a C-terminal flavin-containing reductase domain and an N-terminal oxygenase domain containing a P450- type heme with cysteine as proximal axial ligand to the iron.

Dimerization creates a channel of 30 Å deep with the larger cavity of 15 Å, which is large enough to allow diffusion of both arginine and citrulline. Specifically, human nNOS contains two asymmetric dimers (Figure 5.4) [17]. The distance between the FAD and FMN cofactors is 15.5 Å. [12].

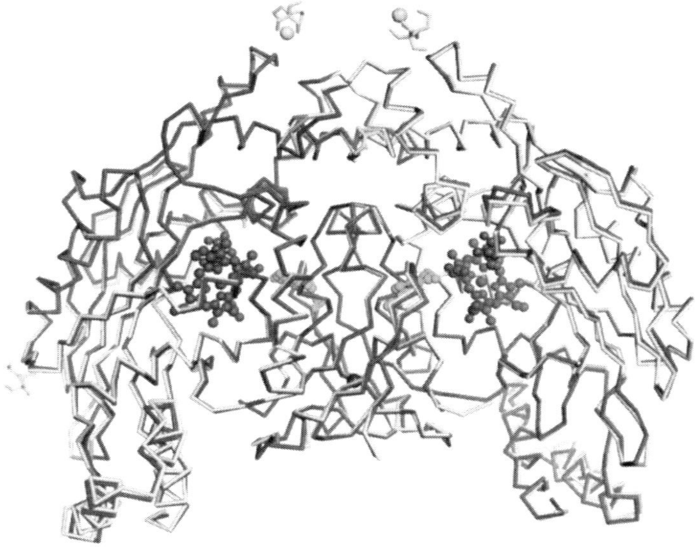

(a)

Figure 5.3. Superimposition of bovine eNOS and human eNOS structures. Heme, tetrahydrobiopterin, and L-Arg are shown as ball-and-stick representations [17]. Copyrights 2014 American Chemical Society.

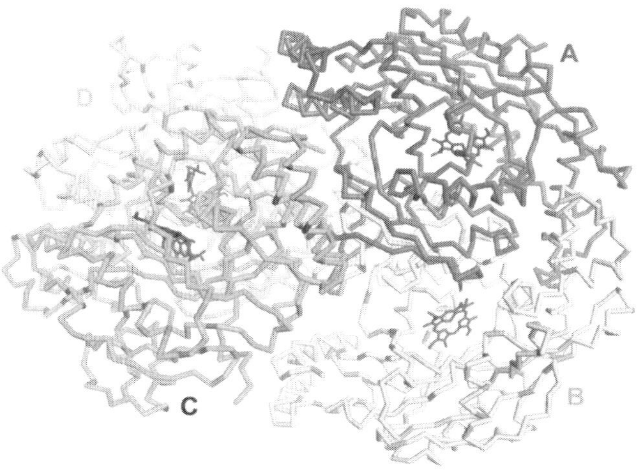

Figure 5.4. Two dimers in the asymmetric unit of human nNOS. Dimer AB packs tightly against dimer CD in a handshake manner [17]. Copyrights 2014 American Chemical Society.

The three-dimensional structures of the oxygenase domains of iNOS, eNOS, and nNOS2 have been determined. All three structures are very similar but not identical to one another [7–17]. The structure of an intact CaM-free rat nNOS reductase with the FMN domain docked onto the NADPH/FAD domain was presented in [3]. According to designed ribbon diagram, the FMN domain docked onto the NADPH/FAD domain in the FMN-shielded conformation. The reductase domain is divided into two different flavin-containing domains: (a) the N terminus, FMN-containing portion, and (b) the C terminus FAD- and NADPH-binding portion [13]. Residues in the vicinity of the FAD isoalloxazine ring were revealed. Extensive hydrogen bonding network with the main chain atoms (carbonyl oxygens of Thr^{1191} and Ala^{1193} and the amide nitrogen of Ala^{1193}) of the polypeptide, and two tightly bound water molecules, W1 and W2, were shown. It was speculated that the water molecule W2 may play a role of general acid/base in the protonation/deprotonation of the N1 atom of the flavin that is necessary during catalysis.

The most important problem, essential for understanding the chemical mechanism of dioxygen activation in the NO production, is the structure of P450-type heme-containing center, which catalyzes the formation of NO from L-arginine, NADPH, and O_2 in a two-step reaction sequence. Positions of heme, H4B, substrate L-arginine on the oxidase domain, and surrounding groups are showed in Figure 5.1 and Figure 5.5 [17].

In the human iNOS oxygenase domain, the iron protoporphyrin-IX complex is pentacoordinate and axially coordinated to the proximal Cys. Reducing agent H4B and substrate L-arginine are located in close proximity to heam, which is buried in the protein interior and makes van der Waals interactions with hydrophobic chains [18].

A position of tetrahydrobiopterin in NOS and in its binding pocket is given in Figure 5.6 [27].

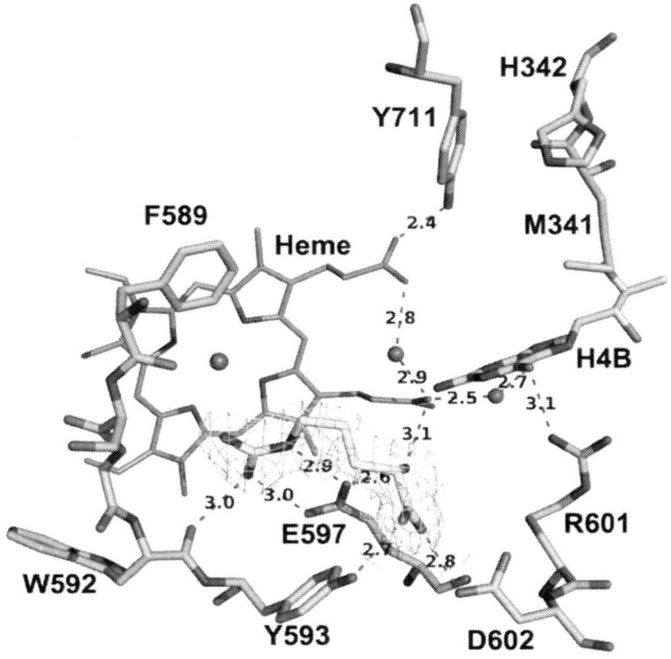

Figure 5.5. The active site of human nNOS with L-Arg bound. The $2F_o - F_c$ electron density for L-Arg is contoured at 1.0σ. The extensive hydrogen-bonding networks involving protein, heme, H_4B, and L-Arg are depicted as dashed lines with distances marked in Å. [17]. Copyrights 2014 American Chemical Society.

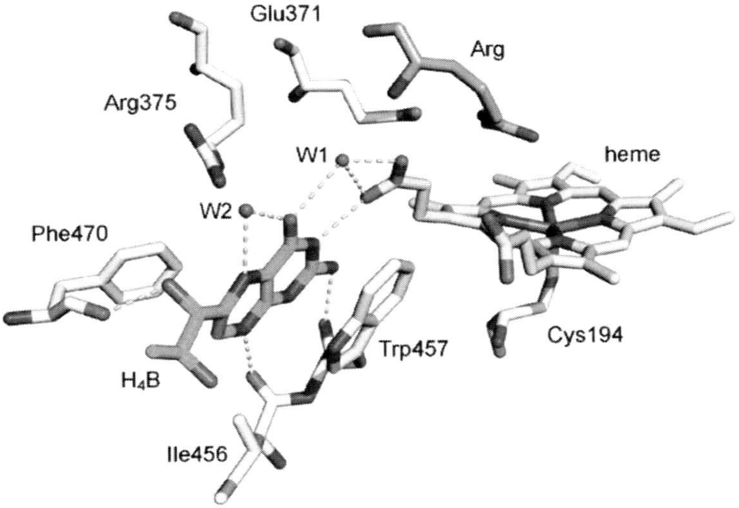

Figure 5.6. Tetrahydrobiopterin in its binding pocket in NOS, with substrate arginine bound (PDB 1nod) and two structural waters W1 and W2 [27]. Copyrights 2010 American Chemical Society.

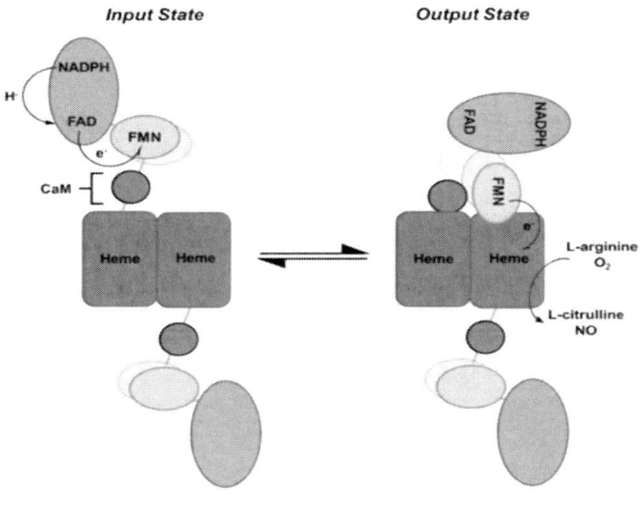

Figure 5.7. Structural organization and electron flow through nitric oxide synthase (NOS). The FAD domain is shown in bright orange, the FMN domain in light orange, the heme domain in red, and the partner protein calmodulin (CaM) in in gray. Ligand (NADPH and CaM) binding is implicated in shifting the conformational equilibrium of nNOS and thus regulating electron transfer during the catalytic cycle. Dimerization of nNOS occurs at an interface between the heme domains [20]. Copyrights 2016 American Chemical Society.

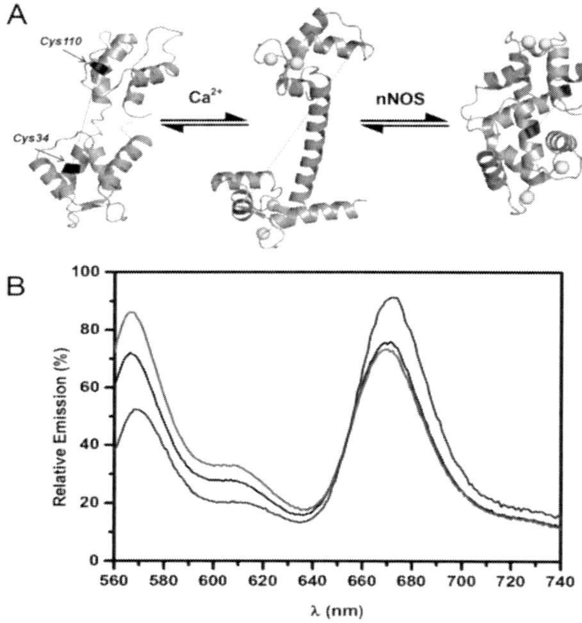

Figure 5.8. Ligand binding and the dynamic landscape of CaM. (A) Structures of apo (PDB_1CLL, shown on the left), Ca^{2+}-bound (PDB_1CFC, shown in the middle) and both Ca^{2+}/nNOS-peptide-bound forms of CaM (PDB_2O60, shown on the right). Divalent calcium ions are shown as yellow spheres, and the nNOS peptide is represented as an orange ribbon. Ligand binding and the dynamic landscape of CaM. The distances between the α-carbon atoms of the two fluorophore labeling sites (Cys34 and Cys110; highlighted in red) are 27, 52.4, and 12.4 Å for the apo, Ca^{2+}-bound, and the Ca^{2+}/nNOS-bound forms, respectively. (B) Normalized fluorescence emission spectra showing ratiometric changes in the donor and acceptor emission [23]. Copyrights 2016 American Chemical Society.

Calmodulin (CaM) regulates a wide range of cellular functions through its reversible Ca^{2+}-dependent binding to target proteins, including NOS. One of the most important observations in the area is the role of CaM in activating the iNOS catalysis. Thus, CaM controls NOS activity by controlling the rates of electron transfer between the two flavin cofactors and between FMN and heme [14–15, 19–26]. First the crystal structure of CaM has been determined to 3.6 A resolution [24]. Later the protein *structure* has been refined up to 1.0 Å resolution [25]. The crystal structure of the CaCaM·FMN complex in four different conformations, each with a different relative orientation, between the FMN domain and the bound CaM have been also reported [26]. The position of CaM *in* NOS and electron flow through the enzyme are indicated in Figure 5.7 [23].

Figure 5.8 describes crystal structure, a ligand (Ca^{2+}) binding, and the dynamic landscape of CaM obtained using time-resolved spectroscopy employing absorbance, Förster resonance energy transfer a flavin analogue (5-deazaflavin mononucleotide [5-dFMN]), and isotopically labeled nicotinamide coenzymes [23]. The distances between the α-carbon atoms of the two fluorophore labeling sites (Cys34 and Cys110) were found as 27, 52.4, and 12.4 Å for the apo, Ca^{2+}-bound and the Ca^{2+}/nNOS-bound forms, respectively.

The structure and dynamics of complexes formed by peptides based on inducible NOS (iNOS) and endothelial NOS (eNOS) with CaM at Ca^{2+} concentrations that mimic the physiological basal and elevated level in mammalian cells were established using fluorescence techniques and nuclear magnetic resonance (NMR) spectroscopy [19]. The Ca^{2+}-dependent binding properties of the CaM-binding domains were investigated for detecting conformational changes using dansyl-labeled CaM proteins. The label fluorescence spectrum is enhanced and shifted when the dansyl moiety becomes embedded in a hydrophobic environment. Detailed information about fluctuations in protein structures was obtained from the measurement of amide proton (NH) hydrogen/deuterium exchange (H/D) rates using NMR spectroscopy. These fluctuations expose some of the NH to the D_2O solvent. The results show the CaM-NOS complexes have similar structures at physiological and fully saturated Ca^{2+} levels, while different dynamics. At 225 nM Ca^{2+}, the CaM-NOS complexes show overall an increase in backbone dynamics. Worm models of CaM–eNOS peptide and CaM–iNOS peptide complexes at 225 nM Ca^{2+} and saturated Ca^{2+} concentrations illustrate their internal dynamics and amide H_2O/D_2O exchange data for CaM–eNOS peptide, and CaM–iNOS peptide complexes were suggested (Figure 5.9).

Thus, the crystallography has paved a highway for detailed research of the mechanism of the NO synthase regulation and purposeful design of inhibitors and activators.

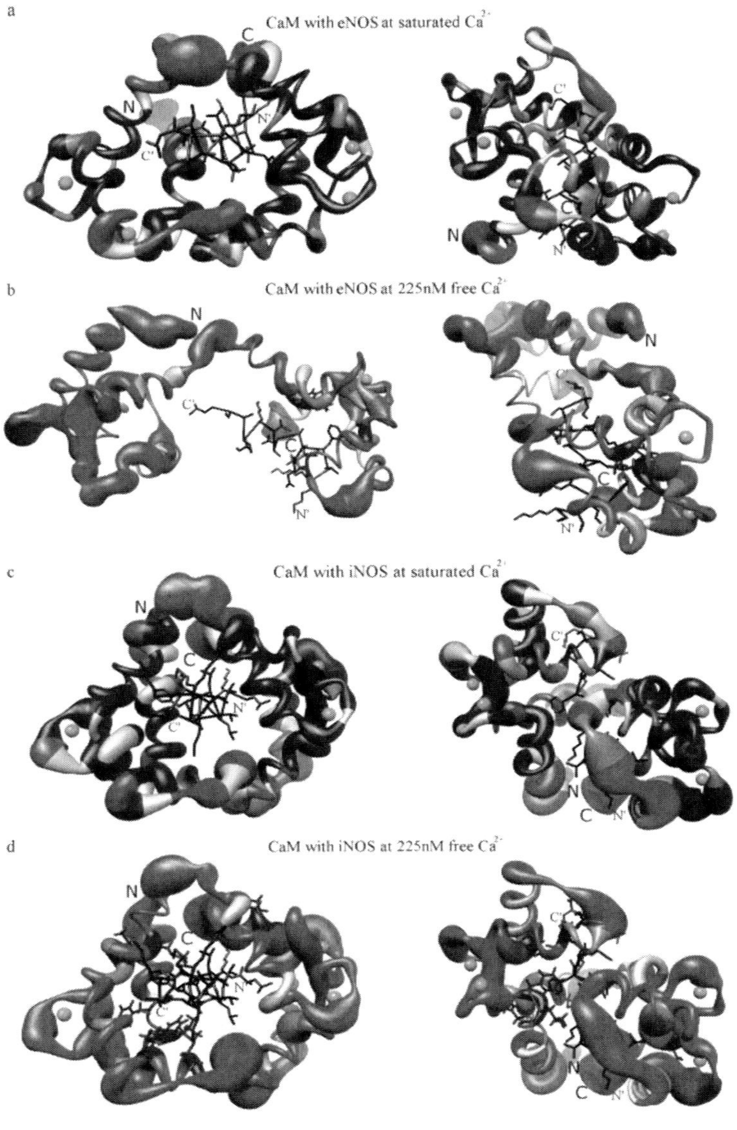

Figure 5.9. Worm models of CaM–eNOS peptide and CaM–iNOS peptide complexes at 225 nM Ca^{2+} and saturated Ca^{2+} concentrations illustrating their internal dynamics and amide H_2O/D_2O exchange data. The worm models were prepared using UCSF Chimera with the render by attribute function. The worm radius ranges from 0.25, corresponding to an S^2 value of 1, to 4, corresponding to an S^2 value of 0 [19]. Copyrights 2016 American Chemical Society.

5.4. Nitric Oxide Synthase Kinetics and Mechanism

To study the kinetics and thermodynamics of the NOS reactions, an arsenal of physical methods, such as steady state assays, stopped-flow spectrophotometry, absorption and ESR and NMR spectroscopies, resolved fluorescence resonance energy transfer (FRET), laser flash photolysis, electrochemical, hydrogen exchange, and potential-metric technics, is widely used. Significant additional information is obtained through the use of mutagenesis and the truncated samples of the enzyme.

The flavin states thermodynamics of the reductase part of mammalian nitric oxide synthase (NOS redox), a FMN, FMNH·, $FMNH_2$, FAD, FADH·, and $FADH_2$ were studied employing *potentiometric titrations* [29]. The experiments have showed the reaction smooth thermodynamic profile which is similar for all three isoforms. The following values of the midpoint potentials were reported: 250 mV (FAD/FADH), -260 mV FADH·/$FADH_2$), -220 mV (FMNH·/$FMNH_2$, -120 mV (FMN/FMNH), −248 mV (Fe^{III} heme/Fe^{II} heme), and −229 mV (Fe^{III}heme/Fe^{II}-CO heme) [29].

The electron flow pathway in NOS turned out to be a lot of steps and complex processes [30, 31]. To monitor electron transfer steps for nNOS, single wavelength kinetic traces at 395, 600, and 750 nm) and curve fitting were used [30]. The formation time of the charge transfer (CT) complex and electron flow from $FADH_2$ to FMN forming $FADH^•$ – $FMN^{•-}$ were estimated as 122 and 38.9 s^{-1}, respectively. The rate constants of the formation of the CT complex (or reduction of FAD) with a rate of 129 s^{-1}, interflavin ET leading to the formation of disemiquinone (38.9 s^{-1}), and disemiquinone $FADH^•$ – FMN transition with a slower rise (35.0 s^{-1}) were measured. The rate constant of formation of $FMNH_2$, the direct electron donor to the heme for wild-type nNOS, was found to be as at 13 s^{-1} at 4°C and much faster at 25°C. A kinetic model linking protein conformational motions, interflavin electron transfer, and electron flux through the endothelial and neuronal NOS flavoprotein domains was proposed [31]. The model includes four kinetic rates: association (k_1 or k_3) and dissociation (k_{-1} or k_{-3}) of the FMN and FNR domains; the FMNH• reduction rate (k_2), and the cytochrome c reduction rate (k_4). The following steps were also considered: the fully-reduced enzyme in the open conformation (species a) reduces cytochrome c and generates species b, which then undergoes conformational closing, interflavin electron transfer, and conformational opening steps [31].

Global kinetic model for NOS catalysis was suggested in [37]. In frame of the model, ferric enzyme reduction (kr) is rate limiting for the biosynthetic reactions, and kcat1 and kcat2 are the conversion rates of the $FeIIO_2$ species to products in the Arg and NOHA reactions, respectively. The ferric heme–NO product complex (FeIIINO) can either release NO (kd) or become reduced (kr) to a ferrous heme–NO complex (FeIINO), which reacts with O_2 (kox) to regenerate the ferric enzyme (See details in [37]).

Another global kinetic model for NOS catalysis in the NOS oxygenase part and a table of corresponding thermodynamic parameters were presented in [32]. In work [33], the electron flux value measured for nNOS with L-agmatine yields was estimated as 10.5 s^{-1} for the heme reduction rate at 25°C. This value is in a good agreement with the 13 s^{-1} estimated for flavin to heme electron transfer derived from laser flash photolysis experiments done at room temperature [34]. Activation parameters for the FMN-heme interdomain electron transfer (IET) kinetics in human iNOS were determined by laser flash photolysis over the temperature range from 283 to 304 K [35]. The best fit to the oxyFMN kinetic data was realized with $\Delta H^{\ddagger}$ = 52,6 kJ mol^{-1} and $\Delta S^{\ddagger}$ = − 17,1 J mol^{-1}K^{-1}, while for the holoenzyme $\Delta H^{\ddagger}$ = 36.9 kJ mol^{-1} and $\Delta S^{\ddagger}$ = − 89.7J mol^{-1}K^{-1}. Note that the distance between donor and acceptor centers (for example between FMN and hemin,) R_{DA} = 15–19 Å. At these distances, the reaction proceeds by a nonadiabatic mechanism and must be described using the Marcus-Levich equation [36]. In such of conditions, the value of $\Delta S^{\ddagger}$ is an apparent parameter. The calculated FMN to heme distance, as measured between the heme iron and the N5 atom of FMN, is in a reasonable agreement with the previously pulse EPR-derived distance of 18.8 Å. [38].

Mapping critical NOS interaction surfaces and direct interactions between the heme domain, the FMN subdomain, and calmodulin were established using *hydrogen-deuterium exchange mass spectrometry (*HDX-MS) and kinetic studies of site-specific interface mutants, respectively [15]. Comparison of HDX-MS results with kinetic analysis of iNOS and calmodulin mutants is given in Figure 5.10.

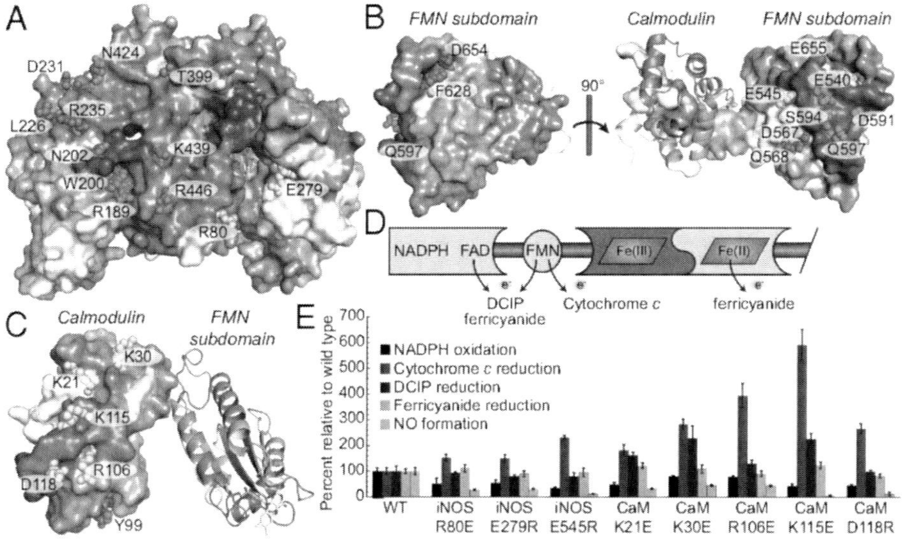

Figure 5.10. Comparison of HDX-MS results with kinetic analysis of iNOS and calmodulin mutants. Previous mutagenesis studies have established residues that are predicted to lie within or outside the interfaces between the iNOS heme domain and the calmodulin-bound FMN subdomain. Residues mutated as part of the current study are shown. (A) Residues mutated in the iNOS heme domain. (B) Residues mutated in the iNOS FMN subdomain. The FMN subdomain is shown in surface representation and calmodulin is shown in ribbon representation. (C) Residues mutated in calmodulin. Calmodulin is shown in surface representation and the FMN subdomain is shown in ribbon representation. (D) Summary of the cofactor specificity of the alternative electron acceptors used in this study. (E) Bar graphs of steady-state rates of NADPH oxidation, cytochrome c reduction, DCIP reduction, ferricyanide reduction, and NO formation for iNOS and calmodulin mutants. Rates were normalized to the rates for wild-type iNOS bound to calmodulin [15]. With permission from PNAS, 2013.

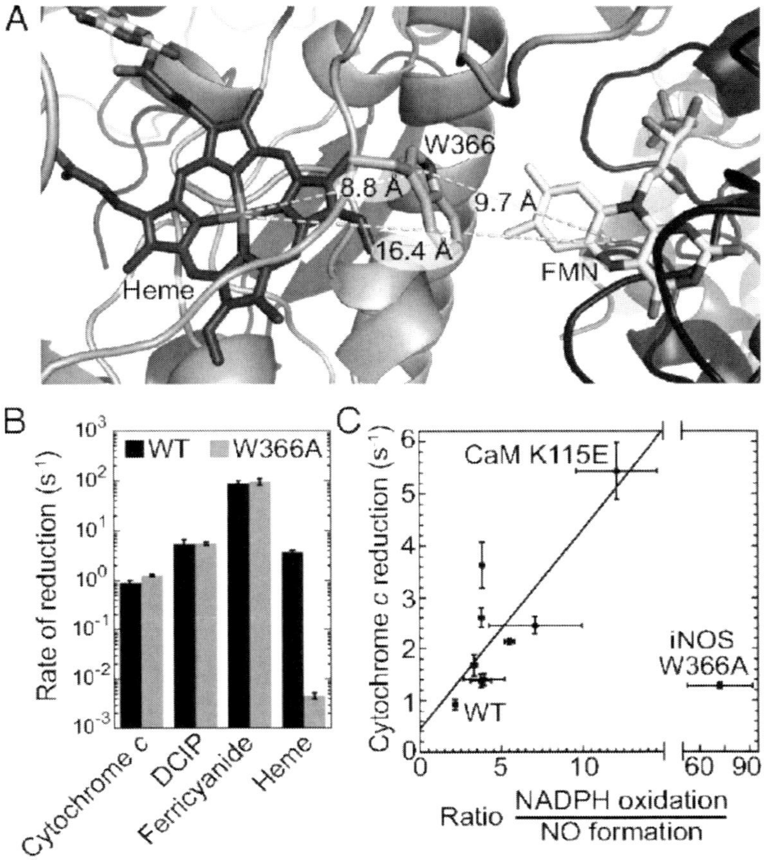

Figure 5.11. iNOS W366 mediates electron transfer from FMN to heme. (A) iNOS output state highlighting W366, heme, and FMN in the iNOS-3 model (B) Comparison of rates of cytochrome *c*, DCIP, ferricyanide, and heme reduction for wild-type iNOS (black) and iNOS W366A (gray). (C) Plot of NOS uncoupling expressed as the ratio of NADPH oxidation to NO formation versus the rate of cytochrome *c* reduction [15]. With permission from PNAS, 2013.

Figure 5.11 indicates a pathway of the electron transfer from FMN to heme iNOS W366 and comparison of rates of cytochrome *c*, DCIP, ferricyanide, and heme reduction for wild-type iNOS and iNOS W366A. Note that ferrocyanicytochrome cde is a hundred times more effective reductant then cytochrome c.

5.5. Conformational Mobility and Catalysis

As was proposed by Smith et al. [15], calmodulin (CaM) binding to nitric oxide synthase (NOS) enables a conformational change, in which the FMN domain shuttles between the FAD and heme domains to deliver electrons to the active site heme center. Later this suggestion was well documented [37–47]. Thus, conformational mobility and electrostatic interactions play a crucial role in processes of electronic transfer between the NOS domains [1].

For probing NOS dynamics and conformational change, several independent fluorescent methods were employed including FAD and FMN fluorescence steady-state emission, external CaM-bound fluorophore fluorescence, and monitoring the distances between the nNOS FAD and FMN by single molecule measurements FRET-based experiments [1]. A constructing a "Cys-lite" variant of the NOS reductase domain that is unreactive toward maleimide fluorophores was also performed. The graph in Figure 5.12 shows kinetic coupling between the nNOS-bound CaM and CPR domain dynamics and reaction chemistry.

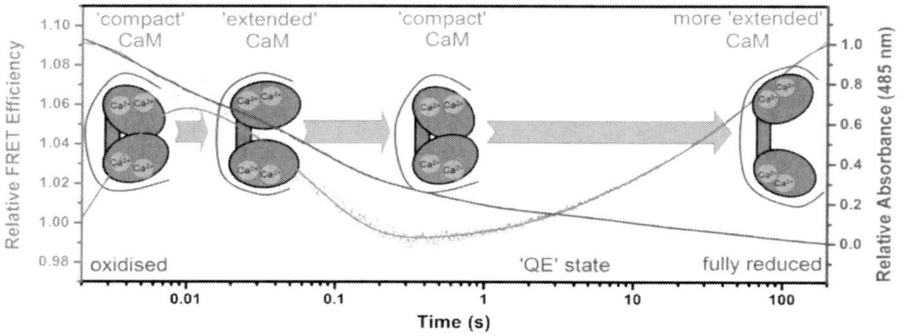

Figure 5.12. nNOS-bound CaM dynamics and reaction chemistry are kinetically coupled. The graph shows the transient reaction chemistry of nNOS (relative absorbance and dynamics of nNOS-bound CaM (relative FRET efficiency, recorded when mixing excess NADPH with nNOS (bound to CaM), under pseudo first-order conditions. The schematic presented here shows the temporal resolved dynamics of nNOS-bound CaM when nNOS is reduced with NADPH [20]. Copyrights 2016 American Chemical Society.

According to three-state model of NOS FMN domain function in electron transfer and heme, reduction proposed in [37], the FMN domain is attached in NOS by two hinge elements (H1 and H2), which allow a shift of shielded (closed) conformation to deshielded (open) non-interacting conformation that allows the FMN domain to interact with electron acceptors such as cytochrome c (Equilibrium A). Equilibrium B provides the FMN–NOSoxy domain interaction required for heme reduction and NO synthesis. The CM binding causes a shift of conformational equilibrium. For example, parameters describing conformational equilibrium A for the 4-electron reduced nNOS flavoproteins $K_{eq}A = 0.1$ and $K_{eq}A = 8\text{-}9$ for process in the absence and presence CM, respectively.

In the FMN domain, a shift from a position adequate to accept electrons from FAD, (FMN-shielded conformation) to a conformation that is able to reach the heme domain (FMN-deshielded conformation) allows corresponding electron transfer and NO synthesis [40]. Stopped-flow and steady-state experiments in combination with mutagenesis revealed that individual charges at Lys^{423}, Lys^{620}, and Lys^{660} in the NOSs, which form a triad of positive charges on the NOSoxy surface, are the most important in heme reduction in nNOS. Conformational equilibrium and electrostatic interactions of the nNOS domains and docking model of a nNOSoxy-FMN-CaM complex were also proposed.

Results on resolved FMN fluorescence lifetimes indicated that neuronal NOS activation by CaM removes constraints favoring a closed "input state," increasing occupation of other states and facilitating conformational transitions [18]. The 90 ps FMN input state lifetime distinguishes it from ~4 ns "open" states in which FMN does not interact strongly with other groups, or 0.9 ns output states in which FMN interacts with ferriheme.

Important conformational effects in the NOS reactions were also demonstrated in [12]. Using the 4-pulse pulsed electron-electron double resonance spectroscopy (PELDOR), fluorescence, and high-pressure techniques up to 600 bar, it was shown that the binding of NADPH and CaM influence interflavin distance relationships, the enzyme global and local conformations and, as a result, provides a chemical reaction. Inter-flavin distances derived from the PELDOR data on two electrons reduced nNOSred were found to be (in Å) 35, 28, 24. This suggests conformational states at low temperature (80 K), with three energy minima, each having a different inter-flavin distance. CaM binding in the full-length protein shifts the equilibrium toward on average slightly longer inter-flavin distances. The observed shortest inter-flavin distances of 20 Å obtained from the PELDOR experiments for nNOSred was in reasonable agreement with reported the inter-flavin distance in the crystal structure of NADP-bound nNOSred 15.5 Å [41].

The analysis of experimental data obtained in [12] revealed the following tendencies: (1) In the absence of C-terminal tail (CT) interaction, a substrate –NOS spends to reduction of volume of the activated complex ($\Delta V‡$ negative) for account restructurings of the enzyme local structure to more compact state; (2) Linkage of CaM by FLnNOS leads to substantial increase of rate and to positive value of $V‡$ for processes testifying to transition to a more friable conformation. Data on pressure dependence of the observed rates High-pressure fit parameters for the analysis of the observed rate constants for steady-state of NADPH oxidation, NO formation, superoxide formation, and high-pressure stopped-flow fit parameters for pre-steady-state flavin reduction and were tabulated [12].

Fluorescence quenching was used to study the binding of CaM labeled with the fluorescent dye Atto532-maleimide to nNOS under high pressure conditions [12]. As pressure is increased, the fluorescence quenching under binding conditions decreased only slightly. The authors suggested the following consequence of the conformational equilibria in nNOS during catalysis: (1) upon binding of NADPH shift toward the population of more closed conformations takes place; (2) in the presence of CaM, the conformational equilibrium shifts toward on average intermediate inter-flavin distances, with conformations compatible with both inter-flavin electron transfer and FMN to heme electron transfer occur; and (3) during catalysis, theFMNsubdomain interacts alternately with the FADsubdomain and hemeoxygenase domain, resulting in NO formation. The key role of CaM in conformational sampling, vectorial electron transfer, and preventing "leakage" of electrons to molecular oxygen or other external electron acceptors was stressed [12].

Single-molecule FRET spectroscopy was used to characterize conformational dynamics of individual molecules of nNOSr in a free state and after binding CaM [42]. Cys residues at two surface sites on the enzyme's FNR and FMN domains were labeled with Cyanine 3 (Cy3) donor and Cyanine 5 (Cy5) acceptor dye molecules followed by the fluorescence measurement efficiency of the singlet singlet energy transfer that depends on the donor acceptor distance. This approach allows to determine how CaM binding influence on conformational states distribution. The associated conformational fluctuation within a distance range of ~40–60 Å for the closed and maximally open states, respectively, were investigated. These results suggested changes in the distance distribution between the NOS domains, shortening the lifetimes of the individual conformational states, and narrowing the distributions of the conformational states. The enzyme fluctuation rates were also estimated.

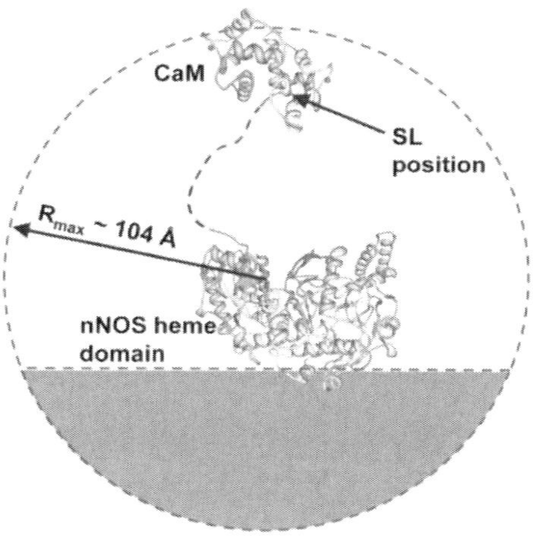

Figure 5.13. Structural model of the open state used in RIDME simulations. The SL position is averaged over the sphere centered at the peptide nitrogen of Tyr706, with the radius of 104 Å. The regions occupied by the oxygenase domain and the gray part of the sphere on the far side of the oxygenase domain (40 Å from the center to the cutoff plane) are excluded from the calculation as in accessible [38].Copyrights 2014 American Chemical Society.

In the work [38], pulsed electron paramagnetic resonance (EPR) study of domain docking in neuronal NOS: the CaM, and output state perspective. The dynamic domain docking in a CaM-bound oxygenase/FMN (oxyFMN) of nNOS was investigated using the relaxation-induced dipolar modulation enhancement (RIDME) technique. The nitroxide spin label (SL) was attached to the protein via previously introduced cysteine at position 110 of CaM (Figure 5.13) and the distance between the label and the ferric heme centers in the oxygenase domain of nNOS was measured by analysis of corresponding magnetic dipole interaction.

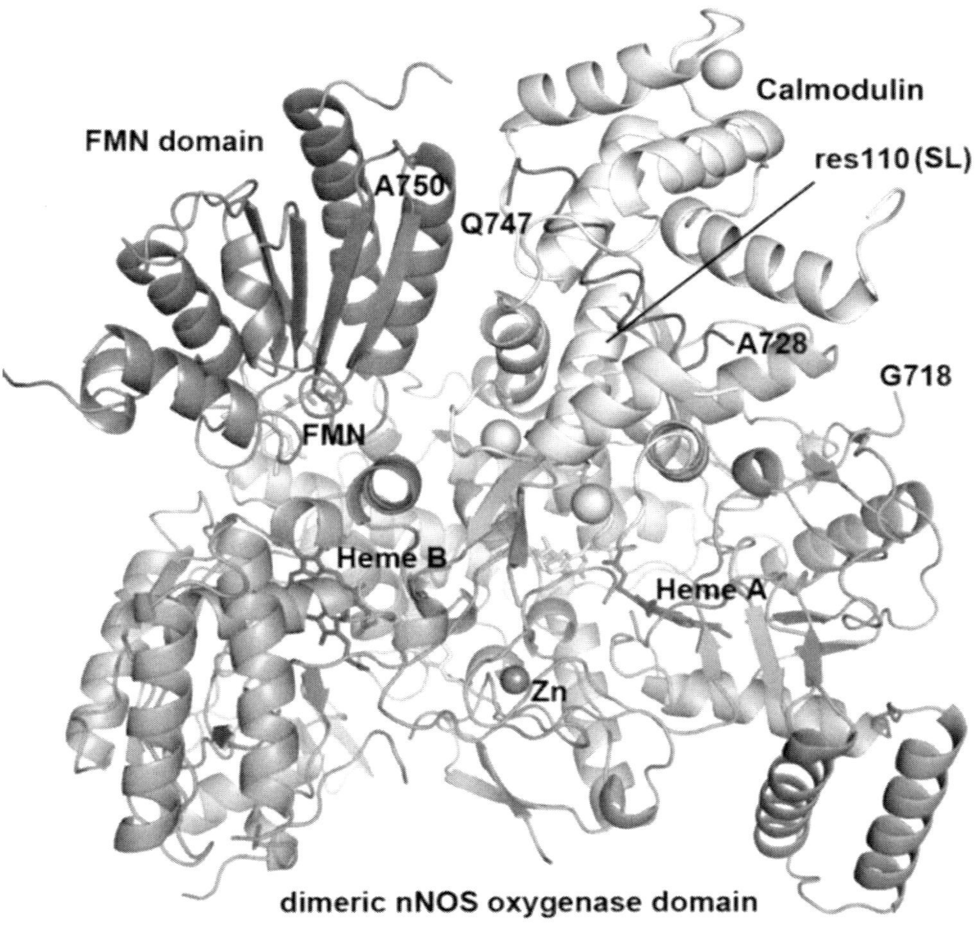

Figure 5.14. Docking model of nNOS FMN domain onto the dimeric oxygenase domain D 4JSH in the presence of CaM. For clarity, the second FMN domain and CaM molecule are not displayed. The SL site (res110) in CaM is labeled, so are the terminal residues in oxygenase and FMN domains that connect with the CaM-binding peptide. The docking model was constructed by carefully cross-checking against the iNOS model in terms of which residues are involved in the docking surface among the proteins [38]. Copyrights 2014 American Chemical Society.

Such an approach allows us to establish the distance between the label and hemin and consequently opens a possibility to keep conformational track up to 60 Å [44]. Primary results of work can be stated as follows. According suggested docking model (Figure 5.14), the SL in the docked state is located at the distances between the Oγ atom of Thr110 and the iron ions of the two heme centers of the oxygenase domain of about 36 and 31 Å, while the distance between these centers is 34 Å.

As it is seen in the Figure 5.15, the dynamic docking complex between the FMN domain and the heme domain can be formed only in the presence Ca^{2+}-enabling formation of the output state of nNOS for NO production.

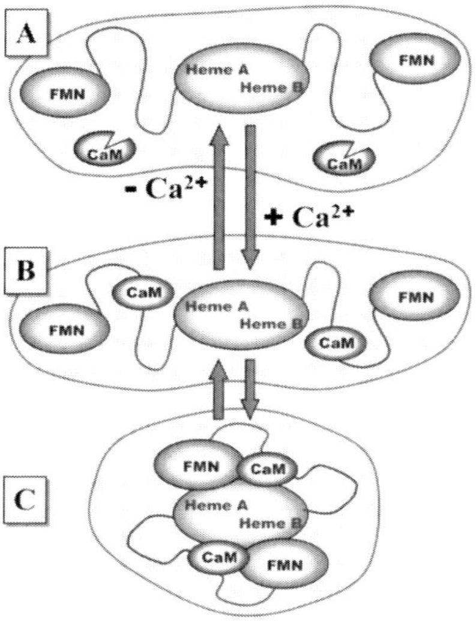

Figure 5.15. Model describing the effect of Ca^{2+} on the structural rearrangements in nNOS. (A) Without Ca^{2+} in solution, no CaM binding to nNOS occurs, and the docking complex between the FMN domain and the heme domain does not form. (B and C) When Ca^{2+} ions are present in solution (up to four Ca^{2+} ions per CaM molecule), CaM binds to the nNOS CaM-binding motif, which facilitates the formation of the IET-competent docking complex between the FMN and heme domains. The CaM-bound nNOS is in equilibrium between the structurally disordered open state (B) and the IET-capable docked state (C) [38]. Copyrights 2014 American Chemical Society.

The crystal structure of the CaCaM·FMN complex of inducible NOS in four different conformations, each with a different relative orientation, between the FMN domain and the bound CaM have been solved [45]. The data demonstrate that the CaM-binding region together with bound CaM forms a hinge and regulates electron transfer from FAD to FMN and from FMN to heme by adjusting the relative orientation and distance among the three cofactors. Thus, the hinge between the two CaM lobes contributes to the fine tuning of both FMN-heme and FMN-FAD alignments for optimal electron transfer between these cofactors.

The above-mentioned experimental data on key role of conformational transitions in NOS have been supported by recent theoretical works with the use of the advanced programs [43–47]. To describe the structural rearrangements and the domain interactions related to the oxy FMN-heme interdomain electron transfer (IET), the molecular dynamics simulations on the iNOS oxyFMN·CaM complex in [Fe(III)][FMNH(-)] and [Fe(II)][FMNH] oxidation states, the pre- and post-IET states have been carried out [43]. The suggested model predicted that an intra-subunit pivot region modulates the FMN domain motion and provides a bottleneck in the conformational sampling that leads a conformational state favorable to the inter domain electron transfer. These dynamic interactions ensure that the FMN domain moves with appropriate degrees of freedom and docks to proper positions at the ends, resulting in efficient IET. Calmodulin binding to NOS enables a conformational change, in which the FMN domain shuttles between the

FAD and heme domains to deliver electrons to the active site heme center. This step was suggested to be the rate-limiting in NOS catalysis [43]. In addition, possible electron transfer routes between the FMN and heme molecules and the motions of the FMN domain were considered. Thus, the function of CaM is to stabilize the ET competent output state long enough for heme reduction, owing to the large domain rearrangements that are necessary for ET, first from FAD to FMN and then FMN to heme.

In recent work [46], a model for the final FMN-to-heme electron transfer step this final electron transfer step for the heme–FMN–calmodulin NOS complex based on the scheme of domains dynamics, using a 105-ns molecular dynamics trajectory, was evaluated. The calculation results indicated that the equilibrated CaM and FMN subdomain model, based the molecular dynamics simulations of the oxy-FMN-CaM in the initial state and after 80 nanoseconds, is closer to chain of the heme domain than in the initial starting model. In the frame of the model, CaM, binding on the a-helical linker between the FMN and oxygenase domain, induces a conformational switch toward the output state to transfer electrons from FMN to the heme group. Such movement and fluctuation mobility of both input and output states are inevitable to avoid severe overlap between CaM and the FAD domain and for electron transfer first from FAD to FMN and then FMN to heme [46]. This provides control FMN subdomain interactions with both its electron donor (NADPH-FAD subdomain) and electron acceptor (heme domain) partner subdomains in nNOS. The suggested bimodal mechanism links a single structural aspect of CaM binding to specific changes in nNOS protein conformational and ET properties that are essential for catalysis.

5.6. BIOPTERIN

5,6,7,8-tetrahydro-L-biopterin, (tetrahydrobiopterin, BH_4) is a naturally occurring cofactor of NOS and other aromatic amino acid hydroxylase enzymes, involving in electron transfer process [48–52]. In the first step of the NOSox reaction, the tetrahydrobiopterin cofactor bound near one of the heme propionate groups acts as a one-electron donor to the P450-type heme active site [51]. The obtained electron oxidized radical is subsequently re-reduced. According to proposed mechanisms for $H_4B^{\cdot+}$, reduction, and regeneration during NOHA oxidation, the cofactor involves three redox processes. The transfer of an electron from H4B to the ferrous-oxy species triggers the formation of a ferric-peroxy species that will capture two protons and eliminate a water molecule to form a compound I-like species that will carry out the hydroxylation of the substrate L-Arginine. The H4B radical is not reduced after NOHA formation and has to receive one electron from the flavin domain to continue the catalytic cycle [51]. Multifrequency 9.5 and 330–416 GHz continuous-wave (cw) EPR and 34 GHz pulse ^{1}H electron nuclear double-resonance (ENDOR) spectra of the radical in $iNOS_{oxy}$, EPR, and ENDOR spectroscopy were used to *establish* the (de)protonation state of the H_4B radical in the oxygenase domain dimer of inducible NO synthase that was trapped by rapid freeze quench [48]. On the basis of obtained data on the ESR magnetic parameters, it was suggested that the cationic $H_4B^{\bullet+}$ is stabilized in protonated form and that the protein environment also prevents further oxidation. It is necessary to emphasize a role of the hydrogen bonding interactions in maintenance of optimum structure of the active center. Several iNOS residues and the heme group form a hydrogen bonding network with the H4B cofactor. Ser112, Ile456, and Trp457 make H-bonds through the main chain carbonyl groups; Arg375 interacts through its side chain. The side chain of Trp457 forms a p-stacking interaction with H4B. In addition, several bsNOS residues and the heme group form a hydrogen bonding network with the H4F cofactor. Thr324 and Trp325 make H-bonds through the main chain carbonyl groups; Arg243 interacts through its side chain. The side chain of Trp325 forms a p-stacking interaction with H4B [51].

The following values of the rate constant have been reported for iNOSoxy: the reduction of the $H_4B^{\cdot+}$ radical in the first step (0.71 s^{-1}), in the second step (8.3 s^{-1}), and the decay of the $Fe^{II}O_2$ species (8 s^{-1}) for the second step [53]. Other data on the NOS kinetics summarized in [49] are the following: flavoprotein-to-oxygenase domain electron transfer rate at 10°C (4 s^{-1}), H_4B radical formation rate during Arg hydroxylation at 10°C (20 s^{-1}), the oxidative decay of its bound H_4B radical (0.6 s^{-1} at 10°C), the rate of ferrous-dioxy decay (12 s^{-1}) the rate of H_4B radical formation (11 s^{-1}), and the rate of Arg hydroxylation (9 s^{-1}). Utilization of 5MeH_4B in place of H_4B led to faster radical formation (51 s^{-1}) and a slower oxidative decay (0.2 s^{-1}) in nNOS Arg hydroxylation reactions compared with H_4B.

5.7. Nitric Oxide Synthase Reaction Computational Studies

Owing to essential biological importance of reactions that involve heme, mechanisms of heme reactions in enzymes like NOS, heme oxygenase (HO), and cytochrome P450s (CYP450s) occupied great deal of attention [47, 54–60]. Experimental investigation in this area preceded corresponding computational studies. For example, spectral experiments on a reaction of NG-hydroxy-L-arginine (L-NHA) and H_2O_2 in the hemin active cite of murine macrophage NOS under anaerobic conditions were performed [54].

On the basis of mechanistic studies of NOS inactivation by amidine analogues of l-arginine and other previous mechanistic results, a new mechanism for NOS-catalyzed l-arginine NG-hydroxylation (the first half of the catalytic reaction) was proposed [55]. The following important aspects of the mechanism were discussed: *the second half of the NOS catalytic reaction, were suggested:* (1) the internal electron transfer between the substrate and heme, on the basis of mechanistic results of NOS inactivation by NG-allyl-l-arginine and the structures of the substrate intermediates; (2) the heme degradation caused by NOS inactivation by amidines, and (3) meso-hydroxylation step during inactivation of NOS by amidines as well as the HO-catalyzed reaction.

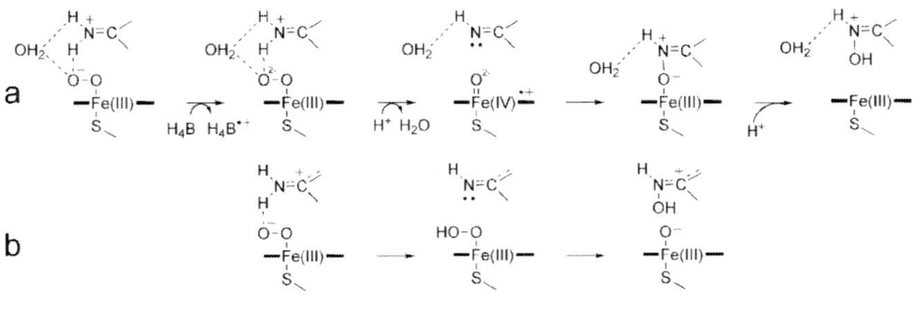

Figure 5.16. Proposed reaction mechanisms mechanism of key steps of the first half-reaction of murine macrophage nitric oxide [57]. Copyrights 2009 American Chemical Society.

Within the last decades, the contribution of the advanced quantum mechanical/molecular mechanical/molecular dynamics (QM/MM/MD) calculations in understanding the NOS detailed mechanism has been raised [43, 57–60]. Though this approach does not give the unequivocal final answers to all questions in this extremely complex area, the calculations pave a way to fruitful discussions about the most probable mechanisms of interest. Mentioned above in a greater degree are concerns to elementary chemical stages, which cannot be studied experimentally, at least nowadays. On the other hand, unexpected predictions of the theory may initiate new experiments

A mechanism of key steps of the first half-reaction of murine macrophage NOS based on the appearance of a ferrous heme-NO complex with a Soret peak at 440 nm and a broad single alpha/beta peak at 578 nm, attributed to a reduced ferric-NO- (nitroxyl) complex, was proposed (Figure 5.16) [57].

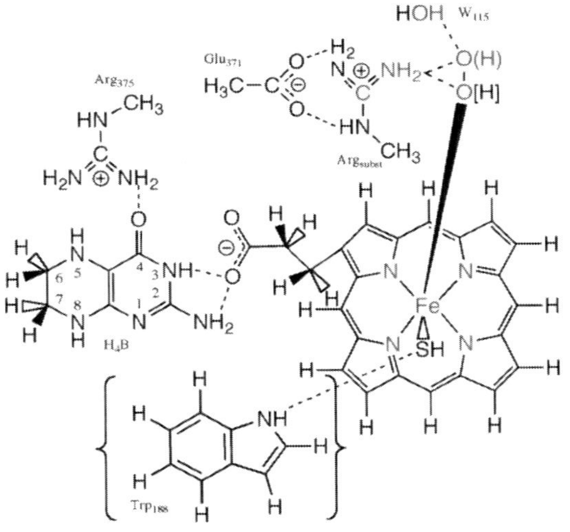

Figure 5.17. Schematic view of the QM region used in the calculations. Numberings as used in the text are marked for H4B [57]. Copyrights 2009 American Chemical Society.

A quantum mechanical/molecular mechanical (QM/MM)13,14 study to decipher the mechanism of the NO first half-reaction has been performed on the base of scheme depicted in (Figure 5.17) [57].

Three alternative mechanistic hypotheses for the arginine hydroxylation process were tested: unprotonated Fe^{III}-OO- ferric-oxy pathway, doubly protonated pathway, and monoprotonated ferric-superoxy pathway via Por+•Fe^{III}-OOH species [57]. The authors suggested that the reaction most likely proceeds after a single proton mechanism. The following elementary steps of key chemical elementary acts have been considered in detail: O-O bond breaking, arginine deprotonation, arginine oxygenation, and product (NHA) forming. Surrounding factors potentially affecting the Por+•Fe^{III}-OOH pathway, namely, role of H4B, Trp188 and crystal water W115, were also taken into consideration. A proposed schematic drawing of the electron flows in the reaction is shown in Figure 5.18.

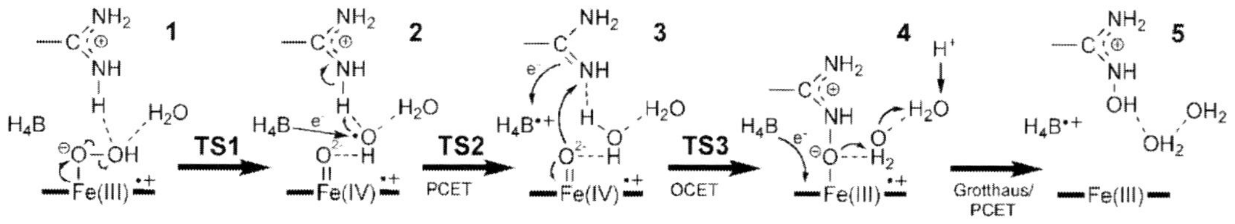

Figure 5.18. Proposed schematic drawing of the electron flows in first half-reaction mechanism of nitric oxide synthase [57]. Copyrights 2009 American Chemical Society.

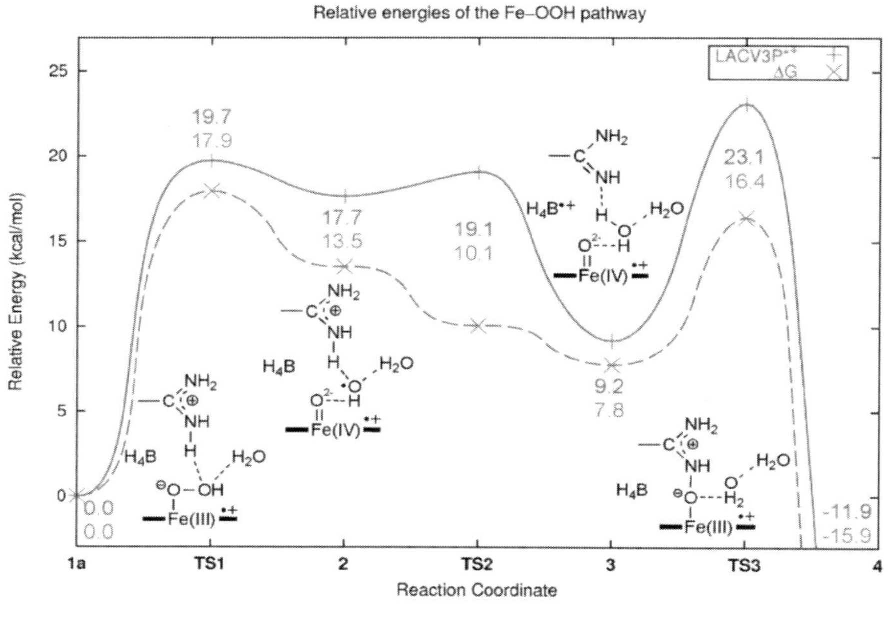

Figure 5.19. The energetics of the NOS first half-reaction. The red solid curve shows the energies calculated by the LACV3P*+ basis set, while the green dashed curve also includes thermal effects (i.e., Gibbs free energy of the QM-only subsystem) [57]. Copyrights 2009 American Chemical Society.

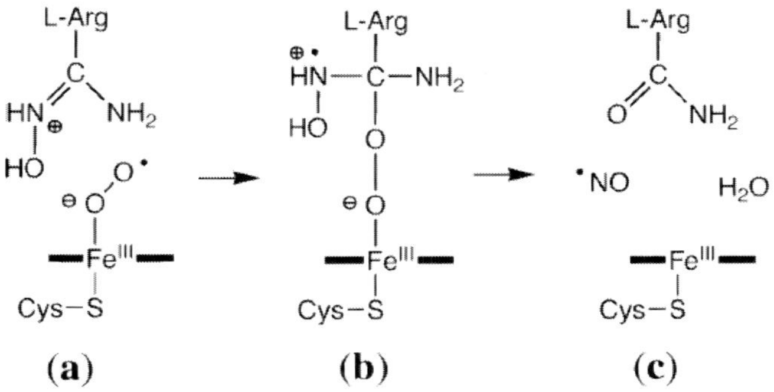

Figure 5.20. Summary of the common features of the proposed mechanisms for the second half-reaction: initial formation of (a) an FeHeme-O2 species, subsequent formation of (b) a tetrahedral "intermediate," and formation of (c) the final products [59]. Copyrights 2009 American Chemical Society.

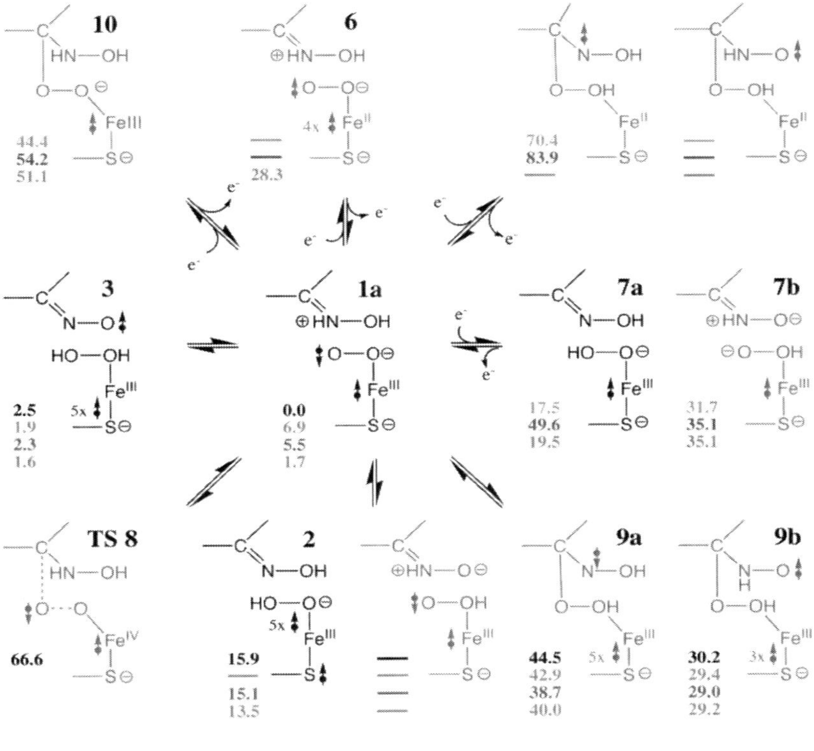

Figure 5.21. Schematic illustration of the structures and relative free energies (kcal mol-[1]) of all initial complexes and tetrahedral intermediate considered in the study [59]. Copyrights 2009 American Chemical Society.

In the Figure 5.18, PCET is the proton coupling electron transfer, and OCET is the oxygen coupling electron transfer. The N-oxygenation transition state TS3 was believed to be the most principle stage of the process, leading to 4, which involves FeO-N formation and occurring by an OCET mechanism. Results of theoretical estimation of the energetics of the first half-reaction nascent from the Por+•FeIII-OOH pathway predicts maximum values of the activation energy for the reaction limiting steps of 16-18 cal/mol that relate to experimental data on corresponded rate constant. The calculated energetic profile of the NOS first half-reaction is given in Figure 5.19.

Employing a combined QM/MM (quantum mechanics/molecular mechanics/DFT) approach, a mechanism of oxygen activation and the conversion of L-arginine into N(omega)-hydroxo-arginine in the active site of the NOS enzymes have been proposed [58]. According to the suggested reaction model, rout in NOS enzymes compound I is first reduced to compound II before the hydroxylation of arginine. The substrate arginine appears to serve as a proton donor in the catalytic cycle to convert the ferric-superoxo into a ferric-hydroperoxo complex and as the substrate that is hydroxylated in the process leading to N(omega)-hydroxo-arginine. The calculations using QM/MM methods also predicted optimized geometries of compound II of NOS enzymes.

A detailed investigation on the key initial step of the second half-reaction of NOS based on advance density functional theory (DFT) has been performed [59]. Intermediate complexes formed via transfer of either *or* both hydrogens of the substrates (Nӧ-hydroxy-L-arginine, NHA), and NHOH group to the Fe-bound O_2 were considered (Figure 5.20).

Starting from the initial structure of the active site with NHA + O_2 bound, the following reaction stages were theoretically considered (Figure 5.21): (1) transfer of a single H/H^+ from NHA to Feheme-O_2; (2) double transfer of H/H^+ from NHA to Feheme-O_2; (3) addition of an electron from H4B; (4) addition of an electron from H4B, without H+/H Transfer; (5) addition of an electron from H4B, with H^+/H transfer; (6) formation of the proposed tetrahedral intermediates, direct formation without electron or H+/H transfer; (7) formation involving transfer of H^+/H from –NH to Feheme-O_2; and (8) formation intermediates with an additional electron from H4B. The author's preferred initial pathway, which involves the simultaneous transfer of both hydrogens of the −NHOH group to the Fe_{heme}−O_2, without an additional electron. This process gives the Fe_{heme}−HOOH species.

In the frame of the proposed reaction mechanism, the most principle step is formation either a shortly living tetrahedral intermediate, in which the Fe_{heme}−O_2 has attached at the guanidinium carbon (C_{guan}) of NHA, that is, forms an Fe_{heme}−O_2−C_{guan} link (Figure 5.22) or, alternatively, corresponding transition state. In the latter case, the elementary act runs on the synchronous mechanism. Schematic illustration of the structures and relative free energies of all initial complexes and tetrahedral intermediates were also considered in [59].

DFT calculations on isolated model systems and hybrid quantum-classical computations of the active sites in the protein environment for NOS wild-type and mutant (Trp 178 Gly) proteins were performed [60]. The calculations on the five-coordinate ferrous heme–thiolate complex were fulfilled on the quintet configuration. In order to analyze how the H-bond interaction of Trp 178 with the proximal cysteine affects NO coordination, a model of wild-type NOS in which the thiolate ligand is H-bonded to a Trp

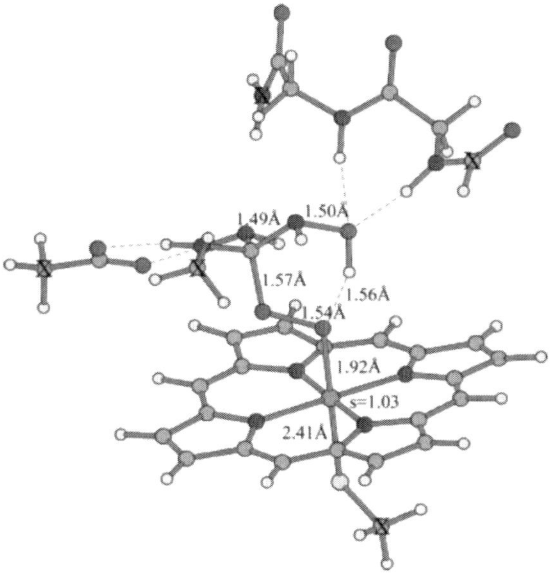

Figure 5.22. Optimized structure and selected spin distribution of the tetrahedral intermediate with the addition of an electron to 1a [59]. Considered in Study [59]. Copyrights 2009 American Chemical Society.

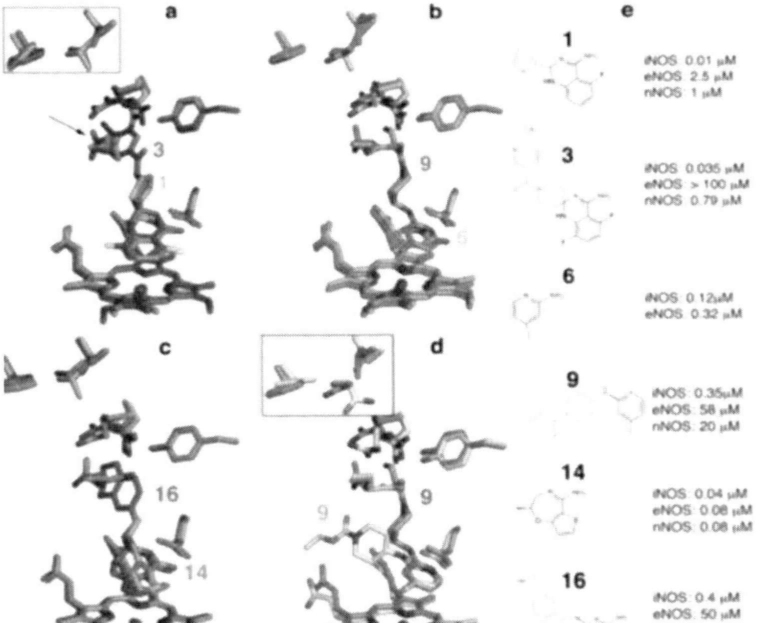

Figure 5.23. NOS inhibitors. X-ray structures of six inhibitors bound to iNOS (a-c) and eNOS (d) representative of three different pharmacophores used in this study. The inhibitor core is shown in black and the "tail" in magenta. IC50 values are shown for all three isozymes. The inhibitor core is shown in black and the "tail" in magenta. Boxed residues represent the isozyme-specific triad of distant residues that modulate the opening of the pocket. IC50 values are shown for all three isozymes [62]. Copyrights 2008 American Chemical Society.

residue was considered. Molecular dynamics simulations on the iNOS oxyFMN·CaM complex models in [Fe(III)][FMNH−] and [Fe(II)][FMNH•] oxidation states, the pre- and post-IET states based on the initial docking model for a human iNOS oxyFMN construct, were carried out. Molecular dynamics simulating of specific residues on the heme, FMN, and CaM domains allowed us to predict the dynamic interacting sites, the FMN domain moving efficient IET, and NO production [43]. In the study, to illustrate the motions of the FMN domain, the heme domains in two conformations at 60 ns of human iNOS oxyFMN·CaM in the pre- and post-IET states were superimposed. Possible electron transfer routes between the FMN and heme molecules and the motions of the FMN domain were discussed.

5.8. Nitric Oxide Synthase Inhibitors

5.8.1. Crystal Structure

Inhibition of the NOS reaction has proved to be a powerful approach for regulation of the enzyme activity and paves the way for a purposeful drug design. NOS inhibitors also improve the effectiveness of antimicrobials. Here we concentrate on advances in determination of crystal structure of the inhibitor-enzymes complexes and corresponding kinetics. Achieves in the area are illustrated by several typical examples.

Seventeen crystal structures of iNOS and eNOS bound to various inhibitors (Figure 5.23), which mimic the binding of the arginine substrate in the active site were determined [62]. These data allowed us to interpret information in the context of a such important part of the enzyme as the heam active center.

Experimental results indicated that binding of the large inhibitors (quinazolines, aminopyridines, and bicyclic thienooxazepines) induces the cascade of conformational changes and opening of the new specificity pocket. As an example, binding of inhibitor 9 in iNOS induces conformational changes resulted in the opening of a new pocket for binding of the inhibitor tail (Figures 5.23 and 5.24). The opening is prevented in eNOS by bulkier residues far from the active site. In contrast, binding mode of inhibitor 9 in iNOS and eNOS does not reveal conformational changes in eNOS [62].

Figure 5.25 illustrates a role of the anchored plasticity on selective enzyme inhibition. Isozyme-specific-induced fit upon inhibitor binding is accompanied by the cascade of conformational changes in human iNOS (hiNOS). Inhibitor binding first induces the Gln-closed to Gln-open conformation and Arg rotation and rotation of second-shell Asn toward third-shell Phe286 and Val305 occurs. In human eNOS (heNOS), bulkier third-shell residues (Ile269 and Leu288) prevent the Asn rotation. In human nNOS (hnNOS), partial rotation of Asn toward third-shell residues Phe506 and bulky Leu525 takes place [62].

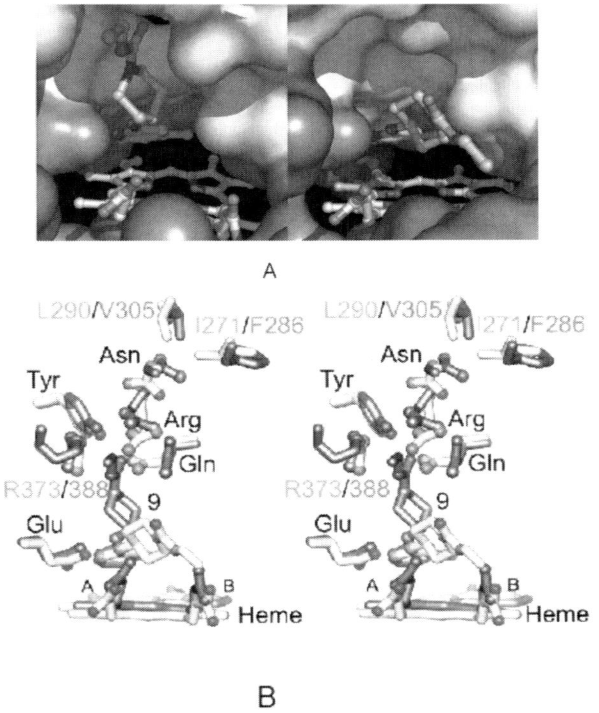

Figure 5.24. Selective aminopyridine compound 9 binding to eNOS versus iNOS. (a) solvent-accessible surfaces for the iNOS (left) and eNOS (right) active sites. The core of compound 9 binds closer and more parallel to the heme in eNOS. In iNOS, side chain rotations of Gln, Arg, and Arg388 open the Gln specificity pocket for binding of the bulky inhibitor tail. (b) Stereo view of the superimposition of bovine eNOS–compound 9 and human iNOS–compound 9 X-ray structures, highlighting the cascade of conformational changes of first-shell and second-shell residues upon inhibitor binding to iNOS [62]. Copyrights 2008 American Chemical Society.

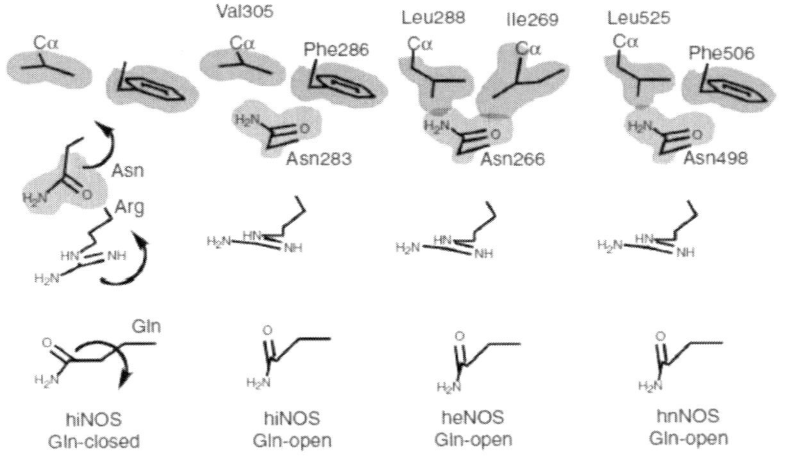

Figure 5.25. Isozyme-specific induced fit upon inhibitor binding. Schematic of the cascade of conformational changes associated with inhibitor binding in the three human NOS isozymes. The van der Waals surfaces for the isozyme-specific triads are shown. Explanation in text and [62]. Copyrights 2008 American Chemical Society.

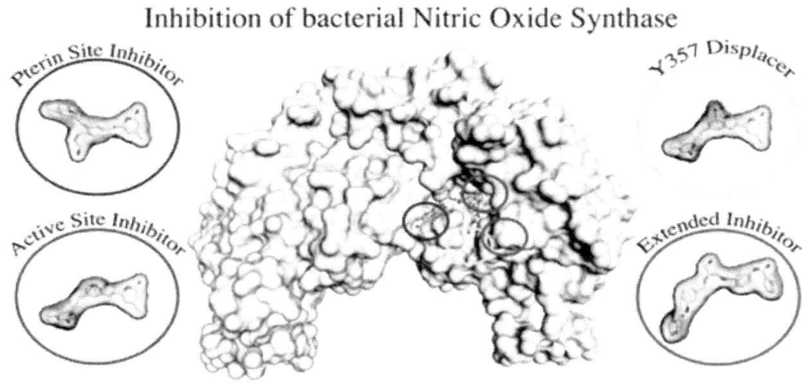

Figure 5.26. Schematic illustration of structure of NOS different inhibitors and its position on the enzyme [63]. Copyrights 2005 American Chemical Society.

Thus, effective binding in iNOS depends not only on interactions within the conserved enzyme active site, but also on interactions 20 Å away from the active site. The inhibitors containing large tails can be bind in this new specificity pocket and block NO production. The authors suggested that the plasticity of distant non-conserved residues can be used for drug design.

Thirty-two crystal structures of inhibitors of were established [63]. The structure of bacterial NOS's different inhibitors and its position on the enzyme is schematically illustrated in Figure 5.26.

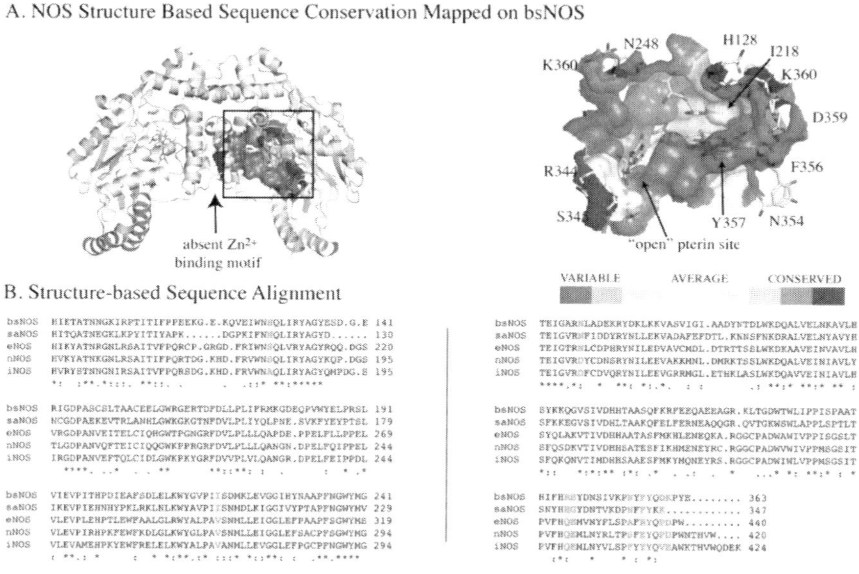

Figure 5.27. (A) Crystal structure of bacterial NOS (bsNOS) shown as a dimer (left) with active site (right) conservation *colored* using the consurf web server. (B) Partial sequence alignment of NOS isoforms based on a structural alignment using Chimera and crystal structures of bsNOS), saNOS human eNOS, human nNOS, and human iNOS) [63]. Copyrights 2005 American Chemical Society.

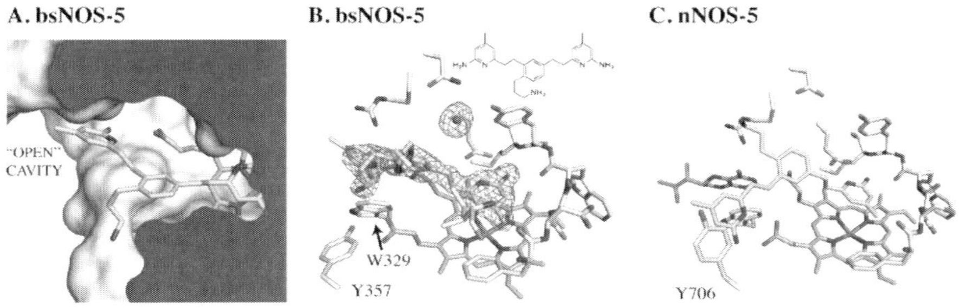

Figure 5.28. (A) Inhibitor 5 bound to bsNOS with the aminopyridine group bound to the heme propionate and exposed to a solvent-accessible surface that is unique to bNOS. (B) Stick representation and $2F_o - F_c$ at 1.0σ of 5 bound to bsNOS. (C) Stick representation of nNOS–5 [63]. Copyrights 2005 American Chemical Society.

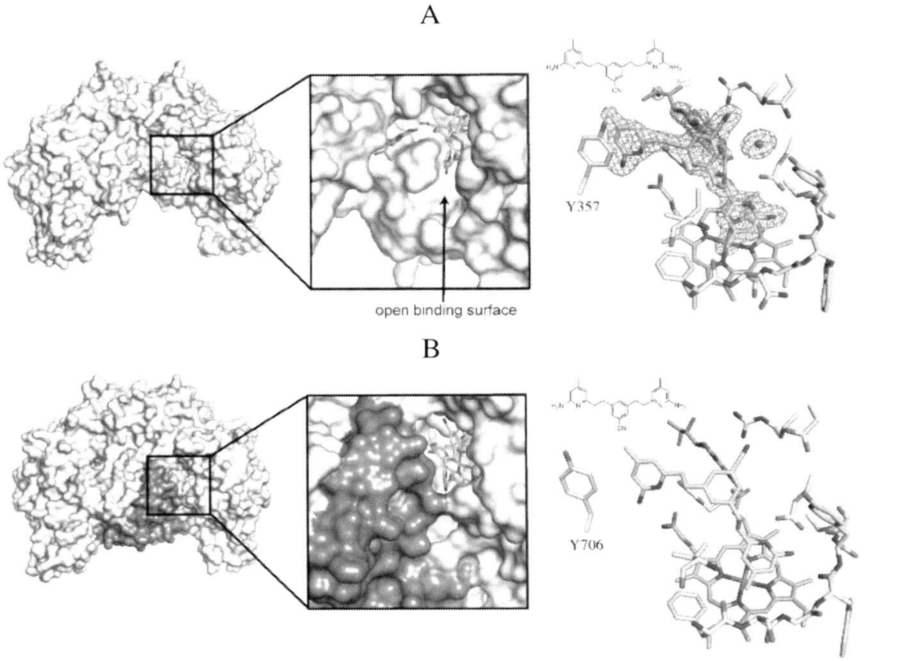

Figure 5.29. Inhibitor 3 induces a conformational change in both (A) bsNOS Y357 and (B) nNOS Y706. Rotation of Y357 in bsNOS opens access to a novel binding surface. [63]. Copyrights 2005 American Chemical Society.

Mutagenesis studies revealed several key residues that unlock access to bacterial NOS surfaces that could provide the NOS inhibitors selectivity (Figures 5.27 and 5.28).

Binding inhibitor 3 to bsNOS Y357 and nNOS Y706 induces conformational change (Figure 5.29).

X-ray crystal structure of complex rat nNOS with N^5-(1-imino-3-butenyl)-L-ornithine (L-VNIO) was reported [64]. It was shown that the amino acid head group of L-VNIO fits into a pocket defined by Gln-478, Tyr-588, Glu-592, and Asp-597 similar to that of L-arginine. The hydrophobic pocket expands slightly because of movement of the Val-567 and Phe-584 side chains. Crystal structures of inhibitors consisting three fragments—an aminopyridine ring, a pyrrolidine, and a tail of various length—and polarity of these *inhibitors*, bound to either wild-type or mutant nNOS and eNOS, were established [65].

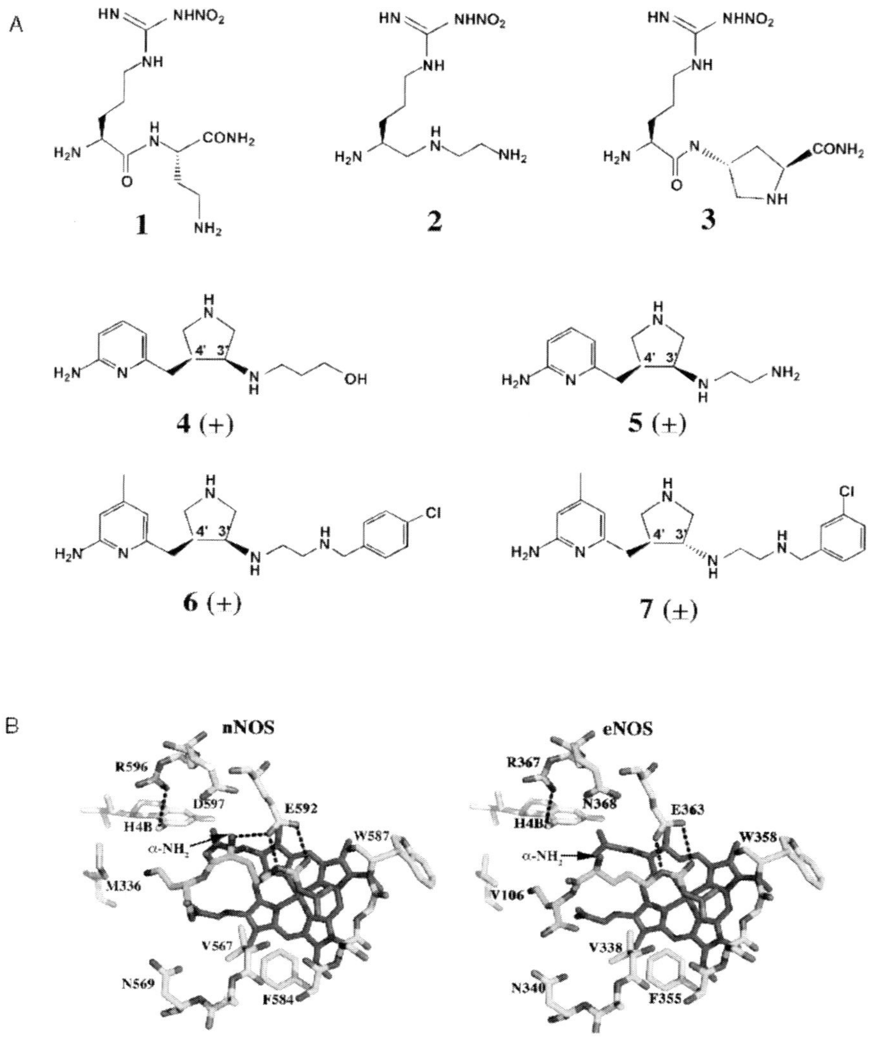

Figure 5.30. Structure of inhibitor 1 complexed to nNOS and eNOS. Ki is the inhibitor equilibrium constant [65]. Copyrights 2005 American Chemical Society.

The aminopyridine ring mimics the guanidinium group of L-arginine and functions to replace the substrate in the NOS active site.

Chemical structures and nomenclature of series of inhibitors shown in Figure 5.30 were presented in the paper [65].

It is interesting that affinity of the inhibitor 1

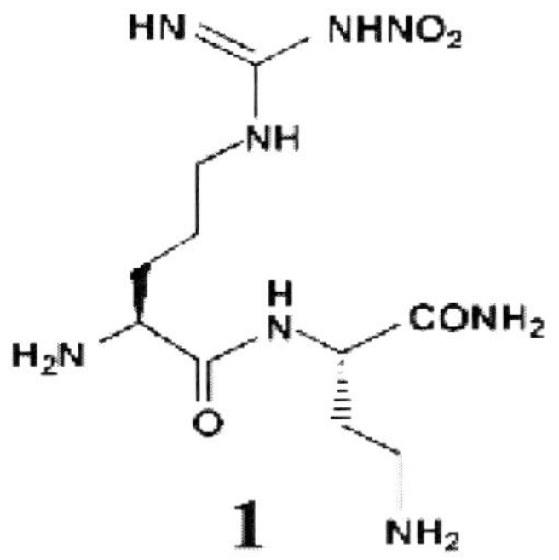

1

for nNOS is of two orders of magnitude higher than that for eNOS (Figure 5.31) A plausible cause of this finding is that, in nNOS, the inhibitor α-amino group enables to optimally interact with both Glu592 and Asp597, while in eNOS, the residue corresponding to Asp597 is Asn368 and, as a result, the inhibitor adopts an extended conformation.

Experimental free energies (ΔG_{exp}) derived from K_i measurements and computed free energies (ΔG_{calc}) were obtained and compared. Although the inhibitors used in [65] are very similar, they span a K_i range of over 10^3, the relative ΔG_{calc} agrees well with ΔG_{exp}. (Figure 5.32).

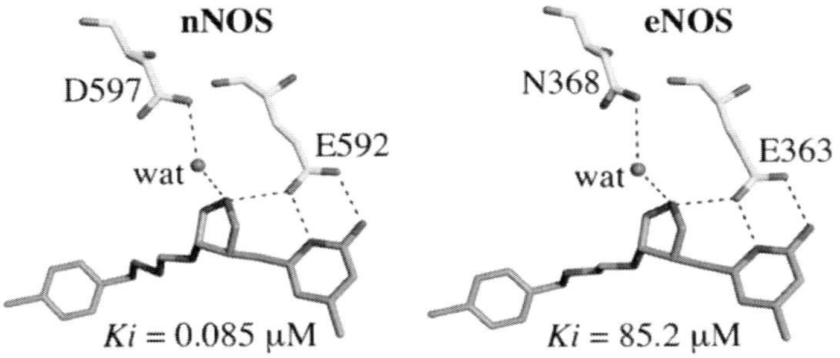

Figure 5.31. (A) Structures of the various inhibitors complexed with either eNOS, nNOS, or both used to construct the plot in panel B. (B) Plot of experimental ΔG_{exp} vs computed ΔG_{calc} using five different crystal structures [65]. Copyrights 2005 American Chemical Society.

Figure 5.32. A series of L-NIO analogues synthesized in [80]. Copyrights 2005 American Chemical Society.

Recent review [77] summarizes principle steps of the history of NOS inhibitor development and more recent advances in developing isoform selective inhibitors using primarily structure-based approaches.

5.9. INHIBITORS KINETICS

The other fruitful avenue of investigation of the NOS regulation molecular mechanism is detection of the inhibition kinetics and thermodynamics. Here we restrict the discussion to data that illustrate the experimental approach and some main results. It is in last decades, a tremendous increase in employing synthetic NOS inhibitors occurred [67–87].

Natural NOS inhibitors can be involved in regulation of the NOS activity in living organism.

LN-Methylarginine, a product of degradation of arginine naturally occurring in organisms, acts as a competitive inhibitor of all NOS isoforms with a Ki value ranging from 10-100 μM [66–67]. Another naturally present inhibitor is N-Dimethylarginine (ADMA) [68]. Following the discovery of synthetic inhibitor L-NAME (LN N_{ω}-Nitro-L-arginine methyl ester hydrochloride), another inhibitor of similar structure, L-NG-nitro arginine, was developed [69, 70]. There have been a number of reports indicating the effectiveness of a variety of classes of inhibitors acting by interfering with the heme moiety. Among them aminoguanidine [71]. S-substituted isothioureas [72], S-methyl-L-thiocitrulline and S-ethyl-L-thiocitrulline, the NOS reaction product analogs [73], indazole and derivative –(7-Nitroindazole (7-NI) and trimethylphenylfluoroimidazole, which binds to the heme group altering both pterin and substrate binding [74, 75], and S-Methyl-L thiocitrulline Inhibitor that binds with heam group [76].

Some inhibitors expose high selectivity related to different NOS. For example, an inhibitor of human inducible nitric-oxide synthase (iNOS) *N*-(3-[Aminomethyl] benzyl)acetamidine1 (400W) showed time-depended saturation kinetics with a maximal rate constant of 0.028 s^{-1} and a binding constant of 2.0 μM. 400W was at least 5000-fold selective for iNOS *versus* eNOS. Inhibition of human neuronal nNOS and endothelial NOS (eNOS) was relatively weaker, rapidly reversible, and competitive with L-arginine, with K_i values of 2 μM and 50 μM [78]. Other selective inhibitors of iNOS such as isothioureas aminoguanidine (16), cyclic amidines, and *N*-iminoethyl- L-ornithine were also found to be about 30-fold more potent against iNOS than eNOS [79].

Initial binding kinetics of novel L-arginine and L-homoarginine analogs were detected [64]. The kinetic studies showed that the inhibitors, L-homoarginine analogs, bind competitively with L-arginine. Binding constants for L-arginine, L-homoarginine, *N*5-(1-imino-3-butenyl)-L-ornithine(L-VNIO), and number of novel inhibitors analogues of L-VNIO were presented [64]. Values of binding constants for L-arginine, L-homoarginine,

L-VNIO, and novel inhibitors related to L-VNIO were tabulated [64]. The following tendency have been drawn: (1) Comparison of the results for L-VNIO *(3-carbon tail).*

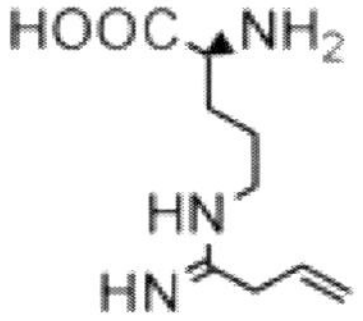

to those for L-ENIPO

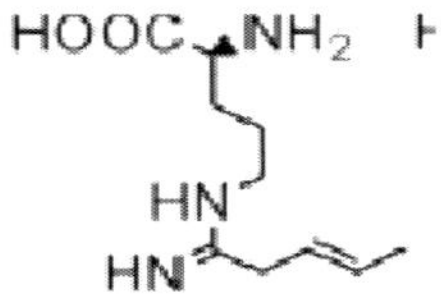

and *L-PrNIO (one carbon tails more)* shows a 5–8-fold increase in the *Ki* values for each of the NOS isoforms; (2) lengthening the tail group weakens eNOS binding relative to the other two isoforms; (3) the branched four-carbon analog, L-2MeNIBO,

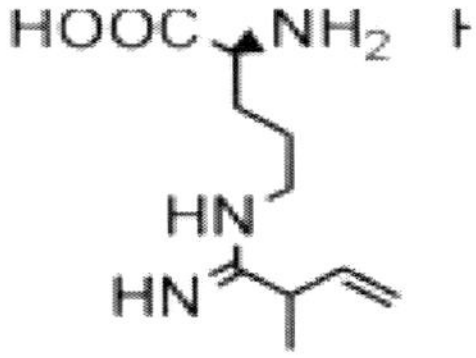

showed a twofold difference in the *Ki* values for nNOS and eNOS or nNOS and iNOS; (4) the *Km* values for homoarginine with the human NOS isoforms are about tenfold higher than the *Km* values for arginine; and (5) the *Ki* values for L-VNIL and L-ENIPL are not higher than the values determined for the corresponding arginine analog inhibitors, L-VNIO, and L-ENIPO. Data on rates of formation and decay of the tight-binding inhibitor complexes with the neuronal and inducible forms of the human NOSs and associated equilibrium constants for L-arginine, L-homoarginine, L-VNIO, and novel inhibitors related to L-VNIO were tabulated [64].

IC50 values for sixteen synthesized de Novo inhibitors of iNOS, eNOS and nNOS isozymes were measured [62, 77]. Anchored plasticity opens doors for selective inhibitor design in nitric oxide synthase

Table 5.1. Binding and Inhibition Constants for Wild-Type and Mutant Human iNOS and eNOS Proteins COMPOUND 9 (Figure 5.23) [62]. Copyrights 2008 American Chemical Society

K_d (μM)[a]			IC_{50} (μM)[b]						
iNOSox WT	iNOSox E377A	iNOSox F286I V305L	iNOS WT	eNOS WT	iNOS Q263A	iNOS Q263N	eNOS Q246A	eNOS Q246N	iNOS Y347F Y373F
0.4	$\gg$100	$\gg$100	0.35	>100	1.0	2.0	56	>100	3.4

The experiments have revealed variety kinetic parameters effects (Table 5.1). For different inhibitors and isozymes, the IC50 values ranged 00.1 mM – 0.4 nNOS for iNOS, 0.32 - >100 mM for eNOS, and 0.08 mM – 0.58 mM for nNOS. This finding reflects fine specificities of the enzyme structure out the conserved residues surrounding heam.

To determine the dual substrate/inactivation properties of L-NIO, serious analogues of amidines as substrates and inactivators of iNOS were synthesized, and its effects on iNOS was investigated (Figure 5.32) [80]. Computer modeling and molecular dynamics simulations were also employed. All these compounds inhibited iNOS and increased inhibitory potency correlated with decreased substituent size. Unlike the methyl amidine (L-NIO), which is the natural inhibitor, the other alkyl groups block binding of O_2 at the heme iron. Compounds 8, 9, and 11 were inactivators; however, no heme was lost, and no biliverdin was formed. No kinetic isotope effect on inactivation was observed with perdeuterated ethyl in compound 8.

Most of the iNOS selective dimerization inhibitors have an substituted imidazole group (pyrimidine imidazole-based or 4-((2-Cyclobutyl-1*H*-imidazo[4,5-*b*]pyrazin-1-yl) methyl)-7,8-difluoroquinolin-2(1*H*)-one (KD7332) and (2,4-difluorophenyl)-6-[2-[4-(1H-imidazol-1-ylmethyl) phenoxy] ethoxy]-2-phenylpyridine, for example), which directly bind to the heme iron in the enzyme's active site and disturb contacts between monomers in the NOS oxidodase domain [81, 82]. Nevertheless, cyclo(dehydrohistidyl-l-tryptophyl) also prevents the dimerization of inducible NOS [83].

The following are typical examples of NOS inhibitors, which are not directly linked to the hemin moiety. [84]. As an example, a series of novel 4-amino pteridine derivatives blocking the pterin binding site of the neuronal isoform of nitric oxide synthase (NOS-I) has been synthesized [84]. A structure-activity relationship of the synthesized inhibitors t substituents in the 2-, 4-, 5-, 6-, 7-, and 8-position of the pteridine nucleus was established. It was shown that bulky hydrophobic substituents in the 6-position, such as phenyl, increased the inhibitory potency of the reduced 4-amino-5,6,7,8-tetrahydropteridines, most probably, due to the allosteric effect. Specific inhibitors of NADPH-dependent flavoprotein of the reductase domain of NO synthases diphenyleneiodonium (DPI), di-2-thienyliodonium (DTI), and iodoniumdiphenyl, (ID), which block the NADPH-dependent flavoprotein of the reductase domain activity with (IC50's 50-150 nM) represent a separate class of inhibitors [85]. Inhibition by DPI was prevented by NADPH, $NADP^+$, or 2'5'-ADP. Cyclo(dehydrohistidyl-l-tryptophyl) prevents the dimerization of inducible NOS [86].

Recently a comprehensive review on the NOS inhibitors was published by Lakshmikirupa Sundaresan et al. [79] A list of thirty-nine inhibitors based on hemin binding sites has been tabulated.

In summary, known inhibitors of the NOS oxygenase domain hemin center can be classified in the following way: as those that (1) replace the substrate arginine, (2) anchor active center arginine-like molecules with tails disturbing structures out the conserved residues surrounding heam, (3) allosterically prevent appropriate protein–protein interactions between the oxygenase domain monomers, and (4) inhibits blocking binding site of pterin and NADPH, which are not directly linked to the hemin moiety. Several inhibitors showed high selectivity regarding NOS isomers.

References

[1] Tobias M, Hedison, Sam Hay, Nigel S. A perspective on conformational control of electron transfer in nitric oxide synthases. *Nitric Oxide.* (2017) 63, 61-67.

[2] Förstermann U, Sessa WC. Nitric oxide synthases: Regulation and function. *Eur Heart J.* (2012) 829–837.

[3] Feng C, Chen L, Li W, Elmore BO, Fan W, Sun X. Dissecting regulation mechanism of the FMN to heme interdomain electron transfer in nitric oxide synthases. *J. Inorg. Biochem.* (2014) 130, 130-40.

[4] Shirran S, Garnaud P, Daff S, McMillan D, Barran P. The formation of a complex between calmodulin and neuronal nitric oxide synthase is determined by ESI-MS. *J. R. Soc. Interface.* (2005) 2, 465-476.

[5] Stuehr DJ, Tejero J, Haque MM. Structural and mechanistic aspects of flavoproteins: Electron transfer through the nitric oxide synthase flavoprotein domain. *FEBS J.* (2009) 276, 3959–3974.

[6] Hollingsworth SA, Holden JK, Li H, Poulos TL. Elucidating nitric oxide synthase domain interactions by molecular dynamics. *Protein Sci.* (2016) 25, 374-82.

[7] Crane BR, Arvai AS, Ghosh DK, Wu C, Getzoff ED, Stuehr DJ, Tainer J. A. Structure of Nitric Oxide Synthase Oxygenase Dimer with Pterin and Substrate (1998) *Science* 279, 2121–2126.

[8] Fischmann TO, Hruza A, Niu XD, Fossetta JD, Lunn CA, Dolphin E, Prongay AJ, Reichert P, Lundell DJ, Narula SK, Weber PC. Structural characterization of nitric oxide synthase isoforms reveals striking active-site conservation. (1999) *Nat. Struct. Biol.* 6, 233–242.

[9] Raman CS, Li H, Martasek P, Kral V, Masters BS, Poulos TL. Crystal Structure of Constitutive Endothelial Nitric Oxide Synthase: A Paradigm for Pterin Function Involving a Novel Metal Center. *Cell* (1998) 95, 939–950.

[10] Andrew PJ, Mayer B. Enzymatic function of nitric oxide synthases. *Cardiovascular Research:* (1999) 111, 521-531.

[11] Plaza C, Pineda SH, Chreifi G, Jing Q, Cinelli MA, Silverman RB, Poulos TL. Structures of human constitutive nitric oxide synthases. *Acta Crystallogr D Biol Crystallogr.* (2014) 70, 2667-2674.

[12] Sobolewska-Stawiarz A, Leferink NGH, l Fisher K, DJ, Hay S, Rigby SEJ, Scrutton NS. Energy Landscapes and Catalysis in Nitric-oxide Synthase. *J. Biol. Chem.* (2014) 289, 11725-11738.

[13] Zhang J, Martàsek P, Paschke R, Shea T, Siler Masters BS, Kim JJ. Crystal structure of the FAD/NADPH-binding domain of rat neuronal nitric-oxide synthase. Comparisons with NADPH-cytochrome P450 oxidoreductase. *J Biol Chem.* (2001)276, 37506-37513.

[14] Piazza M, Futrega K, Spratt DE, Dieckmann T, Guillemette JG. Structure and dynamics of calmodulin (CaM) bound to nitric oxide synthase peptides: *Effects of a phosphomimetic CaM mutation.* (2012) 51, 3651-3661.

[15] Smith BC, Underbakke ES, Kulp DW, Schief WR, Marletta MA. Nitric oxide synthase domain interfaces regulate electron transfer and calmodulin activation. *Proc, Natl. Acad. Sci. U S A.* (2013)110, E3577-86.

[16] Hannibal L, Page RC, Haque MM, Bolisetty K, Yu Z, Misra S, Stuehr DJ. Dissecting structural and electronic effects in inducible nitric oxide synthase *Biochemical Journal* (2015) 467, 153-165.

[17] Li H, Jamal J, Plaza C, Pineda SH, Chreifi G, Jing Q, Cinelli MA, Silverman RB, Poulos TL. Structures of human constitutive nitric oxide synthases. *Acta Crystallogr. D Biol Crystallogr.* (2014) 70, 2667-2674.

[18] Salerno JC, Krishanu Ray K, Thomas Poulos T, Huiying Li H, Dipak K. Ghosh DK. Calmodulin activates neuronal nitric oxide synthase by enabling transitions between conformational states. *FEBS Letters* 587 (2013) 44–47.

[19] Piazza M, Guillemette JG, Dieckmann T. Dynamics of Nitric Oxide Synthase–Calmodulin Interactions at Physiological Calcium Concentrations. *Biochemistry*, (2015) *54,* 1989–2000.

[20] Hedison TM, Leferink NG, Hay S, Scrutton NS. Correlating Calmodulin Landscapes with Chemical Catalysis in Neuronal Nitric Oxide Synthase using Time-Resolved FRET and a 5-Deazaflavin Thermodynamic Trap. *ACS Catal.* 2016 Aug 5; 6(8):5170-5180.

[21] Smith BC, Underbakke ES, Kulp DW, Schief WR, Marletta MA. Nitric oxide synthase domain interfaces regulate electron transfer and calmodulin activation. *Proc. Natl. Acad. Sci. U S A.* (2013) 110, E3577-86.

[22] Babu YS, Bugg CE, Cook WJ. Structure of calmodulin refined at 2.2 Mark A. Wilson and Axel T. Brunger. The 1.0 A Crystal Structure of Ca2.-bound Calmodulin: An Analysis of Disorder and Implications for Functionally Relevant Plasticity. *J. Mol. Biol.* (2000) 301, 1237-1256.

[23] Xu Guohua, Cheng Kai, Wu Qiong, Liu Maili, Li Conggang. Confinement alters the structure and function of calmodulin. *Angewandte Chemie, International Edition* (2017) 56, 530-534.

[24] Kretsinger RH, Rudnick SE, Weissman LJ. Crystal structure of calmodulin. *J Inorg Biochem.* (1986) 28, 289-302].

[25] Wilson MA, Brunger AT. The 1.0 Å Crystal Structure of Ca2.-bound Calmodulin: An Analysis of Disorder and Implications for Functionally Relevant Plasticity. *J. Mol. Biol.* (2000) 301, 1237-1256].

[26] Chuanwu Xia, Ila Misra, Takashi Iyanagi, Jung-Ja P. Kim. Regulation of Interdomain Interactions by Calmodulin in Inducible Nitric-oxide Synthase. *J. Biol. Chem.* (2009) 284, 30708-30717].

[27] Stoll S, Yaser Nejaty Jahromy, Woodward JJ, Ozarowski A, Marletta MA, Britt RD. Nitric Oxide Synthase Stabilizes the Tetrahydro-biopterin Cofactor Radical by Controlling Its Protonation State *J. Am. Chem. Soc.* (2010) 132, 11812–11823.

[28] Mortensen A. and J. Lykkesfeldt. 2013. "Kinetics of acid-induced degradation of tetra- and dihydrobiopterin in relation to their relevance as biomarkers of endothelial function." *Biomarkers* 18:55–62.

[29] Gao YT, Smith SME, Weinberg JB, Montgomery HJ, Newman E, Guillemette JG, Dipak K. Ghosh DK, Roman LJ, Martasek P, Salerno JC. Thermodynamics of Oxidation-Reduction Reactions in Mammalian Nitric-oxide Synthase Isoforms. *J. Biol. Chem.* (2004) 279, 18759–18766.

[30] Li H, Jamal J, Chreifi G, Venkatesh V, Abou-Ziab H, Poulos TL. Dissecting the kinetics of the NADP+-FADH2 charge transfer complex and flavin semiquinones in neuronal nitric oxide synthase *Journal of Inorganic Biochemistry* (2013) 124, 1-10.

[31] Haque MM, Kenney C, Tejero J, Stuehr DJ. A kinetic model linking protein conformational motions, interflavin electron transfer and electron flux through a dual-flavin enzyme-simulating the reductase activity of the endothelial and neuronal nitric oxide synthase flavoprotein domains. *FEBS J.* (2011) 278, 4055-4069.

[32] Haque MM, Tejero J, Bayachou M, Wang Z-O, Fadlalla M, Stuehr DJ Thermodynamic characterization of five key kinetic parameters that define neuronal nitric oxide synthase catalysis. *FEBS J* (2013) 280, 4439–4453.

[33] Tejero J, Hannibal L, Mustovich A, Stuehr DJ. Surface Charges and Regulation of FMN to Heme Electron Transfer in Nitric-oxide Synthase. J Biol Chem. (2010) 285, 27232–27240.

[34] Salerno J. C. Neuronal nitric oxide synthase: Prototype for pulsed enzymology *FEBS Lett.* (2008) 582, 1395–1399.

[35] Li W, Chen L, Fan W, Feng C. Comparing the temperature dependence of FMN to heme electron transfer in full length and truncated inducible nitric oxide synthase proteins. *FEBS Lett.* (2012) 586, 159–162.

[36] Levich VG, Dogonadze R. Quantum mechanical theory of electron transfer in polar media. *Dokl. Acad. Nauk.* (1959) 78, 2148-2121.

[37] Stuehr DJ, Tejero J, Haque MM. Structural and mechanistic aspects of flavoproteins: Electron transfer through the nitric oxide synthase flavoprotein domain. *FEBS J.* (2009) 276, 3959–3974.

[38] Astashkin AV, Elmore BO, Fan W, Guillemette JG, Feng C. Pulsed EPR determination of the distance between heme iron and FMN centers in a human inducible nitric oxide synthase. *J. Am. Chem, Soc.* (2010) 132, 12059–12067.

[39] Li W, Fan W, Chen L, Elmore BO, Piazza M, Guillemette JG, Feng C. Role of an isoform-specific serine residue in FMN-heme electron transfer in inducible nitric oxide synthase *Journal of Biological Inorganic Chemistry* (2012), 17(5), 675-685.

[40] Tejero J, Hannibal L, Mustovich A, Stuehr DJ. *Surface Charges and Regulation of FMN to Heme Electron Transfer in Nitric-oxide Synthase.* (2010) 285, 27232–27240.

[41] Garcin ED, Bruns CM, Lloyd SJ, Hosfield DJ, Tiso M, Gachhui R, Stuehr DJ, Tainer JA, and Getzoff ED. (2004) Structural basis for isozyme-specific regulation of electron transfer in nitric-oxide synthase. *J. Biol. Chem.* 279, 37918–37927.

[42] He Y, Haque Mm, Stuehr DJ, Lu H. Peter Single-molecule spectroscopy reveals how calmodulin activates NO synthase by controlling its conformational fluctuation dynamics. *Proc. Natl. Acad. Sci U S A.* (2015) 112, 11835-11840.

[43] Sheng Y, Zhong L, Guo D, Lau G, Feng C. Insight into structural rearrangements and interdomain interactions related to electron transfer between flavin mononucleotide and heme in nitric oxide synthase: A molecular dynamics study. *J Inorg Biochem.* (2015)153, 186-196.

[44] Likhtenshtein Gertz I. (2016) *Electron Spin in Chemistry and Biology: Fundamentals, Methods, Reactions Mechanisms, Magnetic Phenomena, Structure Investigation.* Springer.

[45] Chuanwu Xia, Ila Misra, Takashi Iyanagi, Jung-Ja P. Kim. Regulation of Interdomain Interactions by Calmodulin in Inducible Nitric-oxide Synthase. *J. Biol. Chem.* (2009) 284, 30708-30717.

[46] Hollingsworth SA, Holden JK, Li H, Poulos TL. Elucidating nitric oxide synthase domain interactions by molecular dynamics. *Protein Sci.* 2016 Feb;25(2):374-382.

[47] Smith BC, Underbakke ES, Kulp DW, Schief WR, Marletta MA. Nitric oxide synthase domain interfaces regulate electron transfer and calmodulin activation. *Proc. Natl. Acad. Sci U S A.* (2013)110, E3577-86.

[48] Stoll S, Nejaty Jahromy Yaser, Woodward JJ, Ozarowski A, Marletta MA, Britt RD. Nitric Oxide Synthase Stabilizes the Tetrahydrobiopterin Cofactor Radical by Controlling Its Protonation State. *J. Am. Chem. Soc*. (2010) 132, 11812-11823.

[49] Wei CC, Wang ZQ, Tejero J, Yang YP, Hemann C, Hille R, Stuehr DJ. Catalytic Reduction of a Tetrahydrobiopterin Radical within Nitric-oxide Synthase. *J. Biol. Chem.* (2008) 283, 11734–11742.

[50] Woodward JJ, NejatyJahromy Y, Britt RD, Marletta MA. Pterin-centered radical as a mechanistic probe of the second step of nitric oxide synthase. *J. Am. Chem. Soc.* (2010) 132, 5105–5113.

[51] Tejero J, Stuehr D. Tetrahydrobiopterin in nitric oxide synthase *IUBMB Life* (2013) 65, 358-365.

[52] Santolini J. The molecular mechanism of mammalian NO-synthases: a story of electrons and protons. *J. Inorg. Biochem.* (2011) 105, 127–14.1.

[53] Wei CC, Wang ZQ, Stuehr DJ. The three nitric-oxide synthases differ in their kinetics of tetrahydrobiopterin radical formation, heme-dioxy reduction, and arginine hydroxylation. *J. Biol. Chem.* (2005) 280, 8929–8935.

[54] Pufahl RA, Wishnok JS, Marletta MA. Hydrogen peroxide-supported oxidation of NG-hydroxy-L-arginine by nitric oxide synthase. *Biochemistry* (1995) 34, 1930-1941.

[55] Zhu Y, Silverman RB. Revisiting Heme Mechanisms. A Perspective on the Mechanisms of Nitric Oxide Synthase (NOS), Heme Oxygenase (HO), and Cytochrome P450s (CYP450s). *Biochemistry* (2008) 47, 2231–2243.

[56] Bernad S, Brunel A, Dorlet P, Sicard-Roselli C, Santolini J. A Novel Cryo-Reduction Method to Investigate the Molecular Mechanism of Nitric Oxide Synthases. *J Phys. Chem. B* (2012) 116, 5595-5603.

[57] Cho K-B, Carvajal MA, Shaik S. First half-reaction mechanism of nitric oxide synthase: the role of proton and oxygen coupled electron transfer in the reaction by quantum mechanics/molecular mechanics. *J. Phys. Chem. B.* (2009)113, 336–346.

[58] de Visser SP. Density functional theory (DFT) and combined quantum mechanical/molecular mechanics (QM/MM) studies on the oxygen activation step in nitric oxide synthase enzymes. *Biochem. Soc. Trans.* (2009) 37, 373–377.

[59] Cho K-B, Gauld JW. Second Half-Reaction of Nitric Oxide Synthase: Computational Insights into the Initial Step and Key Proposed Intermediate *J. Phys. Chem. B* (2005) *109*, 23706–23714

[60] Fernandez MA, Martı MA, Crespo F, Estrin DA. Proximal effects in the modulation of nitric oxide synthase reactivity: A QM-MM study. *J. Biol. Inorg. Chem.* (2005) 10, 595–604.

[61] Pradhan AA, Bertels Z, Akerman S. Targeted Nitric Oxide Synthase Inhibitors for Migraine. *Neurotherapeutics* (2018) 15, 391-401.

[62] Garcin ED, Arvai AS, Rosenfeld RJ, Kroeger MD, Crane BR, Andersson G, Andrews A, Hamley PJ, Mallinder PR, Nicholls DJ, St-Gallay SA, Tinker AC, Gensmantel NP, Mete A., Cheshire DR, Connolly S., Stuehr DJ, Aberg A, Wallace AV, Tainer JA, Getzoff ED. 2008. "Anchored plasticity opens doors for selective inhibitor design in nitric oxide synthase. Structural Studies of Nitric Oxide Synthase Inhibitor Complexes: An Anchored Plasticity Approach for Selective Enzyme Inhibitione." *Nat. Chem. Biol.* 4:700–707.

[63] Holden JK, Dillon Dejam, Lewis MC, He Huang, Soosung Kang, Qing Jing, Fengtian Xue, Silverman RB,Poulos TL. Inhibitor Bound Crystal Structures of Bacterial Nitric Oxide Synthase *Biochemistry* (2015) *54,* 4075–4082.

[64] Bretscher LE, Huiying Li, Poulos TL, Griffith OW. Structural Characterization and Kinetics of Nitric-oxide Synthase Inhibition by Novel N^5-(Iminoalkyl)- and N^5-(Iminoalkenyl)-ornithines *J. Biol. Chem.* (2003) 278, 46789-46797.

[65] Igarashi, Jotaro; Li, Huiying; Jamal, Joumana; Ji, Haitao; Fang, Jianguo; Lawton GR; Silverman RB, Poulos TL. Crystal Structures of Constitutive Nitric Oxide Synthases in Complex with De Novo Designed Inhibitors. *Journal of Medicinal Chemistry* (2009) 52, 2060-2066.

[66] Palmer RMJ, Ashton DS, Moncada S. Vascular endothelial cells synthesize nitric oxide from L-arginine. *Nature* (1988) 333, 664-666.

[67] Olken NM, Marletta MA. NG-methyl-L-arginine functions as an alternate substrate and mechanism-based inhibitor of nitric oxide synthase. *Biochemistry* (1993) 32, 9677-9685.

[68] Colonna, VDG, Bonomo S, Ferrario P, Bianchi M, Berti M, Guazzi M, Manfredi B, Muller EE, Berti F, Rossoni G. Asymmetric dimethylarginine (ADMA) induces vascular endothelium impairment and aggravates post-ischemic ventricular dysfunction in rats. *Eur. J. Pharmacol.* (2007) 557, 178-185.

[68] Colonna, VDG, Bonomo S, Ferrario P, Bianchi M, Berti M, Guazzi M, Manfredi B, Muller EE, Berti F, Rossoni G. Asymmetric dimethylarginine (ADMA) induces vascular endothelium impairment and aggravates post-ischemic ventricular dysfunction in rats. *Eur. J. Pharmacol.* (2007) 557, 178-185.

[69] Palmer RMJ, Ashton DS, Moncada S. Vascular endothelial cells synthesize nitric oxide from L-arginine. *Nature,* (1988) 333, 664-666.

[70] Moore PK, al-Swayeh OA, Chong NW, Evans RA, Gibson A. L-NG-nitro arginine (L-NOARG), a novel, L-arginine reversible inhibitor of endothelium-dependent vasodilatation *in vitro*. *Br. J. Pharmacol.*, (1990) 99, 408-412.

[71] Corbett JA, Tilton RG, Chang K, Hasan KS, Ido Y, Wang JL, Sweetland MA, Lancaster JR, Williamson JR, McDaniel ML. Aminoguanidine, a novel inhibitor of nitric oxide formation, prevents diabetic vascular dysfunction. *Diabetes*, (1992) 41, 552-556.

[72] Garvey EP, Oplinger JA, Tanoury GJ, Sherman PA, Fowler M, Marshall S, Harmon MF, Paith JE, Furfine ES. Potent and selective inhibition of human nitric oxide synthases. Inhibition by non-amino acid isothioureas. *J. Biol. Chem.* (1994) 269, 26669-26676.

[73] Furfine ES, Harmon MF, Paith JE, Knowles RG, Salter M, Kiff RJ, Duffy C, Hazelwood R, Oplinger JA, Garvey EP. Potent and selective inhibition of human nitric oxide synthases. Selective inhibition of neuronal nitric oxide synthase by Smethyl-L-thiocitrulline and S-ethyl-Lthiocitrulline. *J. Biol. Chem.* (1994) 269, 26677-26683.

[74] Mayer B, Klatt P. Werner ER, Schmidt K. Molecular mechanisms of inhibition of porcine brain nitric oxide synthase by the antinociceptiv drug 7-nitroindazole. *Neuropharmacol.* (1994) 33, 1253-1259.

[75] Wolff DJ, Gribin BJ. The inhibition of the constitutive and inducible nitric oxide synthase isoforms by indazole agents. *Arch. Biochem. Biophys.* (1994) 311, 300-306.

[76] Frey C, Narayanan K, McMillan K, Spack L, Gross SS, Masters BS, Griffith OW. L-thiocitrulline. A stereospecific, sheme-binding inhibitor of nitric-oxide synthases. *J. Biol. Chem.* (1994) 269, 26083-26091.

[77] Poulos, Thomas L, Li Huiying. Nitric oxide synthase and structure-based inhibitor design. *Nitric Oxide* (2017), 63, 68-77.

[78] Garvey EP, Oplinger JA, Furfine ES, Kiffi RJ, Laszloi F, Whittlei BJR, Knowlesi RG 1400W Is a Slow, Tight Binding, and Highly Selective Inhibitor of Inducible Nitric-oxide Synthase *in Vitro* and *in Vivo*. *JBC* (1997) Vol. 272, 4959–4963.

[79] Sundaresan L, Giri S Chatterjee S. Inhibitors of Nitric Oxide Synthase: What's Up and What's Next? *Current Enzyme Inhibition* 2016, 12, 81-107, not available.

[80] Tang W, Li H, Poulos TL, Silverman RB. Mechanistic studies of inactivation of inducible nitric oxide synthase by amidines. *Biochemistry* (2015) 54, 2530-2538.

[81] Sennequier N, Wolan D, Stuehr DJ. Antifungal imidazoles block assembly of inducible NO synthase into an active dimer. *J. Biol. Chem.*, 1999, *274*, 930-938.

[82] Ohtsuka M, Konno F, Honda H, Oikawa T, Ishikawa M, Iwase N, Isomae K, Ishii F, Hemmi H, Sato S. PPA250 [3-(2,4-difluorophenyl)-6-[2-[4-(1H-imidazol-1-ylmethyl) phenoxy]ethoxy]-2-phenylpyridine], a novel orally effective inhibitor of the dimerization of inducible nitric-oxide synthase, exhibits an anti-inflammatory effect in animal models of chronic arthritis. *J. Pharmacol Exp. Ther.* (2002) 303, 52-57.

[83] Sohn MJ, Hur GM, Byun HS, Kim WG. Cyclo(dehydrohistidyl-l-tryptophyl) inhibits nitric oxide production by preventing the dimerization of inducible nitric oxide synthase. *Biochem. Pharmacol.* 2008, *75*(4), 923-930.

[84] Frohlich LG, Kotsonis P, Traub H, Taghavi-Moghadam S, Al-Masoudi N, Hofmann H, Strobel H, Matter H, Pfleiderer W, Schmidt HH. Inhibition of neuronal nitric oxide synthase by 4-amino pteridine derivatives: Structure-activity relationship of

antagonists of (6R)-5, 6, 7, 8-tetrahydrobiopterin cofactor. *J. Med. Chem.* (1999) 42, 4108-4121.

[85] Stuehr DJ, Fasehun OA, Kwon NS, Gross SS, Gonzalez JA, Levi R, Nathan CF. Inhibition of macrophage and endothelial cell nitric oxide synthase by diphenyleneiodonium and its analogs. *FASEB J.* (1991) 5, 98-103.

[86] Sohn MJ, Hur GM, Byun HS, Kim WG. Cyclo(dehydrohistidyl-l-tryptophyl) inhibits nitric oxide production by preventing the dimerization of inducible nitric oxide synthase. *Biochem. Pharmacol.* (2008) 75(4), 923-930.

Chapter 6

NITRIC OXIDE AND GUANYLYL CYCLASE

ANNOTATION

Soluble guanylate cyclase (sGC), as an NO receptor, is a key metalloprotein in mediating NO signaling transduction. sGC is activated by NO to catalyze the conversion of guanosine 5′-triphosphate (GTP) to cyclic guanylate monophosphate (cGMP) and, therefore, is a critical component of this signaling pathway. In the NO/cGMP signaling pathways, sGC is important in diverse physiological processes such as vasodilation and neurotransmission. Kinetic and physical methods studies revealed the multistep binding of NO and the decisive involvement of conformational transition in the catalytic activity and regulation of adenylyl cyclase. Efficiency of the NO–sGC–cGMP signaling was demonstrated by the examples of endothelium-dependent relaxation of smooth muscle and vasorelaxation, and inhibition of platelet aggregation in a blood vessel

6.1. INTRODUCTION

Guanylyl cyclase activated by nitric oxide is a key part of the G protein signaling cascade that is initiated by low intracellular calcium levels and inhibited by high intracellular calcium levels [1–15].

Guanylate cyclase catalyzes the reaction of guanosine triphosphate (GTP)

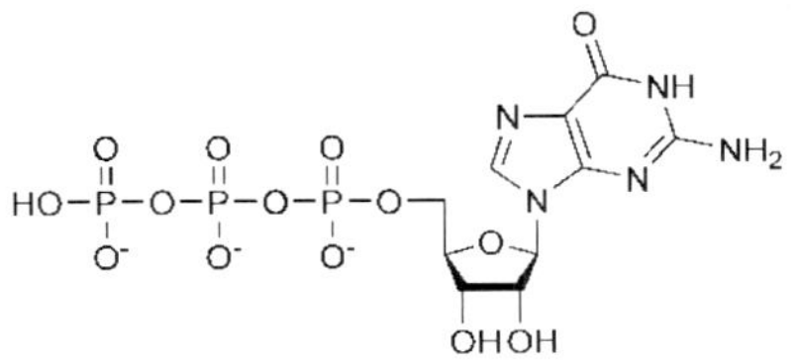

to 3',5'-cyclic guanosine monophosphate (cGMP)

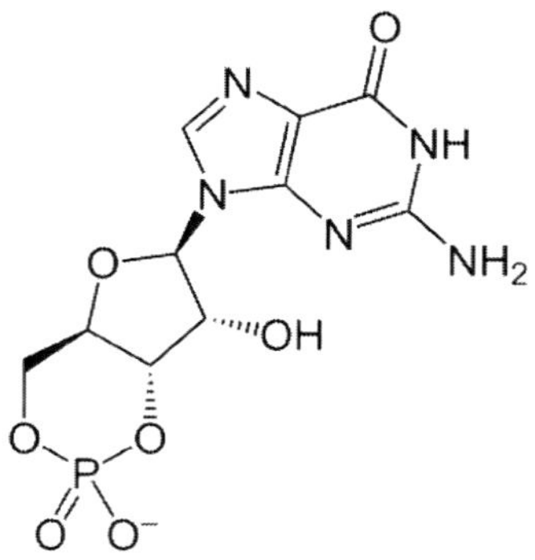

and pyrophosphate [1–9].

6.2. Structure and Action Mechanism

The enzyme is a heterodimer of homologous α and β subunits, each of which is composed of multiple domains. The structure of two homologous subunits α1 and β1 of human sGC is presented in Figures 6.1 and 6.2 [1–4]. The C-terminal domains of both subunits combine to form a heterodimeric catalytic domain. N-terminal HNOX (haem-containing NO/oxygen-binding) domain, a Per/Arnt/Sim PAS domain, and an α-helical region capable of forming CCs (coiled coils) have structural and regulatory function. Only the β subunit can bind haem. Two isoforms of the NO receptor molecule exist: alpha(1)beta(1) and the alpha(2)beta(1) [10]. The alpha(1)beta(1) and alpha(2)beta(1) isoforms were established as the cytosolic and the membrane-associated NO-sensitive guanylyl cyclase, respectively.

The following model for sGC activity modulated by domain–domain interactions that allow the catalytic domains to undergo the transition from an inactive to an active conformation was proposed (Figures 6.3 and 6.4) (1) in the basal state, catalytic domains are constrained in a suboptimal conformation via inhibitory interactions with sGC domains, including the N-terminal βHNOX (and HNOXA) domain; (2) the letter maintained in an inhibited conformation by the αHNOX and αHNOXA domains; and (3) binding of NO and/or activators releases these inhibitory interactions and induces conformational changes transmitted to the catalytic domains via the coiled-coil domain to yield the activated state [4].

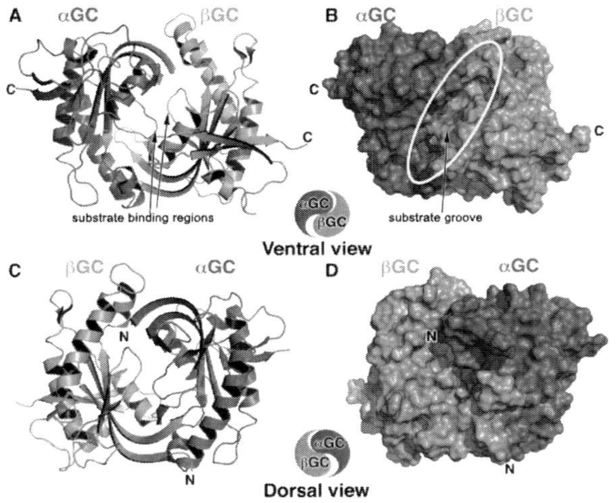

Figure 6.1. Overall structure of heterodimeric wild-type αβGC catalytic domains that resembles the Chinese yin-yang symbol with both subunits arranged in a head-to-tail conformation. (A and B) Ventral face of the heterodimer as a cartoon (A) and a solvent accessible surface representation (B). The deep extended substrate groove (shown by the oval in panel B) bisects the ventral face, which contains the C-termini of both subunits (denoted with C). The αGC subunit is colored blue, and the βGC subunit is colored orange. The substrate binding regions of αGC (residues 523–534) and βGC (residues 470–480) adopt an extended conformation in the absence of substrate and metals (arrows). (C and D) The dorsal face of the heterodimer as a cartoon (C) and a solvent accessible surface representation (D) is flatter than the ventral face and contains the N-termini of both subunits (denoted with N) [4]. Copyrights 2014 the American Chemical Society.

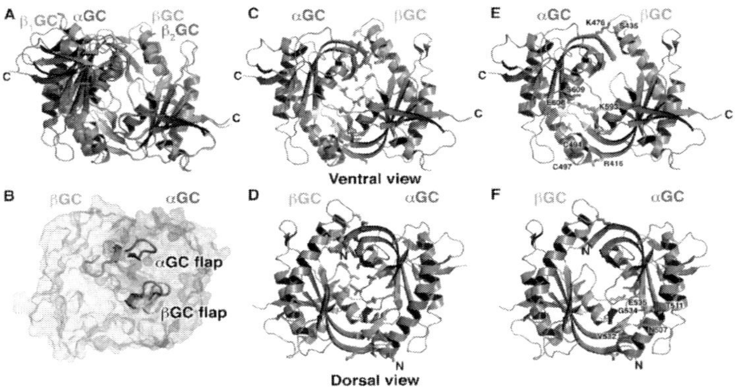

Figure 6.2. Structural determinants of catalytic domain dimerization. (A and B) Homodimeric $\beta_1\beta_2$GC (PDB entry 2WZ1) is colored light green (β_1GC) and dark green (β_2GC). The αβGC heterodimer is colored blue (αGC) and orange (βGC). The ventral view (A) shows that βGC and β_2GC subunits superimpose well, while αGC adopts a conformation different from that of β_1GC in the heterodimer and homodimer, respectively. The dorsal view (B) shows the double flap-wrap conformation of the homodimer flaps. In the heterodimer depicted as a semitransparent solvent accessible surface, only the βGC flap wraps on the αGC subunit while the αGC flap (blue cartoon) is flipped out. (C and D) The core of the heterodimeric interface is formed by numerous hydrophobic interactions between residues from the βGC subunit and the αGC subunit, depicted as sticks. (E and F) Polar interactions (hydrogen bonds and salt bridges) also contribute to the heterodimeric interface. Residues from βGC and αGC participating in these interactions are shown as sticks [4]. Copyrights 2014 the American Chemical Society.

The following model for sGC activity modulated by domain–domain interactions that allow the catalytic domains to undergo the transition from an inactive to an active conformation was proposed (Figures 6.3 and 6.4) (1) in the basal state, catalytic domains are constrained in a suboptimal conformation via inhibitory interactions with sGC domains, including the N-terminal βHNOX (and HNOXA) domain; (2) the letter maintained in an inhibited conformation by the αHNOX and αHNOXA domains; and (3) binding of NO and/or activators releases these inhibitory interactions and induces conformational changes transmitted to the catalytic domains via the coiled-coil domain to yield the activated state [4].

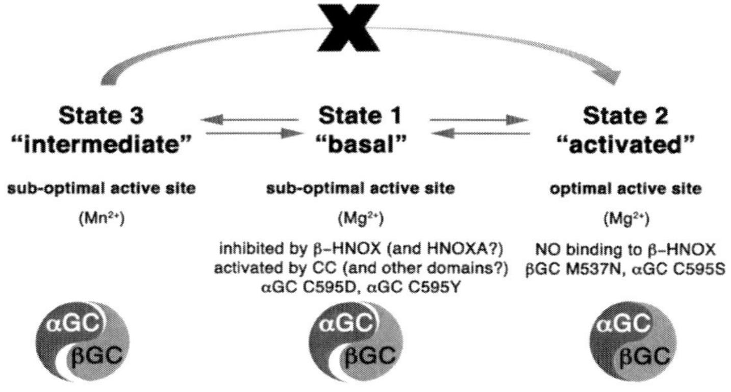

Figure 6.3. Model for domain–domain interactions that influence the conformation of the heterodimeric GC interface to modulate the activity of sGC. In state 1 ("basal"), competing domain–domain interactions yield sGC with low activity. While some sGC domains (including the coiled coil) promote conformational changes of the catalytic subunits, the regulatory βHNOX domain (and possibly the HNOXA domain) inhibits activity via direct binding to αGC. Mutations in the αGC subunit (Cys595Asp and Cys595Tyr) also prevent conformational changes in the active site. In state 2 ("activated"), binding of NO to βHNOX removes the inhibition and allows further conformational changes of the catalytic domains to yield fully active sGC. Mutations in catalytic subunits (βGC Met537Asn and αGC Cys595Ser) also yield an "activated" phenotype. In state 3 ("intermediate"), Mn^{2+} allows catalysis with a nonoptimal conformation of the catalytic domains. However, the enzyme is now locked in an intermediate state and cannot be further activated by NO and/or activators. The catalytic domains (represented as a blue and orange yin-yang) are in an inactive conformation (poor alignment) in states 1 and 3, and an active conformation in state 2 (perfect alignment of yin and yang) [4]. Copyrights 2014 the American Chemical Society.

The reaction of cyclic GMP formation of guanylyl cyclase is initiated by binding the NO molecule. The binding which causes heam a scission of the Fe–His bond and additional covalent and conformational changes in the HNOX domain (Figure 6.5) [2]. These effects lead to the enzyme full activation and are accompanied by transmission to the catalytic domains through a combination of direct contacts and long-range allosteric interactions. Guanylyl cyclase may be inactivated by oxidation, nitrosylation, and haem loss. cGMP activates cGMP-dependent protein kinase, resulting in protein phosphorylation, decreased cytosolic calcium levels to regulate ion transport, myosin light dephosphorylation, and relaxation.

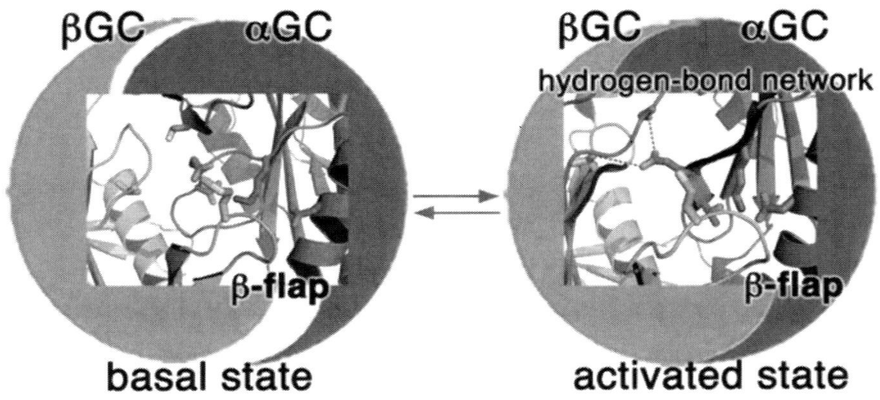

Figure 6.4. Proposed "basal" and "activated" states of the sGC catalytic subunits [4]. Copyrights 2014 the American Chemical Society.

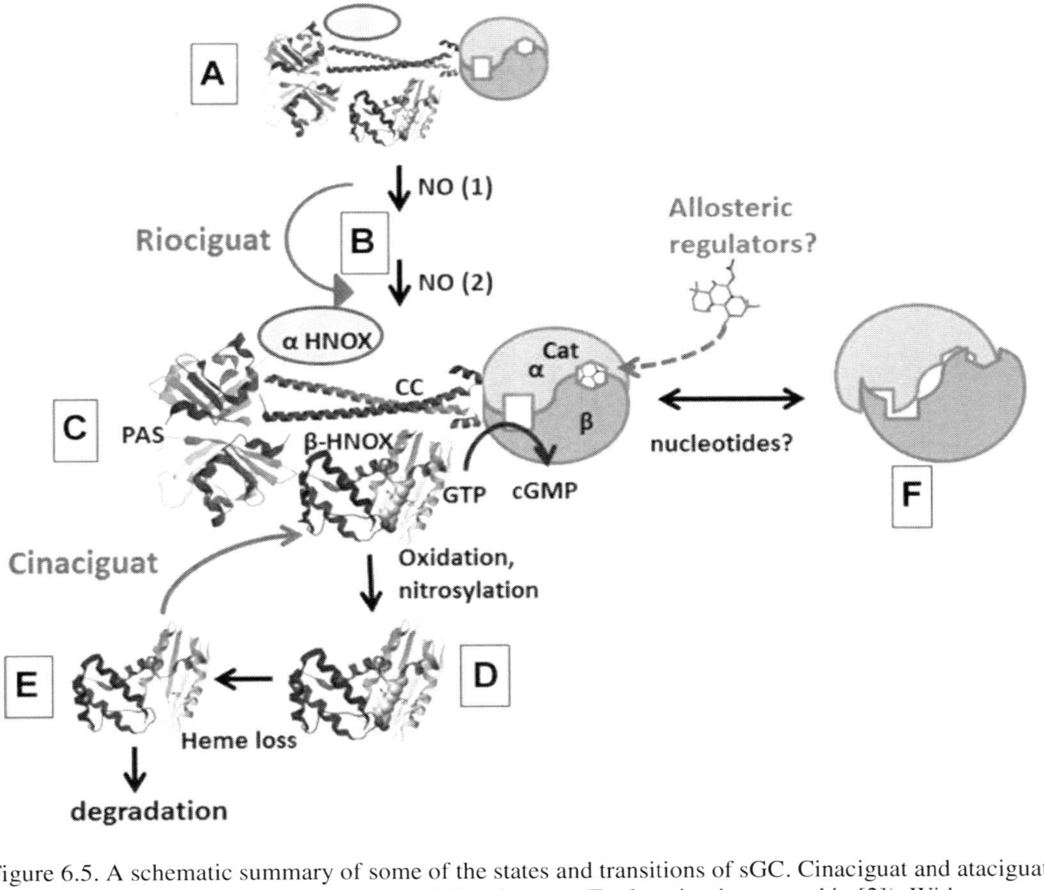

Figure 6.5. A schematic summary of some of the states and transitions of sGC. Cinaciguat and ataciguat are the direct NO- and haem-independent sGC activators. (Explanation in text and in [2]). With permission from the Portland Press, 2014.

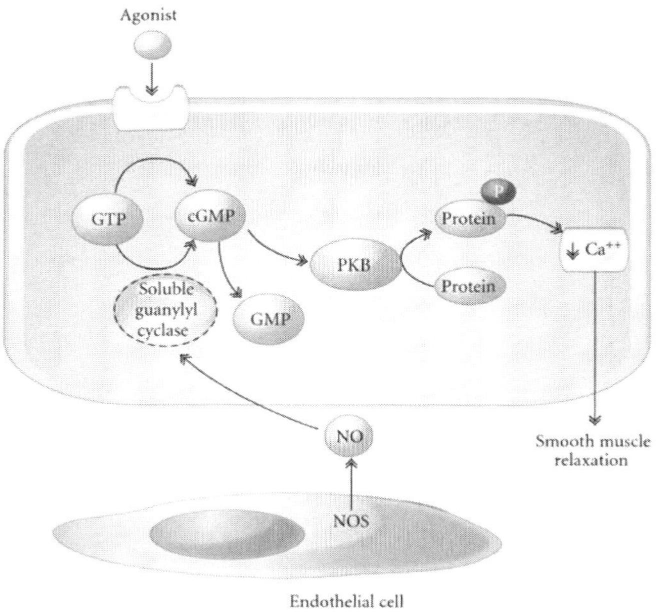

Figure 6.6. Simplified role of NO (nitric oxide) stimulating soluble guanylyl cyclase smooth muscle relaxation. PKB (protein kinase B), NOS (nitric oxide synthase) [17]. With permission from Hindawi, 2012.

The ubiquitous second messenger cAMP regulates a wide array of functions, from bacterial transcription to mammalian memory. This messenger is synthesized by six adenylyl cyclase (AC) families. A potential bacterial model of GTP binding to human sAC and a for substrate recognition during catalysis was suggested in [5]. According to the model, the following reaction steps occur: (1) ATP first binds with an ion B in a loose conformation; (2) Adjustment of ion B and binding of ion A initiate the S_{N2} reaction with in-line geometry; (3) The α-phosphate moves towards the 3′OH, and elongation of the bond to the leaving group is supported by a shift of the AMP/cAMP molecule toward Lys334; (4) After break of the bond to PP_i, the concomitantly formed cAMP can shift the ribose and base back to their previous positions (see further details in [5]).

Several recent studies supported the decisive involvement of the conformational transition in the catalytic activity and regulation of adenylyl cyclase, in creation of inactive, "open" conformation and active, "closed" conformation; in particular, hydrogen/deuterium exchange mass spectrometry (HDX-MS) was employed to probe the NO-induced conformational changes of sGC [13]. Results revealed the process in the following steps: (1) NO binding to the heme-NO/O2-binding (H-NOX) domain perturbs a signaling surface implicated in Per/Arnt/Sim (PAS) domain interactions; (2) the perturbation elicits striking conformational changes in the junction between the PAS and helical domains; and (3) the conformational changes propagate as perturbations throughout the adjoining helices and delineate an allosteric pathway linking NO binding to activation of the catalytic domain.

Thus, NO binding stimulates the catalytic domain by contracting the active site pocket. The catalytic domain exchange rate differences induced by NO were mapped.

Two mechanisms of the allosteric pathway bridging the sensor and output domains were proposed [13]: The junction between PAS and helical domains, caused by conformational changes, can indicate interdomain pivoting that relieves inhibitory contacts between H-NOX and catalytic domains (A), and the PAS-helical junction can mediate remote allosteric effects via long s via long-range conformational changes propagated through the helical domains (B) [13]. Models of the catalytic domain of sGC in "inactive" or "active" conformations (Figure 6.8) were constructed using the crystal structure o of the catalytic domains of "inactive" sGCs (2WZ1, 3ET6) and of "active" adenylate cyclase (1CJU) [7]. Each model was submitted to six independent molecular dynamics simulations of about 11 s. Docking of YC-1, sGC activator, and calculation of absolute binding free energies with the linear interaction energy method revealed a potential high-affinity binding site on the "active" structure.

Soluble GC is the receptor for NO in vascular smooth muscle in the cardiovascular system (Figures 6.6 and 6.7) [17].

NO generated by endothelial NO synthase (eNOS) from L-arginine activates sGC in adjacent vascular smooth muscle cells to increase cGMP levels and induce relaxation.

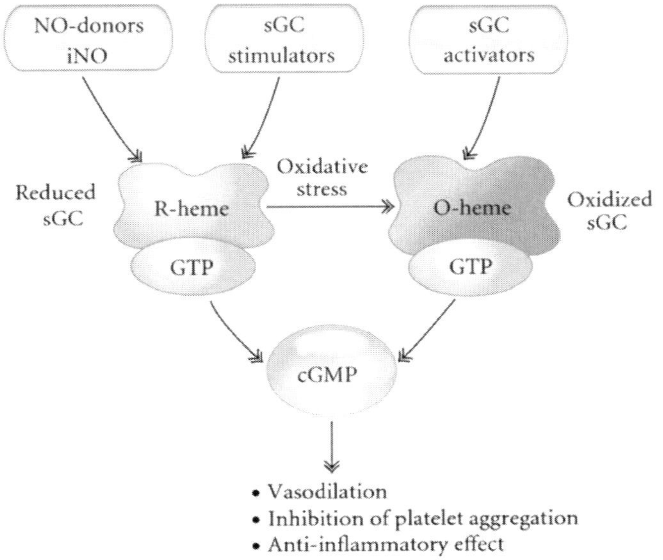

Figure 6.7. Role of NO, inhaled NO, and sGC stimulators in stimulating the reduced heme of sGC and the role of sGC activators in stimulated oxidized sGC to stimulate cGMP leading to vasodilation, inhibition of platelet aggregation, and an anti-inflammatory effect in the vascular bed [17]. With permission from Hindawi, 2012.

sGC can be activated by sodium azide (NaN_3), sodium nitrite ($NaNO_2$), hydroxylamine (NH_2OH), nitroglycerin ($C_3H_5N_3O_9$), and sodium nitroprusside ($Na_2[Fe(CN)_5NO]$). These nitrogen-containing compounds were able to activate sGC, causing an increase in cGMP, and vascular relaxation.

6.3. Kinetics and Chemical Mechanism

Direct kinetic measurements of the multistep binding of NO to sGC was carried out [11]. The following values of the rate constant were found at 4ºC: (1) NO binds to sGC to form a six-coordinate, nonactivated, intermediate with (k(on) > 1.4×10^8 $M^{-1}s^{-1}$; (2) subsequent release of the axial histidine, the rate of which also depends on NO concentration (k = 2.4×10^5 $M^{-1}s^{-1}$); and (3) NO binding to the isolated heme domain of sGC (k = 7.1×10^8 $M^{-1}s^{-1}$). Thus, sGC acts as a fast, specific, and efficient trap for NO. Cleavage of the iron-histidine bond provides the driving force for activation of sGC. NO can be also oxidized to NO_2 or trapped by hemoglobin or myoglobin

Models for NO-binding and activation of NO-sensitive GC are presented in [12]. Binding first NO molecule to heam forms six-coordinated state of 420 nm that converts into the active state of the enzyme in the presence of the reaction product (Mg 2+/cGMP/ pyrophosphate). Formation of proposed dinitrosyl-heme of 399 nm can lead to either the enzyme activation or inactivation

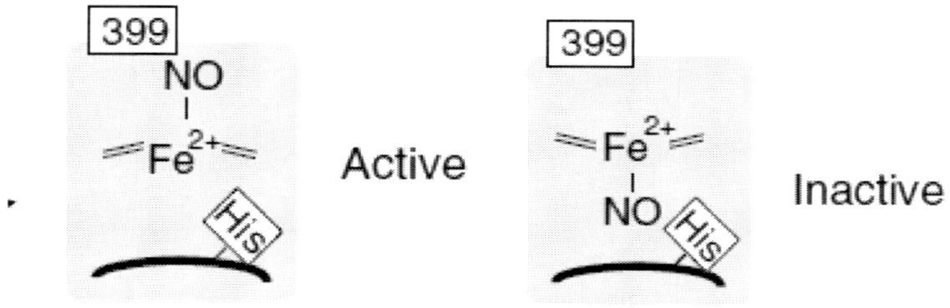

Remarkably, the conversion of the nonactivated into the activated conformation of the enzyme required the simultaneous presence of NO and the reaction products. The intracellular redox equilibrium of the two sGC redox states, NO-sensitive reduced and NO-insensitive oxidized sGC can be shifted by reactive oxygen species to the oxidized (ferric haem) state and by reductases to the reduced (ferrous haem) form [15]. Disequilibrium towards the oxidized NO-unresponsive enzyme can be associated with oxidative stress.

6.4. Signaling : Nitric Oxide–Soluble Guanylate Cyclase–Cyclic Guanylate Monophosphate Signal Transduction Pathway

Guanylyl cyclase activated by NO is a key part of the G protein signaling cascade that is, in turn, activated by low intracellular calcium levels and inhibited by high intracellular calcium levels [15]. The NO–sGC–cGMP signal transduction pathway can lead to tissue protection or cellular dysfunction and death. NO–sGC–cGMP signaling can be compromised either by reducing the bioavailability of NO via chemical interaction of NO with $^{\bullet}O_2^-$ or altering the redox state of sGC itself, though oxidative stress or the action of peroxynitrite can have an effect on the NO–sGC–cGMP signaling. NO activation of sGC converts GTP to cyclic GMP, mediating important physiological and tissue protective effects [15]. Transformation of cGMP to GMP is catalyzed by several phosphodiesterase (PDE) families. Excessive amounts of NO produced under pathological conditions associated with increased inflammation and oxidative stress. Peroxynitrite and other reactive nitrogen and oxygen species induce cell damage via lipid peroxidation, inactivation of enzymes, and other proteins by oxidation and nitration and activation of matrix metalloproteinases (MMP) and the nuclear enzyme poly(ADP-ribose) polymerase (PARP), which leads to cellular dysfunction. The NO–sGC–cGMP signal transduction pathway and potential drug targets were discussed in detail [15]. During the NO–sGC–cGMP signaling in a blood vessel, NO diffuses into both the vessel lumen and the vessel wall, activating sGC. L-arginine is converted in the endothelium monolayer by the endothelial nitric oxide synthase (eNOS). Via the direct activation of sGC, haem-dependent sGC stimulators and haem-independent sGC activators increase the cellular cGMP concentration. As a result, both vasorelaxation and inhibition of platelet aggregation occur in a blood vessel.

Effects of nitrovasodilators producing NO on endothelium-dependent relaxation of smooth muscle were evaluated [14]. It was found that endothelium-dependent vasodilators or nitrovasodilators produce NO, which activates sGC to form cGMP. In turn, cGMP activates cGMP-dependent protein kinase, resulting in protein phosphorylation, which decreased cytosolic calcium levels, myosin light chain dephosphorylation, and relaxation. ANF-1 and ANF-2, receptors of atrial natriuretic factors atriopeptins are coupled to particulate guanylyl cyclase activation or phospholipid metabolism [14]. Cyclic GMP can regulate cation channels, cyclic GMP-dependent protein kinases, and cyclic nucleotide phosphodiesterases, some isoforms of phosphodiesterase. The related downstream processes and their effects were also taken into consideration.

Important aspects of sGC activity, such as involving in NO signaling transduction [15], regulation by nucleotides [16], and the enzyme NO-independent stimulators and activators of soluble guanylate cyclase: discovery and therapeutic potential [17], were discussed.

REFERENCES

[1] Purohit R, Weichsel A. Montfort WR. Crystal structure of the Alpha subunit PAS domain from soluble guanylyl cyclase. *Protein Science* (2013) 22, 1439-1444.

[2] Gileadi, O. Structures of soluble guanylate cyclase: Iimplications for regulatory mechanisms and drug development. *Biochem. Soc. Trans.* (2014) 42, 108-113.

[3] Allerston CK, von Delft F, Gileadi O. Crystal structures of the catalytic domain of human soluble guanylate cyclase. *PLoS One.* (2013) 8, e57644.

[4] Seeger F, Quintyn, R, Tanimoto A, Williams GJ. Tainer JA, Wysocki VH, Garcin ED. Interfacial Residues Promote an Optimal Alignment of the Catalytic Center in Human Soluble Guanylate Cyclase: Heterodimerization Is Required but Not Sufficient for Activity. *Biochemistry* (2014), 53, 2153-2165.

[5] Kleinboelting S, van den Heuve J, Steegborn C. Structural analysis of human soluble adenylyl cyclase and crystal structures of its nucleotide complexes—Implications for cyclase catalysis and evolution. *FEBS J* (2014) 281, 4151-4164.

[6] Steegborn C. Structure, mechanism, and regulation of soluble adenylyl cyclases—Similarities and differences to transmembrane adenylyl cyclases. *Biochim. Biophys. Acta* (2014) 1842, 2535-2547.

[7] Martin E, Berka V, Tsai AL, Murad F. Soluble guanylyl cyclase: The nitric oxide receptor. *Methods Enzymol.* (2005) 396:478-492.

[8] Papapetropoulos A, Hobbs AJ, Topouzis S, Papapetropoulos A, Hobbs AJ, Topouzis S. Extending the translational potential of targeting NO/cGMP-regulated pathways in the CVS. *Br J Pharmacol.* (2015) 172, 1397-414.

[9] Schmidt HH, Schmidt PM, Stasch JP. NO- and haem-independent soluble guanylate cyclase activators. *Handb Exp. Pharmacol.* (2009) 191, 309-339.

[10] Russwurm M, Wittau N, Koesling D. Guanylyl cyclase/PSD-95 interaction: Targeting of the nitric oxide-sensitive $\alpha_2\beta_1$ guanylyl cyclase to synaptic membranes. *J. Biol. Chem.* (2001) 276, 44647-44652.

[11] Zhao Y, Brandish PE, Ballou DP. Marletta basis for nitric oxide sensing by soluble guanylate cyclase. *Proc. Natl. Acad. Sci USA* (1999) 96, 14753-14758.

[12] Russwurm M, Koesling D. NO activation of guanylyl cyclase, *EMBO J.* (2004) 23, 4443-4450].

[13] Underbakke ES, Iavarone AT, Chalmers MJ, Pascal BD, Novick S, Griffin PR, Marletta MA. Mapping NO-Induced Confor-mational Changes to Models of Catalytic Domain Activation. *Structure* (2014) 22, 602-611.

[14] Ferid Murad. Nitric Oxide and Cyclic GMP in Cell Signaling and Drug Development. *N Engl. J. Med.* (2006)355, 2003-2011.

[15] Pan Jie, Zhong Fangfang, Tan Xiangshi. Soluble guanylate cyclase in NO signaling transduction. *Reviews in Inorganic Chemistry* (2013) 33, 193-205.

[16] Sürmeli NB, Müskens FM, Marletta MA. The influence of nitric oxide on soluble guanylate cyclase regulation by nucleotides: Role of the pseudosymmetric site. *J. Biol. Chem.* (2015) 290, 15570-15580.

[17] Nossaman B, Pankey E, Kadowitz P. Stimulators and activators of soluble guanylate cyclase: Review and potential therapeutic indications. *Critical care research and practice* (2012), 2012, 290805.

Chapter 7

REACTIVE NITROGEN SPECIES

ANNOTATION

Nitric oxide, a free radical of modest reactivity, is able to form other reactive intermediates, having effects on biological molecules activity and injury, and on the normal and pathological functions of the entire organisms by radical and signaling mechanisms. The most important component of the reactive nitrogen species (RNS) system is peroxynitrite, which is produced in reaction with superoxide by the body in response to a variety of toxicologically relevant molecules including environmental toxins, as well as in reperfusion injury and inflammation. Connection of RNS with reactive oxygen species (ROS) and inflammation has been well documented.

7.1. INTRODUCTION

Peroxynitrite is the first and the most important RNS, which is derived from nitric oxide (•NO) and superoxide ($O_2^{\bullet-}$), while superoxide is produced via the enzymatic activity of nitric oxide synthase (NOS) and NADPH oxidase, respectively. Another way for the RNS production is reaction of NO with atmospheric oxygen. Peroxynitrite is formed in macrophages by the diffusion-limited reaction of superoxide and NO. This highly reactive component of RNS involves to biological oxidative processes directly and via the spontaneous decomposition for reactive radical OH and nitric dioxide NO_2. The reactive species participate in oxidizing, nitrating, or modifying sites, such as protein thiols and tyrosine residues and nucleic acids. These reactive intermediates can trigger nitrosative damage on biomolecules. RNS has been shown to have a direct role in cellular signaling, vasodilatation, and immune response. Nitrosation of thiol-containing compounds, such as glutathione and thiol proteins, can also make important contributions to the oxidative stress dysfunction.

7.2. Peroxinitrite

7.2.1. General Properties

In this section, the role peroxynitrite in nitrosative stress is illustrated by several typical examples. Among the processes affected on the cytotoxic effect of peroxynitrite are [1–5]: (1) direct inhibition of mitochondrial respiratory chain enzymes and of membrane Na^+/K^+ ATP-ase activity, (2) affecting calcium channeling, (3) initiation of oxidation of lipids and other biological molecules, (4) inactivation of glyceraldehyde-3-phosphate dehydrogenase and of membrane sodium channels, (5) DNA breakage, and (6) promoting chemical and conformational modifications in cytochrome c and other injuring oxidative processes.

In 1990, Beckman et al. described peroxynitrite anion formed from the diffusion limited biradical reaction between NO and the superoxide anion [1].

Peroxynitrite

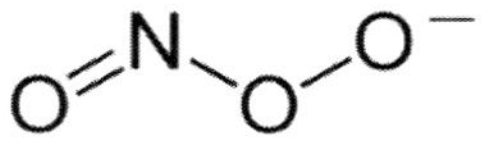

is an anion with a strong oxidant and nitrating reactivity. The molecule can also react as nucleophilic agent to form nitroso-peroxycarbonate ($ONOOCO_2^-$), for example. In chemistry, peroxynitrite is prepared by the reaction of hydrogen peroxide with nitrite:

$$H_2O_2 + NO^- \rightarrow ONOO^- + H_2O$$

In biology, $ONOO^-$ arises on the reaction of the free radical superoxide with the free radical nitric oxide,

$$O_2^- + {}^{\bullet}NO \rightarrow ONO^-$$

which proceeds as the diffusion-controlled reaction with rate constant 1.6×10^{10} M^{-1} s^{-1} [3].

In the biological system, this reaction can compete with the O_2^- dismutation,

$$O_2^- + O_2^- + 2H^+ \rightarrow H_2O_2 + H_2O$$

which occurs catalytically in the presence of enzyme superoxide oxidase dismutase (SOD) with rate constant k = (1–2) 10^9 M^{-1} s^{-1} [4]. Anionic peroxynitrite is stable in basic solutions while its conjugate acid (peroxynitrous acid, HNO_3, pK_a 6.8) is unstable.

7.2.2. Physical Properties

To analyze the peroxynitrite anion experimental optical spectrum, the electronic excited states of the two planar stable conformers was calculated by means of multiconfiguration self-consistent-field (MCSCF) methods using General Atomic and Molecular Electronic Structure System GAMESS [5]. The following main results were described: (1) excitation energies from the ground state equilibrium geometry is obtained by MCSCF solutions for either the A' or A" symmetries' four anion states; (2) two ^{2}A' and two ^{2}A" states are attributed to NO (^{2}II) + O_2^-(^{2}II), the ground state is ^{2}A' and the only strong electronic transition is 1 ^{2}A'+2 ^{2}A'; (3) three excited states predicted to be dissociative to NO and O_2; and (4) in vacuum excitation energy for the *cis-cis* conformer was calculated as 324.3 nm.

By absorption measurements, the maximum extinction coefficient at 240 nm is 770 M^{-1} cm-1 for ONOOH and 1670 M^{-1} cm^{-1} at 300 nm [6]. The absorption spectrum of the fundamental ν_1 band of the *cis-cis* isomer of HOONO was observed at 302,4 nm [7]. The infrared absorption spectra of matrix-isolated *cis, cis*-peroxynitrous acid (HOONO and DOONO) in argon have been detected and compared with CCSD(T) predictions [9]. The experimental matrix frequencies of six fundamental vibrational modes for *cis, cis*-HOONO have been assigned and found to be in reasonable agreement with the *ab initio* anharmonic force field calculations. The following fundamental frequencies for the deuterated isotopomer, *cis*- cis DOONO, were reported: (a′ modes) ν1 = 2447.2±0.6,ν2 = 1595.7±0.7, ν3 = 1089.1±0.4, ν4 = 888.1 ± 0.4, ν5 = 786.6 ± 0.5, ν6 = 613.9±0.9; and (a″ mode) ν8 = 456.5 ± 0.5.

7.2.3. Basic Chemical Reactions

Primary chemical reactions of ONOOH are shown in Figure 7.1. Thermodynamic values for ONOOH/$ONOO^-$ are presented in Table 7.1 and Table 7.2 [10], and kinetic values were listed in [2] correspondingly [10].

Various mechanisms of the dissociation HOONO and related details of its electronic and vibration structure were discussed [10–18]. Values of rate constant of reactions related to RNS with SIN-1 (3-morpholinosydnonimine), a nitric oxide/superoxide/peroxynitrite donor, DAN (2,3-diaminonaphthalene), and NAT (2,3-naphthotriazole) were reported in [2].

In aqueous solution, peroxynitrous acid dissociates on ions

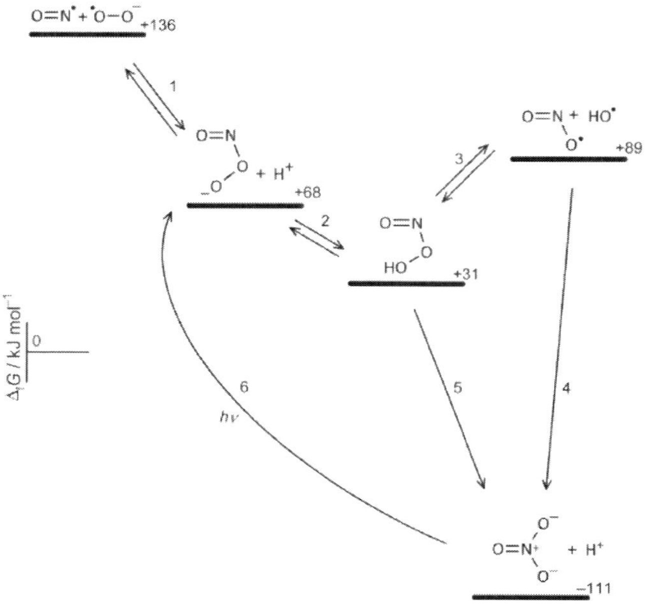

Figure 7.1. Reactions of ONOOH/ONOO$^-$, with Gibbs energies of formation [10]. Koppenol et al. *Dalton Trans.* (2012) 41, 13779-13787, with permission from the Royal Chemical Society, 2012.

Table 7.1. Thermodynamic Values for ONOOH/ONOO$^-$ [10]. With Permission from the Royal Chemical Society, 2012

	Gibbs energy (kJ mol^{-1}) $\Delta_f G°$	Enthalpy (kJ mol^{-1}) $\Delta_f H°$	Entropy (J mol^{-1} K^{-1}) $S°$
$ONOOH_g$ [119]	+44 ± 2	−16 ± 2	+271 ± 3
$ONOOH_{aq}$ [7,54]	+31 ± 1	−62 ± 7	+(16 ± 2) × 10
$ONOO^-_{aq}$ [54,120]	+68 ± 1	−44.8	+91 ± 3
	$\Delta_{rxn}G°$	$\Delta_{rxn}H°$	$\Delta_{rxn}S°$
Ionisation			
$ONOOH_{aq} \rightarrow ONOO^-_{aq} + H^+_{aq}$ [7,55]	+37	+17 ± 7	−(7 ± 2) × 10
Hydration			
$ONOOH_g \rightarrow ONOOH_{aq}$ [7,54,119]	−13 ± 2	−47 ± 2	−113 ± 6
Isomerisation			
$ONOOH_{aq} \rightarrow NO_3^-{}_{aq} + H^+_{aq}$ [7,54,120]	−142 ± 2		
Homolysis			
$ONOOH_g \rightarrow NO_2^{\cdot}{}_g + HO^{\cdot}{}_g$ [119,120]	+42 ± 4		
$ONOOH_{aq} \rightarrow NO_2^{\cdot}{}_{aq} + HO^{\cdot}{}_{aq}$ [7,54,120,121]	+58 ± 4		
$ONOO^-_{aq} \rightarrow NO^{\cdot}{}_{aq} + O_2^{\cdot -}{}_{aq}$ [54,120,121]	+68 ± 2		

$$ONOOH \rightarrow NO_3^- + H^+$$

with the rate constant and $k = 5.7 \times 10^{-1}$ s^{-1}, and pK = 6.8. Peroxynitrous acid and peroxinitrite can react by several mechanisms [14, 15]. The homolytic cleavage of peroxynitrous acid

$ONOOH \rightarrow \bullet NO_2 + \bullet OH \bullet$

producing two strongly oxidizing/hydroxylating and nitrating species occurs at 37°C pH 7.4, with a half-life of 0.8 s in phosphate buffer ($k_{app} = 0.9\ s^{-1}$) [14].

The decay of peroxynitrous acid proceeds through the first-order isomerization of ONOOH to HNO_3 with rate constants k_{iso}, = 1.0 x s^{-1}, and dissociation on $NO_2^{\cdot}$ and $HO^{\cdot}$ occurs with rate constant 1.2 s^{-1} at 25 C [10]. The kinetic hydrogen isotope effect on isomerization and for dissociation of HOONO was determined [10]. The experimental isotope effect (IE) on isomerisation $k_H/k_D = 1.6$ was attributed to a bending vibration of the OH group against the fairly stiff. N–O moiety initiates the intramolecular shift of the OH (secondary IE). For dissociation of HOONO, the IE was found to be $k_H/k_D = 3.3$, a value commonly found for acid/base equilibria. Hemolysis of ONOOH at 355 nm and pH 4.0–5.5 generates $NO^{\cdot}$ and $HO_2^{\cdot}$ radicals, which recombine with a rate constant of $(1.2 \pm 0.2) \times 10^{10}\ M^{-1}\ s^{-1}$ [18]. When NO_3^-, the product of the ONOOH isomerization, is photolyzed, the $ONOO^-$ formed is rapidly protonated with rate constant of $(1.7 \pm 0.8) \times 10^{10}\ M^{-1}\ s^{-1}$.

The formation of ONOOH/$ONOO^-$ in water has been studied using the pulse radiolysis of nitrite and nitrate solutions. The following overall rate constants were detected or calculated: (1) reaction$^{.bul.}$OH with $^{.bul.}NO_2$ generating ONOOH and NO_3^- + H^+ (1.0) $\times 10^{10}$ $M^{-1}\ s^{-1}$); (2) reaction of $^{.bul.}NO_2$ with $O^{.bul.-}$(3-4) $\times 10^9\ M^{-1}\ s^{-1}$); (3) homolysis of $ONOO^-$ into $^{.bul.}NO_2$ and $O^{.bul.-}$ (0.9-3.5) $\times 10^{-6}\ s^{-1}$); and (4) $ONOO^-$ decomposition at pH 13 (25°C) 1.3 $\times 10^{-5}\ s^{-1}$. The activation parameters for the decomposition of $ONOO^-$ at pH 14 were detected to be $A = 8 \times 10^{10}\ s^{-1}$ and $E_a = 21.7$ kcal/mol [16]. The yield of spin adducts of spin trapping of $ONOO^-$ decomposition products with 4-pyridyl-l-oxide-*N-tert*-butylnitrone (4-POBN)–ethanol did not exceed 4%. It was concluded that $HO^{\cdot}$ does not play a significant role in this process [17].

7.2.4. Reactions with Biologically Important Molecules

Peroxynitrous acid peroxynitrite is involved in key reactions of significant biochemical and physiological importance [14–48]. Two key subcellular compartments where peroxynitrite is generated are the mitochondria and the phagosome (Figure 7.2) [15]. $^{\cdot}$NO fluxes that can either arise by readily diffusing from extramitochondrial compartments or be formed locally by NOS-dependent (102) or independent-pathways cause peroxynitrite formation. Peroxynitrite will be either detoxified with the rate constant $7 \times 10^7\ M^{-1}\ s^{-1}$ (pH 7.4, 25°C) by peroxiredoxins (Prx) by reaction

$ONOOH + PrxS^- \rightarrow NO_2^- + PrxSOH$

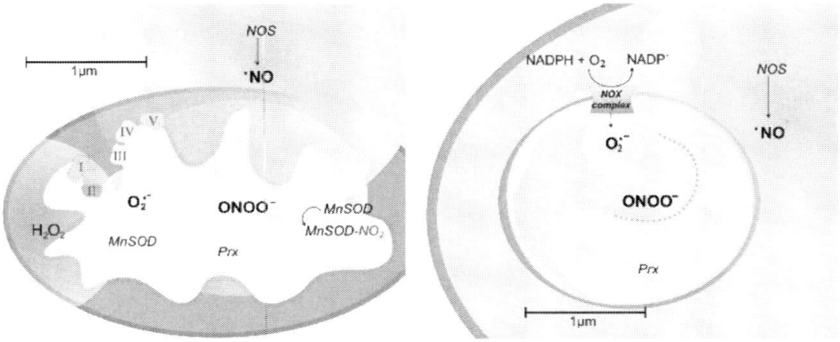

Figure 7.2. Peroxynitrite formation and reactions: • See details in text [15]. Copyrights 2009 American Chemistry Society.

or will trigger oxidative modification events intramitochondrially, such as the nitration and inactivation of MnSOD. Peroxynitrite formation inside a phagosome occurs during phagocytosis of an invading microorganism (e.g., bacteria, parasite) by a neutrophil or macrophage *via* the assembly and activation of NADPH oxidase (NOX), which generates $O_2^{•}$. Peroxynitrite and its secondary radicals can react with the pathogen plasma membrane and intracellular components and act as a cytotoxic effector molecule. The bactericidal and parasiticidal potency of peroxynitrite can be partially or totally neutralized by effective antioxidant mechanisms.

Figure 7.3 lists kinetic data on peroxynitrite scavenging, k is the rate constant (k) of target reactions with peroxynitrite, concentration [T] is the target concentration.

Peroxynitrite reactions with heme proteins are diverse. For instance, horseradish peroxidase and prostaglandin endoperoxide H synthase-1 are oxidized by two electrons, yielding peroxidase compound I and nitrite (Figure7.4), while myeloperoxidase and chloroperoxidase oxidations are one-electron reactions yielding peroxidase compound II and nitrogen dioxide constants on the order of 10^4 M^- s^{-1}. Methemoglobin and metmyoglobin catalyze the isomerization of peroxynitrite to nitrate (route 3) with rate constants on the order of 10^4 M^{-1} s^{-1}. Catalase and oxidized (Fe^{III}) cytochrome *c* do not react with peroxynitrite at any detectable rate.

Peroxynitrite is known to modify several types of biomolecules, including proteins, lipids, and DNA [14]. Peroxynitrite is involved in several basic biochemically relevant reactions, such as two-electron oxidation of thiols, homolysis, nucleophilic addition to CO_2, evolution to radicals, and reaction with transition metal centers. Peroxynitrite also can act as a mediator of superoxide radical and nitric oxide-dependent oxidative and cytotoxic processes [14]. Protonation weakens the O–O bond in ONOOH and leads to homolytic cleavage to hydroxyl radicals (•OH) and nitrogen dioxide ($^•NO_2$). The homolytic cleavage occurs with a k_4 of 4.5 s^{-1} at 37°C. Peroxinitrous acid can rapidly react with CO_2 by two ways [19]: radical mechanism

$$ONOO^- + CO_2 \rightarrow {}^\bullet NO_2 + CO_3^{\bullet -} \quad k = 3.0 \times 10^4\ M^{-1}\ s^{-1}$$

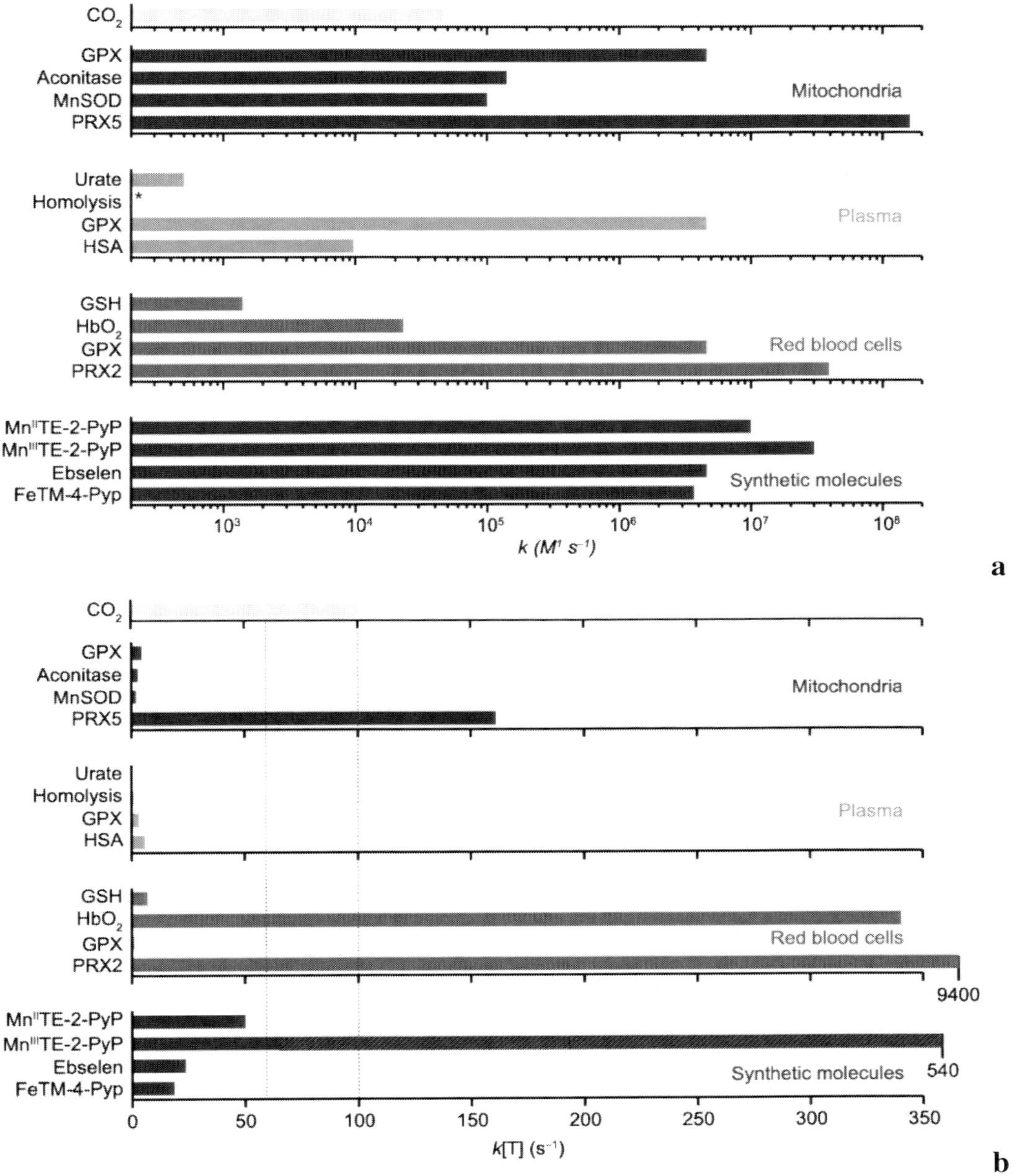

Figure 7.3. Peroxynitrite scavenging at a glance. Quantitative assessment of reactivity toward peroxynitrite is presented through second-order rate constants (k) selected targets (panel a) and through k[T] (panel b). These allow the comparison of reactivities as in simple competition kinetics. All rate constants in panel a are at pH = 7.4 and either determined at 37°C or extrapolated to that temperature for comparison assuming an activation energy of 10 kcal/mol (See details in [15]). Copyrights 2009 American Chemistry Society.

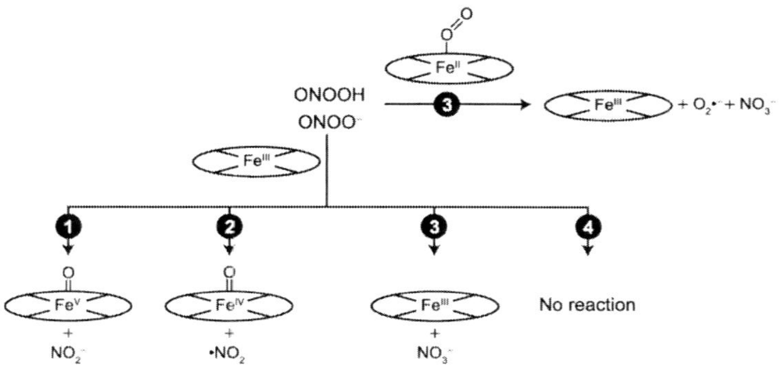

Figure 7.4. Reactions of peroxynitrite with heme proteins [15].

and nucleophilic mechanism

$$ONOO^- + CO_2 \rightarrow ONOOCOO^- \rightarrow NO_2^{\bullet} + CO_3^{\bullet -}$$

Chemical-induced dynamic nuclear polarization (CIDNP) is a powerful method for probing the spin multiplicity of intermediates involved in a reaction, the spin density distribution, and structure of radical or radical ion intermediates [20–22]. CIDNP is a non-Boltzmann nuclear spin state distribution produced in thermal or photochemical reactions, usually from colligation and diffusion, or disproportionation of radical pairs, and detected by NMR spectroscopy as enhanced absorption or emission signals. The decay of $O^{15}NOO^-$ in an aqueous solution was investigated by CIDNP [23]. Data on ^{15}N CIDNP during decomposition of peroxynitrite and the peroxynitrite–CO_2 adduct and during reactions with L-tyrosine, 4-hydroxyphenylacetic acid, and phenylacetic acid at pH 5.25 were discussed in the frame of radical pair model of CIDNP:

Radical precursors (S,T pair) or free radical encounters (F pair)

$$\Downarrow$$

$$\text{cage products} \Leftarrow [R_1^{\bullet}, R_2^{\bullet}]^{S,T,F} \Rightarrow R_1^{\bullet} + R_2^{\bullet} \xrightarrow[R_1^{\bullet},\, R_2^{\bullet}]{\text{Solvent}} \text{escape products}$$

During decomposition of peroxynitrite and the peroxynitrite–CO_2 adduct at pH 5.25, the ^{15}N NMR spectra of $^{15}NO_3^-$ showed enhanced absorption. This finding unequivocally indicated the formation of $^{15}NO_2^{\bullet}$ in radical pairs [$^{15}NO_2^{\bullet}$, $HO^{\bullet}$] and [$^{15}NO_2^{\bullet}$, $CO_3^{\bullet -}$]. A radical pathway of formation of 3-nitrotyrosine the reaction of peroxynitrite with l-tyrosine also took place, because the ^{15}N NMR signal of the nitration product 3-nitrotyrosine exhibited emission.

The rate constant (in M^{-1} s^{-1}) of peroxynitrite-selected reactions of biochemical relevance derived from review [14]. determined as: Tyrosine 0 (Tyr oxidation and nitration), Tryptophan 40, Methionine 360 (Methionine sulfoxide formation), Uric acid

500 (oxidation), Glutathione 1400 (glutathione disulfide), Cysteine 5900 (cysteine disulfide), Human serum Albumin 9700 (sulfenic acid derivative), Oxyhemoglobin $2.3x10^4$ (peroxynitrite to nitrate), Mn SOD > 10^4, (enzyme nitration at Tyr-34, CO_2 5.8 $x10^4$ (Nucleophilic addition), MnP > 10^7, Boronate-based Compounds >10^6 (oxidation), Aconitase 1.4 $x10^5$, (oxidation and disruption) and Peroxiredoxins 10^6–10^7 (cysteine residue)

Peroxynitrite possesses both an oxidant and nucleophile reactivity. First, as an oxidant, it can promote one- and two-electron oxidations by direct reactions with biomolecular targets. The redox potentials of peroxynitrite at pH 7 (E_0) for the $ONOO^-/{}^\cdot NO_2$ and ONOO-/NO_2^- pairs have been estimated as 1.4 and 1.2 V, respectively. An example of two-electron oxidations corresponds to the reaction of peroxynitrite with thiols, which yields the sulfenic acid derivative (and nitrite) following equation [14]:

$$ONOOH + RS^- \rightarrow NO_2^- + RSOH$$

For cysteine and the single thiol group of albumin (Cys-34) an apparent second-order rate constant $k_{app.} = 5.9x10^3\ M^{-1}\ s^{-1}$ at pH 7.4 and 37°C.

Peroxynitrite can be involved in oxidation of the dithiol dihydrolipoic acid (DHLA) (pK_{SH} 10.7) at alkaline pH with stoichiometry for the reaction of two thiols oxidized per peroxynitrite [25, 26]. Nitrite and LA-thiosufinate were formed as reaction products. It was found that the value of the reaction apparent rate constant are: $k_{app} = {}^o250\ M^{-1}\ s^{-1}$ per thiol, at pH 7.4 and 37ºC. It is interesting that an increase in the thiol pK_{SH} value correlated with a decrease of k_{app} for the reaction with peroxynitrite at pH 7.4. Protein stabilization of the transition state was suggested to be achieved by (a) a relatively static charge distribution around the cysteine that provide a cationic environment that stabilizes the reacting thiolate, the transition state, and the anionic leaving group; (b) a dynamic set of polar interactions that stabilize the thiolate in the resting enzyme and increases the nucleophilicity of the attacking sulfur; and c) facilitating the correct positioning of the substrate [25]. Peroxynitrite can also promote one-electron oxidations directly, for example, oxidation of cytochrome c_2 or secondarily through the homolysis of peroxynitrite. The reaction between peroxynitrous acid (ONOOH) and cytochrome c2 occurs with a second-order rate constant of 2.3 x $10^5\ M^{-1}\ s^{-1}$ [27]. The activation enthalpy, free energy, and entropy were found as +10.8 kcal mol^{-1}, +11.8 kcal mol^{-1}, and -3.15 cal $mol^{-1}\ K^{-1}$. The second path way of the oxidation is through the homolysis of peroxynitrite.

Peroxynitrite can act as secondary oxidants, which interact with transition metal centers (Me) of metalloproteins (e.g., Mn-SOD and hemeproteins) or metal complexes (e.g., Mn-porphyrins (MnP) [15, 28]:

$$ONOO^- + Me^n + X \rightarrow ONOO\text{-}Me^n\ X \rightarrow {}^\cdot NO_2 + {}^\cdot O\text{-}Me^{n+}X \rightarrow {}^\cdot NO_2 + O = Me^{(n+1)}) + X$$

In the frame of this scheme, metal-based Lewis adducts yield NO_2 and the oxyradical-metal complex, which in turn rearranges to a strongly oxidizing oxo-metal complex A following catalytic cycle involving peroxinitrite and manganese ions was suggested [29, 30].

The reaction

$$Mn^{3+} + e^- \rightarrow Mn^{2+}$$

is carried out by flavoenzymes, glutathione, and electron transport chain complexes.

The reaction

$$Mn^{2+} + ONOO^- \rightarrow Mn^{4+} = O + NO_2^-$$

proceeds with the rate constant $k = 10^6$–10^7 M^{-1} s^{-1}

$Mn^{4+} = O$ and $ONOO^-$ are reduced by reductants, such as ascorbate, glutathione, and uric acid in reactions:

$$Mn^{4+} = O + e^- + 2H^+ \rightarrow Mn^{3+} + H_2O$$

$$ONOO- + 2\ e^- + 2H^+ \rightarrow NO_2^- + H_2O$$

The last reaction is catalyzed by synthetic molecules (MnP) that are effective against peroxynitrite *in vitro* and *in vivo*.

NO-derived species responsible of the inactivation of mitochondrial aconitase. Disassembly of the cubane [4Fe-4S] cluster by peroxynitrite via oxidative attack leads to an inactive [3Fe-4S] enzyme by the reaction:

$$[4Fe\text{-}4S]^{2+} + ONOO^- \rightarrow [3Fe\text{-}4S]^{1+} + Fe^{3+} + NO_2^-$$

yielding the inactive [3Fe-4S] enzyme (k = 1.1 x 10^5 M^{-1} s^{-1}) [31]. Iron released from aconitase can further propagate intramitochondrial oxidative damage by metal-mediated formation of oxidizing and nitrating species. Carbon dioxide enhanced the $ONOO^-$-dependent inactivation via reaction of $CO_3^{\cdot-}$ with the [4Fe-4S] cluster (k = 3 x 10^8 M^{-1} s^{-1}) [31].

The reaction between ascorbic acid and peroxinitrite is of special interest from the point of view of biochemistry and physiology [32–35]. A rate constant of the reaction between ascorbic acid and HOONO at ambient conditions was found to be k = 235 M^{-1} s^{-1} [32]. Relative antioxidant activities of seventeen compounds measured using Pyrogallol Red toward radicals, derived from peroxynitrous acid, were reported [36]. For example, peroxynitrite reacts with ebselen [2-phenyl- 1,2-benzisoselenazol-3(2H)- one] with a rate constant k = 2x 10^6 M^{-1} s^{-1}. Reactions thiols and ascorbic acid with peroxynitrous acid occurred with k = (1- 6) x10^3 M^{-1} s^{-1} (pH 7.4, 37°C) and k = 235 M^{-1} s^{-1}, respectively. Ascorbic acid has been recognized as a nitrosation inhibitor [37]. The role of ascorbate in affecting the kinetics of nitrosation was investigated in detail in a model dynamic system ascorbate and SIN-1 (3-morpholinosydnonimine), which produces NO reactive species NO_2 and peroxynitrite as more reactive compound [2]. DAN (2,3-diaminonaphthalene) was used as a probe for N-nitrosation reaction. Reactions of peroxynitrite and NO_2 with ascorbate run the rate constants (in $M^{-1}s^{-}1$) 2.3 x 10^5 and 3.5x 10^5, respectively [38]. Kinetics of the reaction of ascorbate and hydroxyl radical adduct of DMPO (5,5-dimethyl-1-pyrroline N-oxide) was investigated by means of stopped-flow electron spin resonance method [39]. The experiments evaluated the reaction rate constant k = 5:2 x$10^3$$M^{-1}$ s^{-1}.

Peroxynitrite can trigger lipid peroxidation in membranes, liposomes, and lipoproteins by abstracting a hydrogen atom from polyunsaturated fatty acids with resulting products lipid hydroperoxyradicals, conjugated dienes, and aldehydes [40]. In review [41], a free radical one-electron mechanism of unsaturated lipid oxidation and nitration by peroxynitrite was discussed. The following chemical processes were taken in consideration: (1) rapid reaction peroxynitrite with CO_2 to yield$^{\cdot}NO_2$ and$^{\cdot}CO_3$, which causing inhibition of lipid oxidation; (2) homolization of ONOOH to NO_2 and $^{\cdot}OH$; (3) abstract H to unsaturated fatty acids, initiating lipid oxidation; and (4)$^{\cdot}NO$ termination reactions with lipid derived radicals (LOO), and reaction $^{\cdot}NO_2$ lipid derived radicals leading to the formation of nitro-fatty acids. Nitrosoperoxolinolenate, hydroxylnitroso-peroxolinolenate, and hydroperoxonitrosoperoxolinolenate are formed from the NO_2 reaction with a carbon-centered radical, likely via a caged radical reorganization of unstable alkyl peroxynitrite intermediates (LOONO).

Nitrogen dioxide coming from ONOOH homolysis can mediate oxidation and nitration of unsaturated fatty acids under anaerobic or aerobic conditions via NO_2 reaction with unsaturated fatty acids through a radical pathway yielding nitrovinyl, nitroallyl, or nitrohydroxy derivatives. At greater oxygen tensions, the formed oxidants react with unsaturated fatty acids enhancing lipid oxidation processes yielding isomerized lipid derivatives. Alternatively, ONOOH homolysis can induce formation of nitro-epoxy products [41].

Low molecular mass selenols react with peroxynitrite with second-order rate constants k_2 in the range 5.1 × 10^5-1.9 × 10^6 M-1s^{-1} and much slower with selenides, including selenosugars (k_2 = 2.5 × 10^3 $M^{-1}s^{-1}$), diselenides k_2 = (0.72-1.3) × 10^3 M-1s^{-1}), sulfides (k_2 = 2.1 × 10^2 $M^{-1}s^{-1}$), and tiols, in particular [42].

Peroxynitrite, as a strong oxidant, can damage both DNA nucleobases and sugar-phosphate backbone [43–48]. The mechanism of sugar damage by peroxynitrite, including an abstraction a hydrogen atom from the deoxyribose moiety and resulting in the opening of the sugar ring and the generation of DNA strand breaks, was suggested in [43]. The general scheme of peroxynitrite-induced G oxidation and nitration products was presented [46]. Among the four nucleobases, guanine is the most reactive with peroxynitrite due to its low reduction potential. The reaction is suggested to occur by the generation of radical intermediates that can recombine with the $^{\cdot}NO_2$ formed during peroxynitrite degradation. The major product of guanine oxidation is 8-oxoguanine and can further react with peroxynitrite, yielding cyanuric acid and oxazolone.

With respect to peroxynitrite modifying DNA, guanine (G) is the most readily modified nucleobase. Specifically, it was found that a single guanine in the loop of the hairpin is most susceptible to a modification (Figure 7.5) [45]. Upon exposure to peroxynitrite, G is converted to three major products: 8-oxoG, 8-nitroguanine, 5-guanidino-4-nitroimidazole. Piperidine treatment leads to conversion of 8-nitroguanine to a strand break.

The peroxynitrite-derived radicals rapid reaction with 8-oxoG, and bimolecular rate constants of 8×10^8 M^{-1} s^{-1} and 5×10^6 M^{-1} s^{-1} have been reported for $CO_3^{\cdot -}$ and $^{\cdot}NO_2$, respectively [47]. As derived from electron-transfer equilibria in aqueous solution, direct electron transfer from 8-oxo-7,8-dihydro-2′-deoxyguanosine to the oxidizing guanosine$^{\cdot}$ occurs with a bimolecular rate constant of 4.6×10^8 M^{-1} s^{-1}[48].

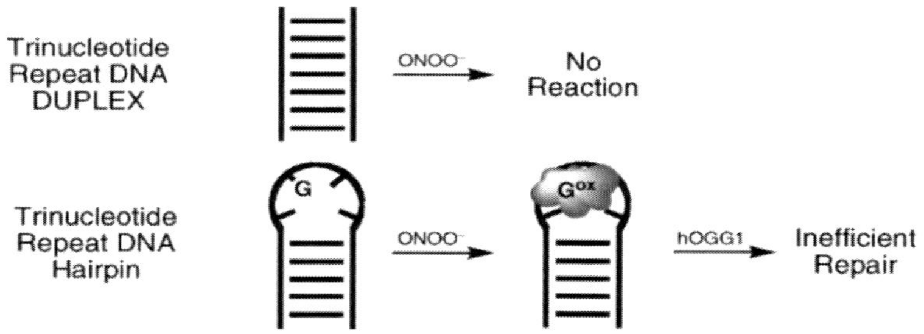

Figure 7.5. Schematically illustration that unpaired guanine in the loop of the hairpin represents a hot spot for damage [45]. Copyrights 2009 American Chemistry Society.

7.2.5. Nitrosation

Nitrosation, caused by peroxinitrite and other RNS, is a process of converting organic compounds into compounds containing the R-NO functionality [49–54]. Forming S-nitroso-proteins (SNO-proteins) through S-nitrosylation of allosteric and active-site cysteine thiols within proteins is the main mechanism of the cellular signal transduction. S-nitrosylation of proteins has been demonstrated to affect a broad range of functional parameters including enzymatic activity, subcellular localization, protein–protein interactions, protein stability, and transnitrosylation. A growing literature suggests that over three thousand proteins are S-nitrosylated in cell:

Nitrosation of thiol-containing compounds, such as glutathione and thiol proteins, can make important contributions to the oxidative stress dysfunction. For example, S-nitrosylation of enzyme peroxiredoxin 2 promotes oxidative stress-induced neuronal cell death in SNO-Prx2 (Figure 7.6) [49].

Evidences for transnitrosylation reactions between hemoglobin/anion exchanger 1, thioredoxin/caspase-3, X-linked inhibitor of apoptosis/ caspase-3, (GAPDH-HDAC2/SIRT1/DNA-PK, and Cdk5/dynamin related protein 1 (Drp1) have been discussed in review [52]. As another example, in sickle cell anemia erythrocyte membranes, hemoglobin (HbS; Glu6Val Hb) is impaired both in its ability to form SNO-sickle hemoglobin (SNO-Hb, Metal-to-Cys transfer of the NO group) and to transnitrosylate AE1 (Cys-to-Cys transfer of the NO group) (Figure 7.7) that resulted in decrease in levels of red blood cells (RBC) [54].

Dysregulation of hemoglobin's nitrosylase activity in sickle cell anemia was investigated in RBCs from patients with sickle cell anemia [54]. In this system, the aberrant intramolecular transfer of NO from heme iron nitrosyl to Cys (impaired Metal-to-Cys nitrosylase activity) results in deficient formation of SNO-HbS. Additionally, aberrant docking of S-nitrosylated SNO-HbS to the membrane protein AE1 disrupts transnitrosylative transfer of the NO group to the membrane (impaired Cys-to-Cys nitrosylase activity). It was suggested that decreased levels of membrane SNO were associated with the decreased ability of RBCs to effect hypoxic vasodilation [54].

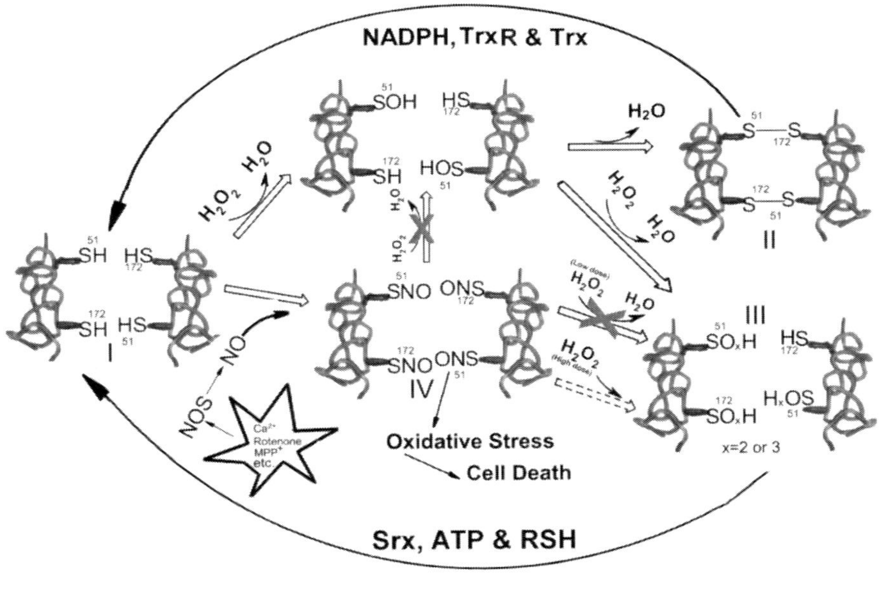

Figure 7.6. Reaction mechanism of SNO-Prx2 contribution to oxidative stress and neuronal cell death in human neurodegenerative disorders. TrxR, thioredoxin reductase; Trx, thioredoxin; Srx, sulfiredoxin [49]. With permission from PNASUSA, 2007.

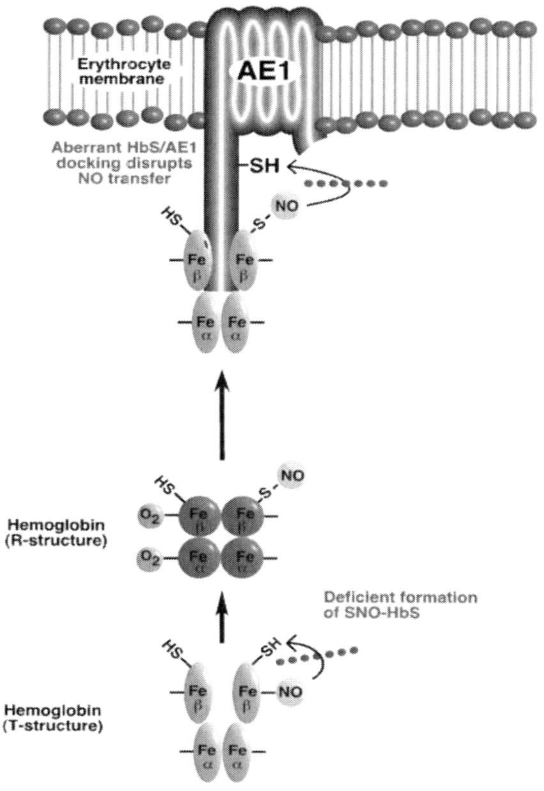

Figure 7.7. Schematic summary of the proposed defects in NO trafficking in the sickle erythrocyte, which inhibit generation of vasodilatory NO bioactivity at the RBC membrane. As shown in the Figure at the bottom, the allosterically (O_2/redox) regulated intramolecular transfer of NO from heme to thiol is impaired in HbS. This defect is likely to represent differences between HbS vs. HbA in heme redox potential (and in P_{50} upon polymerization), and thus in the ability to support *S*-nitrosylation. Note that some portion of NO groups bound to β-heme will redistribute to α-heme ("recapture") rather than thiol, that one oxygen is omitted from R-state SNO-Hb to provide for the possibility of β-heme oxidation ($Fe^{2+} \rightarrow Fe^{3+}$) coupled to SNO-Hb formation, and that O_2/redox-regulation of Hb is represented as a simplified two-state (R/T) model. In addition, transfer of NO groups from SNO-HbS to a cysteine thiol within the CDAE1 is deficient, shown on the left, as a result of aberrant binding of HbS and CDAE1. As shown on the right, oxidative (disulfide) coupling of HbS to CDAE1, which may be facilitated by aberrant binding, results in the formation of cross-linked AE1-hemichrome complexes that preclude NO group transfer from SNO-HbS to the RBC membrane [54]. With permission from PNAS USA, 2007.

7.2.6. Examples of Biochemical and Physiological Effects of Peroxynitrite

Peroxiredoxins, catalase, peroxidases, methemoglobin, myoglobin, metal-porphyrins, hydralazine, imidazole-based thiourea, selenourea derivatives, mercaptoethylguanidine, guanidine, and synthetic compounds are catalytic scavengers of peroxynitrite [55–65]. The fast and selective oxidation of alpha-tocopherol by low concentrations of peroxynitrite in synaptosomes (nerve ending particles) and mitochondria was reported [57]. This result suggests that vitamin E may play an important role in preventing membrane oxidation induced by peroxynitrite. Effects of O_2 on the reactivity of NO, NO_2, and $ONOO^-$/ONOOH for nitration of tyrosine and nitrosation of glutathione (GSH) and morpholine (MOR) were investigated [58]. Two important conclusions have been made: generation of ·OH from $ONOO^-$/ONOOH was suppressed under anaerobic conditions and the reactivity of $ONOO^-$/ONOOH and ·OH generation from $ONOO^-$ were reversibly controlled by the O_2 concentration.

Results of work [59] indicated that Se deficiency induced oxidative damage in the intestinal tracts of chickens and that low levels of GSH-Px and high contents of NO may exert a major role in the injury of the intestinal tract induced by Se deficiency. It was found that oxidative stress in *Salmonella typhimurium* derived by the aerobically functioning flavohemoglobin, Hmp, which catalyses the reaction between oxygen and NO to produce relatively inert nitrate, affected peroxynitrite stress [60]. The authors hypothesized that Hmp-expressing cells, in the absence of NO, generate reactive oxygen species, the toxicity of which is exacerbated by peroxynitrite *in vitro* and in macrophages.

A role of oxidative inactivation of NO and peroxynitrite formation in the vasculature related to hypercholesterolemia, hyperglycemia, turbulent blood flow, inflammatory mediators, and cigarette smoke components was discussed in [61]. S-nitrosylation of peroxiredoxin 2 (Prx), effective antioxidant enzyme, promotes oxidative stress-induced neuronal cell death in Parkinson's disease [62]. This process inhibits the Prx protective function by interrupts the normal redox cycle and results in the accumulation of cellular peroxides, inducing oxidative stress and eventually induces neuronal cell death. Specifically, the S-nitrosylation targets the redox-active Cys-51 and Cys-172 residues critical for Prx2 activity. Formation of the nitrosylation product SNO-Prx2, resulting in Prx2 dysfunction, therefore, provides a mechanistic link between nitrosative/oxidative stress and neurodegeneration.

In neurodegenerative diseases, several chaperones in the cytosol or endoplasmic reticulum (ER) ameliorate the accumulation of misfolded proteins triggered by oxidative or nitrosative stress, or of mutated gene products [63]. It was shown that, in brains manifesting sporadic Parkinson's or Alzheimer's disease, the accumulation of polyubiquitinated proteins and activation the unfolded protein response are induced by S-nitrosylation a critical cysteine thiol of protein-disulfide isomerase (PDI). S-nitrosylation also abrogates protein-disulfide isomerase (PDI)-mediated attenuation of neuronal cell death triggered by ER stress, misfolded proteins, or proteasome inhibition. The ER 1111 withstands relatively mild insults through the expression of stress proteins or chaperones, such as glucose-regulated protein and PDI. S-nitrosation of proteins is also increased in diabetes, when the retinal mitochondria become dysfunctional [65] during ischemia and reperfusion [66], in many chronic inflammatory diseases [67], neurodegeneration in murine [68], and a modification of connexin (Cx)43 hemichannels in astrocytes [69].

7.3. OTHERS COMPOUNDS RELATED TO NITRIC OXIDE EFFECTS ON OXIDATIVE AND NITROSATIVE STRESSES

7.3.1. Nitrite and Nitrate

Though nitrite and nitrate, two endogenous oxides of nitrogen, are known as relatively stable and chemically inertness compounds *in vitro*, nevertheless, they are toxic *in vivo* [70–72]. Sodium nitrite (NO_2^-), a reducing agent, is formed in the body via the oxidation of nitric oxide (NO) or through the reduction of nitrate (NO_3^-). Thermodynamics of some reactions involving nitrous acid are shown below:

Half-reaction	E^0 (V)
$NO_3^- + 3 H^+ + 2 e^- \rightleftharpoons HNO_2 + H_2O$	+0.94
$2 HNO_2 + 4 H^+ + 4 e^- \rightleftharpoons H_2N_2O_2 + 2 H_2O$	+0.86
$N_2O_4 + 2 H^+ + 2 e^- \rightleftharpoons 2 HNO_2$	+1.065
$2 HNO_2 + 4 H^+ + 4 e^- \rightleftharpoons N_2O + 3 H_2O$	+1.29

Reduction of nitrite to NO occurs in blood and tissues and proceeds through several enzymatic and non-enzymatic pathways [72]. Nitrite is involved in a series of important reactions:

Deoxyhaemoglobin/myoglobin

$NO_2^- + Fe^{2+} + H^+ \rightarrow {}^\bullet NO + Fe^{3+} + OH^-$

Xanthine oxidoreductase

$NO_2^- + Mo^{4+} + H^+ \rightarrow {}^\bullet NO + Mo^{5+} + OH^-$

Ascorbate

$NO_2^- + H^+ \rightarrow HNO_2$

$2 HNO_2 + Asc \rightarrow 2 {}^\bullet NO + dehydroAsc + 2 H_2O$

Polyphenols (Ph-OH)

$NO_2^- + H^+ \rightarrow HNO_2$

$Ph\text{-}OH + HNO_2 \rightarrow Ph\text{-}{}^\bullet O + {}^\bullet NO + H_2O$

Nitrite oxidation

Haemoglobin

$4 NO_2^- + 4 HbO_2 + 4H^+ \rightarrow 4NO_3^- + 4 Met\text{-}Hb + 2 H_2O + O_2$

The effect of sodium nitrite ($NaNO_2$) on human erythrocytes was studied under *in vitro* conditions [72]. Incubation of erythrocytes in the sodium nitrite solution resulted in a decrease in the levels of reduced glutathione, total sulfhydryl, and amino groups and an increase in hemoglobin oxidation and aggregation, lipid peroxidation, protein oxidation, and hydrogen peroxide levels. Suggesting the induction of oxidative stress was supported by finding that activities of all major erythrocyte antioxidant defense enzymes were decreased in $NaNO_2^-$-treated erythrocytes. The *in vivo* stable nitrite can inhibit mitochondrial oxidative phosphorylation through modulating levels of NO, changing the activity of heme proteins [71]. It is interesting that in mitochondria, nitrate is more effective than that unstable peroxynitrite. In these conditions, more stable nitrate is reduced to nitrite. In human erythrocytes, sodium-nitrite-induced oxidative stress causes membrane damage and alters major metabolic pathways [72]. In particular, the incubation of erythrocytes with 0.1-10.0 mM $NaNO_2$ at 37°C for 30 min.

7.3.2. Nitric Dioxide

Nitrogen dioxide (NO_2), a toxic-free radical gas, is a major component of both indoor and outdoor air pollution. NO_2 is a strong oxidizer and exists in equilibrium with dinitrogen tetroxide $2\ NO_2 \rightleftharpoons N_2O_4$. The equilibrium is characterized by $\Delta H = -57.23$ kJ/mol. Hydrolyses of both compounds gives nitric acid and nitrous acid [73]:

$$2\ NO_2/N_2O_4 + H_2O \rightarrow HNO_2 + HNO$$

In biological systems, NO_2 is formed by the spontaneous decomposition of peroxynitrite and from the metabolism of nitrite by heme peroxidase enzymes [74–76]. This reactive species participates in oxidizing, nitrating, or modifying sites such as protein thiols and tyrosine residues and nucleic acids. Nitric dioxide initiates lipid peroxidation for production of free radicals and also oxidizes tyrosine to 3-nitrotyrosine. NO_2 as a substrate for the mammalian peroxidase and lactoperoxidase provides an additional pathway contributing to cytotoxicity or host defense associated with increased NO [74]. NO_2 is an air toxicant capable of inducing lung damage to the respiratory epithelium [75, 76]. NO_2 has been documented to cause pulmonary injury in both animal and human studies. To understand the mechanisms of the signaling pathways leading to epithelial cell death by NO2, the cells in were exposed to continuous gas-phase NO_2. The experiments indicated that NO_2-induced cell death is limited to cells localized in the leading edge of the wound. Potential cell-signaling mechanisms include the mitogen-activated protein kinase, c-Jun N-terminal kinase, and the Fas/Fas ligand pathways

7.4. Nitric Oxide and Inflammation

Inflammation is a part of the complex biological response of human and animal tissues to pathogens, damaged cells, or irritants and provides a protective response involving immune cells, blood vessels, and molecular mediators [77–85]. In inflammatory processes, NO functions as an effector molecule toward the infectious organisms and can influence many aspects of the inflammatory cascade in both acute and chronic inflammatory diseases In particular, NO induces vasodilatation in the cardiovascular system, is involved in immune responses by cytokine-activated macrophages, contributes to the regulation of apoptosis, is involved in the pathogenesis of inflammatory disorders of the joint, gut, and lungs, and acts as a potent neurotransmitter at the neuron synapses. A connection between NO production, inflammation, and nitroxide production was discussed in details in .

Nitric oxide plays a double role in the pathogenesis of inflammation giving an anti-inflammatory effect under normal physiological conditions and being a pro-inflammatory mediator that induces inflammation due to overproduction in abnormal situations [78–89]. Comprehensive analysis of the metabolism of NO in living cells in the normal state and pathology indicated that most of physiological fluids, including blood, normally contain nitrite and nonthiolate nitroso compounds in concentrations less than 100 nM, while its concentration in blood is dramatically increased in case of inflammatory diseases [78].

One of possible mechanisms of inflammation is that NO is rapidly oxidized to reactive nitrogen oxide species (RNOS), which can S-nitrosate thiols to modify key signaling molecules, such as kinases and transcription factors, and also inhibits several key enzymes in mitochondrial by RNOS. This leads to a depletion of ATP and cellular energy [81]. A combination of these interactions may explain the multiple actions of NO in the regulation of immune and inflammatory cells. The effects of both NO and superoxide in immune and inflammation regulation can be exerted through interaction with cell signaling systems like cGMP, cAMP, G-protein, JAK/STAT, or MAPK-dependent signal transduction pathways and modulation of the expression of multiple other mediators of inflammation [82].

An alternative mechanism of inflammation suggested in [83] was developed and modified in [78]. A scheme of supposed mechanism of appearance in the blood plasma of nitrite and RNO in inflammation is given in Figure 7.8. The proposed scheme elucidates a key role of dinitrosyliron complexes (DNIC) in the process. The results of investigations, obtained in [78], allowed to formulate the following conclusions: (1) living tissues in the norm have a mechanism preventing oxidation of NO to nitrite; (2) the content of nitrite and nitrate is determined not only by the intensity of NO synthesis but also by the intensity of decay of donor compounds; and (3) in inflammatory diseases in blood, plasma nitrite and nonthiolate nitrosocompounds are appeared. An important regulatory/ modulatory role of NO in a variety of inflammatory conditions was illustrated by several typical examples. The content of nitro and nitrosocompounds and dinitrosyl iron complexes (DINC) in the plasma of healthy donors and more than a hundred patients suffering from various inflammatory diseases were measured using a new enzymatic sensor [78]. It was found that amount of RS-NO and RNO is neglected for healthy people and ranged from 0.5 to 20 mM for sick patients though variated for different diseases within each group. The concentration of DINC in the plasma from the healthy and sick patients were found to be similar (4-20 mM). The authors concluded that appearance of nitrite and RNO in plasma in concentrations higher than 150 nM unequivocally presents a sign of inflammation, and the index of the content of NO + RNO in plasma is the most sensitive and specific index of the presence of nonspecific inflammatory process.

Protein kinase C (PKC) is a family of ten isoenzymes that play a crucial role in cellular signal transduction. The role of PKC isoenzymes in the pathogenesis and as a potential drug target in inflammation has been discussed in detail [84]. The effects of presenting iNOS, as an example of an inflammatory gene regulated by the pleiotropic PKC signaling pathway, was also considered. Nitric oxide inhibits cell proliferation via inhibition of polyamine synthesis, prevents the proliferative response following cytokine exposure, and protects nephrotoxic serum-induced glomerulonephritis, autoimmune tubular interstitial nephritis, and experimental allergic encephalomyelitis [85]. This molecule also plays a significant role in many forms of immune injury.

Nitric oxide exhibits multiple effects [86]: (1) regulates inflammatory erythema and oedema; (2) has cytotoxic action against micro-organisms; (3) has cytoprotective properties from reperfusion injury; (4) mediates some destructive effects of pro-inflammatory cytokines, such as interleukin, and regulates the synthesis of several inflammatory mediators and functions of inflammatory cells; (5) regulates the synthesis of several inflammatory mediators; (6) functions of inflammatory cells; and (7) suppresses lymphocyte proliferation and increased concentrations of nitrite. Constitutive NO synthase is activated in neutrophils in response to inflammatory stimuli. In the inflamed joint, NO regulates the synthesis of several inflammatory mediators and functions of inflammatory cells. In addition, NO mediates some destructive effects of pro-inflammatory cytokines such as interleukin-1.

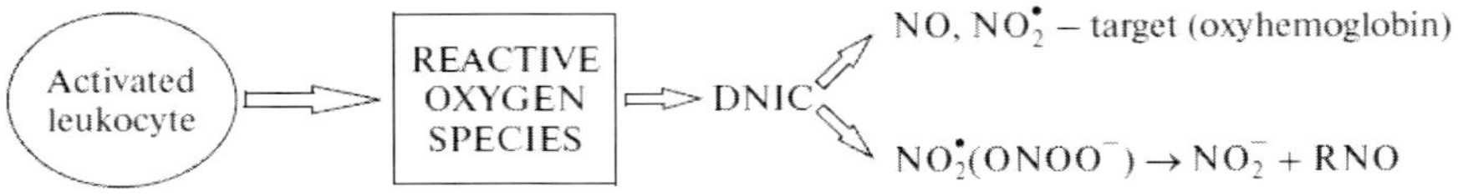

Figure 7.8. Supposed mechanism of appearance in blood plasma of nitrite and RNO in inflammation [78]. With permission from Pleiades Publishing, LTD.

Selective NO biosynthesis inhibitors and synthetic arginine analogues are proved to be important therapeutic means in the management of inflammatory diseases used for the treatment of NO-induced inflammation [87]. The undesired effects of NO can be due to its impaired production, including vasoconstriction, inflammation, and tissue damage. Pro-inflammatory factors including NO can be associated with osteoarthritis [88]. NO is involved in the degradation of matrix metalloproteinases, inhibits the synthesis of both collagen and proteoglycans, prevents the nuclear localization of the transcription factor nuclear factor-κB, and helps to mediate apoptosis [89].Carbon monoxide (CO), nitric oxide (NO) and hydrogen sulfide (H_2S) are able to penetrate cellular membranes and are produced endogenously in the body [90] . In the cardiovascular system, for example, these molecules act as vasodilators, promote angiogenesis, protect tissues against damage (e.g., ischemia–reperfusion injury), and exhibit both pro- and anti-inflammatory effects.

The oxidative stress of inflammation in aggressive forms of cancer was proposed to drive a continuous process of DNA adducts and crosslinks, as well as post-translational modifications to lipids and proteins [91]. It was concluded that the RNS may restore the cancer cells to an apoptosis-permissive and growth-inhibitory state. The pivotal role played by oxidative and nitrative stress in cell death, inflammation, and pain and its consequences for toxicology and disease pathogenesis was described in review [92]. Physiological disorders including neurodegenerative/neuropsychiatric and cardiovascular diseases, and liver and skin carcinogenesis, were considered. Mechanisms of reactive oxygen and nitrogen species drive sustained cell proliferation, cell death including both apoptosis and necrosis, and formation of nuclear and mitochondrial DNA mutations stimulation of a pro-angiogenic environment initiated by reactive oxygen and nitrogen species were discussed.

The hypothesis that an inflammatory microenvironment in *p53−/−NOS2+/+* C57BL6 mice with an enhanced level of $NO^{\bullet}$ accelerates spontaneous tumor development was tested [93]. A dose-dependent model of $NO^{\bullet}$-mediated modulation of tumorigenesis in p53-deficient mice was examined. The experiments on $NO^{\bullet}$-mediated modulation of tumorigenesis in p53-deficient mice indicated that an increase in $NO^{\bullet}$ production under inflammatory conditions can inhibit apoptosis, increase proliferation, and modulate the immune profile, giving rise to an internal milieu conducting to the tumor growth. The rapid tumor development in C. *parvum*–treated *p53−/−NOS2+/+* mice was suggested as genetic evidence of a role of NO.

The matrix metalloproteinases (MMPs) with a zinc ion in the active site regulate cell function, growth and division, host defense, extracellular matrix (ECM) synthesis, morphogenesis, wound healing, tissue repair, skeletal formation, apoptosis, cleavage of transmembrane proteins, and bioactive molecules [94]. These enzymes with combination with nitric oxide were found to increase in inflammation states and in cancer. A study on the mechanism of IL-2/α-CD40 immunotherapy showed that the inhibition of metastasis was a result of induction of iNOS expression by stromal macrophages and a reduction in MMP-9 expression and activity within the tumor. The authors concluded that NO mediates between the tumor and stroma resulting in the regulation of MMP-9.

References

[1] Beckman JS, Beckman TW, Chen J, Marshall PA, Freeman BA. Apparent hydroxyl radical production by peroxynitrite: Implications for endothelial injury from nitric oxide and superoxide. *Proc. Natl. Acad. Sci. USA* (1990) 87, 1620-1624.

[2] Teh-Min Hu, Yu-Jen Chen. Nitrosation-modulating effect of ascorbate in a model dynamic system of coexisting nitric oxide and superoxide. *Free Radical Research* (2010) 44, 552-562.

[3] Nauser T, Koppenol WH. The Rate Constant of the Reaction of Superoxide with Nitrogen Monoxide: Approaching the Diffusion Limit. *J. Phys. Chem. A* (2002) 106, 4084-4086.

[4] Fridovich I. Superoxide radical and superoxide dismutases. *Annu. Rev. Biochem.* (1995) 64, 97-112.

[5] Krauss M. Electronic structure and spectra of the peroxynitrite anion. *Chemical Physics Letters* (1994) 222, 13-15.

[6] Logager T, Sehested K. Formation and decay of peroxynitrous acid: A pulse radiolysis study. *J. Phys. Chem.* (1993) 97, 6664.

[7] Bean BD, Mollner AK, Nizkorodov SA, Nair G, Okumura M, Sander SP, Peterson KA, Francisco JS. Cavity Ringdown Spectroscopy of *cis-cis* HOONO and the HOONO/$HONO_2$ Branching Ratio in the Reaction OH + NO_2 + M. *J. Phys. Chem. A* (2003)107, 6974-6985.Lo WJ, Lee YP, Tsai IYM, S. Beckman JS.

[8] Ultraviolet absorption of cis-cis and trans-perp peroxynitrous acid (HOONO) in solid argon. *Chemical Physics Letters* (1994), 229, 357-361.

[9] Zhang X, Nimlos MR, Ellison GB, Varner ME, Stanton JF. Infrared absorption spectra of matrix-isolated *cis,cis*-HOONO and its ab initio CCSD(T) anharmonic vibrational bands. *J. Chem. Phys.* (2006) 124, 084305/1-084305/7.

[10] Koppenol WH, Bounds PL, Nauser T, Kissner, Rüegger H. Peroxynitrous acid: controversy and consensus surrounding an enigmatic oxidant. *Dalton Trans.* (2012) 41, 13779-13787.

[11] Berski S, Latajka Z, Gordon AJ. Electron localization function and electron localizability indicator applied to study the bonding in the peroxynitrous acid HOONO. *Journal of Computational Chemistry* (2011) 32, 1528-1540.

[12] Grubb MP, Warter ML, Xiao H, Maeda S, Morokuma K, North S. No Straight Path: Roaming in Both Ground- and Excited-State Photolytic Channels of $NO_3 \rightarrow NO + O_2$. *Science* (2012) 335, 1075-1078.

[13] Contreras R, Galván M, Oliva M, Safont VS, Andrés J, Guerra D, Aizman A. Two state reactivity mechanism for the rearrangement of hydrogen peroxynitrite to nitric acid. *Chem. Phys.* (2008) 457, 216–222.

[14] Radi R. Peroxynitrite, a Stealthy Biological Oxidant. *J. Biol. Chem.* (2013) 288, 26464–26472.

[15] Ferrer-Sueta G, Radi R. Chemical Biology of Peroxynitrite: Kinetics, Diffusion, and Radicals. *ACS Chem. Biol.* (2009) 161–177.

[16] Merényi G, Lind J, Goldstein S, Czapsk Gi, Mechanism and Thermochemistry of Peroxynitrite Decomposition in Water. *J. Phys. Chem. A*, (1999) 103, 5685–5691.

[17] Pou S, Nguyen SY, Gladwell T, Rosen GM. Does peroxynitrite generate hydroxyl radical? *Biochim. Biophys. Acta, Gen. Subj.* (1995) 1244, 62–68.

[18] Sturzbecher-Hohne M, Nauser T, Kissner R, Koppenol WH. Photon-Initiated Homolysis of Peroxynitrous Acid. *Inorganic Chemistry* (2009), 48, 7307-7312.

[19] Lymar SV, Hurst JK. Rapid reaction between peroxonitrite ion and carbon dioxide: Implications for biological activity. *J. Am. Chem. Soc.* (1995) 117, 8867–8868.

[20] Kaptein R, Oosterhoff LJ. Chemically induced dynamic nuclear polarization III: Anomalous multiplets of radical coupling and disproportionation products. *Chem. Phys. Lett.* (1969) 214–216.

[21] Closs GL. A mechanism explaining nuclear spin polarizations in radical combination reactions. *J. Am. Chem. Soc.* (1969) 91, 4552–4554.

[22] Likhtenshtein GI. *Electron Spin Interaction in Chemistry and Biology. Fundamentals, methods, reactions mechanisms, magnetic phenomena, structure investigation,* SPRINGER Publisher, 2016.

[23] Lehnig M. Radical Mechanisms of the Decomposition of Peroxynitrite and the Peroxynitrite–CO_2 Adduct and of Reactions with L-Tyrosine and Related Compounds as Studied by ^{15}N Chemically Induced Dynamic Nuclear Polarization. *Arch. Biochem. Biophys.* (1999) 368, 303–318.

[24] Koppenol WH, Moreno JJ, Pryor WA, Ischiropoulos H, Beckman JS. Peroxynitrite, a cloaked oxidant formed by nitric oxide and superoxide. *Chem. Res. Toxicol.* (1992) 5, 834–842.

[25] Trujillo M, Radi R. Peroxynitrite reaction with the reduced and the oxidized forms of lipoic acid: New insights into the reaction of peroxynitrite with thiols. *Arch. Biochem. Biophys.* (2002) 397, 91–98.

[26] Ferrer-Sueta G, Manta B, Botti H, Radi R, Trujillo M, Denicola A. Factors affecting protein thiol reactivity and specificity in peroxide reduction. *Chem. Res. Toxicol.* (2011) 24, 434–450.

[27] Thomson L, Trujill M, Telleri R, Radi R. Kinetics of cytochrome *c*2_oxidation by peroxynitrite: Implications for superoxide measurements in nitric oxide-producing biological systems. *Arch. Biochem. Biophys.* (1995) 319, 491–497.

[28] Crow JP. Manganese and iron porphyrins catalyze peroxynitrite decomposition and simultaneously increase nitration and oxidant yield: Implications for their use as peroxynitrite scavengers in vivo. *Arch Biochem Biophys.* (1999) 371, 41–52.

[29] Ferrer-Sueta G, Hannibal L, Batinic´-Haberle I, and Radi R. Reduction of manganese porphyrins y flavoenzymes and submitochondrial particles: A catalytic cycle for the reduction of peroxynitrite. *Free Radic. Biol. Med.* (2006) 41, 503–512.

[30] Batinic-Haberle I, Reboucas JS, Spasojevic I. Superoxide dismutase mimics: Chemistry, pharmacology, and therapeutic potential. *Antioxid. Redox Signal.* (2010) 13, 877–918.

[31] Tortora V, Quijano C, Freeman B, Radi R, Castro L. Mitochondrial aconitase reaction with nitric oxide, *S*-nitrosoglutathione, and peroxynitrite: Mechanisms and relative contributions to aconitase inactivation. *Free Radic. Biol. Med.* (2007) 42, 1075–1088.

[32] Squadrito GL, Jin X, Pryor WA. Stopped flow kinetic study of the reaction of ascorbic acid with peroxynitrite. *Arch. Biochem. Biophys.* 322, (1995) 53–59.

[33] Kuzkaya N, Weissmann N, Harrison DG, Dikalov S. Interactions of peroxynitrite, tetrahydrobiopterin, ascorbic acid, and thiols: Implications for uncoupling endothelial nitric-oxide synthase. *J Biol Chem.* (2003) 278, 22546–22554.

[34] Guidarelli A, Fiorani M, Cantoni O. Enhancing effects of intracellular ascorbic acid on peroxynitrite-induced U937 cell death are mediated by mitochondrial events resulting in enhanced sensitivity to peroxynitrite-dependent inhibition of complex III and formation of hydrogen peroxide. *Biochem J.* (2004) 378: 959–966.

[35] Sakurai Yasuhiro, Sanuki Hodaka, Komatsu-Watanabe Rushiru, Ideguchi Tomoko, Yanagi Naoki, Kawai Kiyoshi, Kanaori Kenji, Tajima Kunihiko. Kinetic investigation of reaction of ascorbate and hydroxyl radical adduct of DMPO (5,5-dimethyl-1-pyrroline N-oxide) studied by stopped-flow ESR. *Chemistry Letters* (2008) 37, 1270-1271.

[36] Balavoine GGA, Geletii YV. Peroxynitrite Scavenging by Different Antioxidants. Part I: Convenient Assay. *Nitric oxide: Biology and Chemistry* (1999) 3, 40–54.

[37] Suzuki H, Iijima K, Scobie G, Fyfe V, McColl KE. Nitrate and nitrosative chemistry within Barrett's oesophagus during acid reflux. *Gut* (2005) 54, 1527–1535.

[38] Teh-Min Hu, Yu-Jen Chen. Nitrosation-modulating effect of ascorbate in a model dynamic system of coexisting nitric oxide and superoxide. *Free Radical Research* (2010) 44, 552–562.

[39] Sakurai Yasuhiro, Sanuki Hodaka, Komatsu-Watanabe Rushiru, Ideguchi Tomoko, Yanagi Naoki, Kawai Kiyoshi, Kanaori Kenji, Tajima Kunihiko. Kinetic investigation of reaction of ascorbate and hydroxyl radical adduct of DMPO (5, 5-dimethyl-1-pyrroline N-oxide) studied by stopped-flow ESR. *Chemistry Letters* (2008) 37, 1270–1271.

[40] Denicola A, Radi R. Peroxynitrite and drug-dependent toxicity. *Toxicology* (2005) 208, 273–288.

[41] Rubbo H, Trostchansky A, Valerie B. O'Donnell VB. Peroxynitrite-mediated lipid oxidation and nitration: Mechanisms and consequences. *Archives of Biochemistry and Biophysics* (2009) 484, 167–172.

[42] Storkey C, Pattison DI, Ignasiak MT, Schiesser CH, Davies MJ. Kinetics of reaction of peroxynitrite with selenium- and sulfur-containing compounds: Absolute rate constants and assessment of biological significance. *Free radical biology & medicine* (2015), 89, 1049–56.

[43] Szabó C, Ohshima H. DNA damage induced by peroxynitrite: Subsequent biological effects. *Nitric Oxide*. 1997 Oct;1(5):373–385.

[44] Pacher P, Beckman JS, Liaudet L. Nitric oxide and peroxynitrite in health and disease. *Physiol Rev*. (2007), 87, 315–424.

[45] Jarem DA, Wilson NR, Delaney S. Structure-Dependent DNA Damage and Repair in a Trinucleotide Repeat Sequence. *Biochemistry* (2009) *48,* 6655–6663.

[46] Niles JC, Wishnok JS, Tannenbaum SR. Peroxynitrite-induced oxidation and nitration products of guanine and 8-oxoguanine: structures and mechanisms of product formation. *Nitric Oxide.* (2006) 14, 109–121.

[47] Shafirovich V, Dourandin A, Huang WD, Geacintov NE. The carbonate radical is a site-selective oxidizing agent of guanine in double-stranded oligonucleotides. *J. Biol. Chem.* (2001) 276, 24621–24626.

[48] Steenken S, Jovanovic SV, Bietti M, Bernhard K. The trap depth (in DNA) of 8-oxo-7,8-dihydro-2′-deoxyguanosine as derived from electron-transfer equilibria in aqueous solution. *J. Am. Chem. Soc.* (2000) *122*, 2373–2374.

[49] Fang J, Nakamura T, Cho DH, Zezong Gu Z, Lipton SA. S-nitrosylation of peroxiredoxin 2 promotes oxidative stress-induced neuronal cell death in Parkinson’s disease. *PNAS* (2007) 104, 18742–18747.

[50] Jianguo Fang, Tomohiro Nakamura, Dong-Hyung Cho, Zezong Gu, Lipton SA. S-nitrosylation of peroxiredoxin 2 promotes oxidative stress-induced neuronal cell death in Parkinson’s disease. *PNAS* (2007) 104, 18742–18747.

[51] Lancaster JR, Jr. How are nitrosothiols formed de novo in vivo? *Archives of Biochemistry and Biophysics* (2017) 617, 137–144.

[52] Nakamura T, Lipton SA. Emerging role of protein-protein transnitrosylation in cell signaling pathways. *Antioxidants Redox Signal.* (2013) 18, 239–249.

[53] Rizza S, Montagna C, Di Giacomo G, Cirotti C, Filomeni G. S-Nitrosation and Ubiquitin-Proteasome System Interplay in Neuromuscular Disorders. *International Journal of Cell Biology* (2014) 2014, 428764, 10 pages.

[54] Pawloski JR, Hess DT, Stamler JS. Impaired vasodilation by red blood cells in sickle cell disease. *Proc. Nat. Acad. Sci. USA* (2005) 102, 2531–2536.

[55] Gebicka L, Didik J. Oxidative stress induced by peroxynitrite. *Postepy Biochem.* (2010) 56, 103–106.

[56] Krishna P. Bhabak, Kandhan Satheeshkumar, Subramaniam Jayavelu Govindasamy Mugesh.Inhibition of peroxynitrite- and peroxidase-mediated proteintyrosine nitration by imidazole-based thiourea and selenourea derivatives. *Org. Biomol. Chem.* (2011) 9, 7343–7350.

[57] Vatassery GT, Smith WE, Quach HT. Alpha-tocopherol in rat brain subcellular fractions is oxidized rapidly during incubations with low concentrations of peroxynitrite. *Nutr.* (1998) 128, 152–715.

[58] Kikugawa K, Hiramoto K, Ohkawa T. Effects of oxygen on the reactivity of nitrogen oxide species including peroxynitrite. *Biol. Pharm. Bull.* (2004) 27, 17–23.

[59] Yu, Jiao; Yao, Haidong; Gao, Xuejiao; Zhang, Ziwei; Wang, Jiu-Feng; Xu, Shi-Wen. The Role of Nitric Oxide and Oxidative Stress in Intestinal Damage Induced by Selenium Deficiency in Chickens. *Biological Trace Element Research* (2015) 163, 144–153.

[60] McLean S, Bowman LA, Poole RK. Peroxynitrite stress is affected by flavohaemoglobin, Hmp, derived oxidative stress in *Salmonella Typhimurium* and is relieved by nitric oxide. *Microbiology* (2010) 3556–3565.

[61] Subelzu N, Bartesaghi S, de Bem A, Radi R. Oxidative inactivation of nitric oxide and peroxynitrite formation in the vasculature. *Pictures ACS Symposium Series. Oxidative Stress* (2015) 2, 91–145.

[62] Fang J, Nakamura T, Cho D-H, Gu Z, Lipton SA. S-nitrosylation of peroxiredoxin 2 promotes oxidative stress-induced neuronal cell death in Parkinson's disease. *Proc. Nat. Acad. Sci. USA* (2007) 104, 18742–18747.

[63] Uehara, Takashi; Nakamura, Tomohiro; Yao, Dongdong; Shi, Zhong-Qing; Gu, Zezong; Ma, Yuliang; Masliah, Eliezer; Nomura, Yasuyuki; Lipton, Stuart A S-Nitrosylated protein-disulphide isomerase links protein misfolding to neurodegeneration *Nature (London, United Kingdom)* (2006),441, (7092), 513–551.

[64] Anand P, Stamler JS. Enzymatic mechanisms regulating protein S-nitrosylation: implications in health and disease. *J.Mol.Med. Berl.,* (2012) 90, 233–244.

[65] Kowluru RA. *Diabetic retinopathy, oxidative stress and antioxidants.* (2005) 3, 209–218.

[66] Espey G, Miranda KM, Wink DA, Colton CA, Pluta M, Hewett SJ. Nitric oxide and the NMDA receptor in ischemia and reperfusion injury: is NO protective or injurious? *Oxidative Stress and Disease (Free Radicals in Brain Pathophysiology)* (2000), 5523–5539.

[67] Dilek F, Ozceker D, Ozkaya E, Guler EM, Yazici M, Tamay Z, Kocyigit A, Guler N. Elevated Nitrosative Stress in Children with Chronic Spontaneous Urticaria. *International Archives of Allergy and Immunology* (2017) 172, 33–39.

[68] Caruso G, Fresta CG, Martinez-Becerra F, Antonio L, Johnson RT, de Campos RPS, Siegel JM, Wijesinghe MB, Lazzarino G, Lunte SM. Carnosine modulates nitric oxide in stimulated murine RAW 264.7 macrophages. *Mol. Cell. Biochem.* (2017) 431, 197–210.

[69] Retamal MA, Cortés CJ, Reuss L, Bennett MV, Sáez JC. RS-nitrosylation and permeation through connexin 43 hemichannels in astrocytes: Induction by oxidant stress and reversal by reducing agents. *Proc. Nat. Acad. Sci. USA* (2006) 103, 4475–4480.

[70] Lundberg JO, Weitzberg E, Gladwin MT. The nitrate-nitrite-nitric oxide pathway in physiology and therapeutics. *Nat Rev Drug Discov.* (2008) 7:156-167.

[71] Vatassery GT, SantaCruz KS, DeMaster EG, Quach HT, Smith WE. Oxidative stress and inhibition of oxidative phosphorylation induced by peroxynitrite and nitrite in rat brain subcellular fractions. *Neurochem. Int.* (2004) 45, 963–970.

[72] Ansari FA, Ali SN, Mahmood R. Sodium nitrite-induced oxidative stress causes membrane damage, protein oxidation, lipid peroxidation and alters major metabolic pathways in human erythrocytes. *Toxicology In Vitro* (2015) 29, 1878–1886.

[73] Altshullern AP. Thermodynamic Considerations in the Interactions of Nitrogen Oxides and Oxy-Acids in the Atmosphere. *Journal of the Air Pollution Control Association* (2012) 6, 97–100.

[74] Drew B, Leeuwenburgh C. Aging and the role of reactive nitrogen species. *Annals of the New York Academy Science* (2002) 959, 66–81.

[75] Mirowsky JE, Dailey LA, Devlin RB. Differential expression of pro-inflammatory and oxidative stress mediators induced by nitrogen dioxide and ozone in primary human bronchial epithelial cells. *Inhal Toxicol.* (2016) 28, 374–382.

[76] Persinger RL, Poynter ME, Ckless K, Janssen-Heininger YM. Molecular mechanisms of nitrogen dioxide induced epithelial injury in the lung. *Toxicol. In Vitro.* (2015) 29, 1878–1886.

[77] Kaji N, Horiguchi K, Iino S, Nakayama S, Ohwada T, Otani Y, Firman; Murata T, Sanders KM, Ozaki H, Masatoshi H. Nitric oxide-induced oxidative stress impairs pacemaker function of murine interstitial cells of Cajal during inflammation. *Pharmacological Research* (2016), 111, 838–848.

[78] Titov VYu, Osipov AN, Kreinina MV, Vanin AF. Features of the Metabolism of Nitric Oxide in Normal State and Inflammation. *Biophysics* (2013), 58, 676–688.

[79] Sharma JN, Al-Omran A, Parvathy SS. Role of nitric oxide in inflammatory diseases. *Inflammopharmacology*. (2007) 15, 252–259.

[80] Guzik TJ, Korbut R, Adamek-Guzik T. Nitric oxide and superoxide in inflammation and immune regulation. *J Physiol. Pharmacol.* (2003) 54, 469–487.

[81] Coleman JW. Nitric oxide in immunity and inflammation. *Int. Immunopharmacol.* (2001) 1, 1397–406.

[82] Guzik TJ, Korbut R, Adamek-Guzik T. Nitric oxide and superoxide in inflammation and immune regulation. *J. Physiol. Pharmacol.* (2003)54, 469–487.

[83] Iyengar R, Stuehr DJ, Marletta MA. Macrophage synthesis of nitrite, nitrate, and N-nitrosamines: Precursors and role of the respiratory burst. *Proc; Natl. Acad. Sci. U S A* (1987) 84, 6369–6373.

[84] Leppanen T, Tuominen RK, Moilanen E Protein Kinase C and its Inhibitors in the Regulation of Inflammation: Inducible Nitric Oxide Synthase as an Example *Basic & Clinical Pharmacology &* Toxicology (2014) 114, 37–43.

[85] Blantz RC, Munger K. Role of nitric oxide in inflammatory conditions. *Nephron* (2002) 90, 373–378.

[86] Moilanen E, Vapaatalo H. Nitric oxide in inflammation and immune response. Renal and mesenteric vasoconstriction. *Ann. Med.* (1995) 27, 359–67.

[87] Cirino G, Distrutti E, and John L. Wallace JL. Nitric Oxide and Inflammation. *Inflammation & Allergy - Drug Targets,* (2006) 5, 115–119.

[88] Abramson SB. Nitric oxide in inflammation and pain associated with osteoarthritis *Arthritis Research & Therapy (*2008) 10, (Suppl 2) S2.

[89] Tripathi P[1], Tripathi P, Kashyap L, The role of nitric oxide in inflammatory reactions. *Pathogens and Disease* (2007) 51, 443–452.

[90] Lo Faro ML, Fox B, Whatmore JL, Winyard PG, Whiteman M. Hydrogen sulfide and nitric oxide interactions in inflammation. *Nitric Oxide* (2014) 41, 38–47.

[91] Grimm EA, Sikora AG, Ekmekcioglu S. Molecular Pathways: Infammation-Associated Nitric-Oxide Production as a Cancer-*Supporting Redox Mechanism and a Potential Therapeutic Target. Clin Cance.r Res.* (2013) 19, 5557–5563.

[92] Roberts RA, Smith RA, Safe S, Szabo C, Tjalkens RB, Robertson FM. Toxicological and pathophysiological roles of reactive oxygen and nitrogen species. *Toxicology* (2010) 276, 85-94.

[93] Husain SP, He P, Subleski J, Hofseth LJ, Trivers GE, Mechanic L, Hofseth AB, Bernard M, Schwank J, Giang Nguyen, Mathe LM, Djurickovic D, Haines D, Weiss J, y Back T, Gruys E, Laubach VE. Wiltrout RH, Harris CC. Nitric Oxide Is a Key Component in Inflammation-Accelerated Tumorigenesis. *Cancer Res.* (2008) 68, 7130–7136.

[94] O'Sullivan S, Medina C, Ledwidge M, Radomski ML, Gilmer JF. Nitric oxide-matrix metaloproteinase-9 interactions: Biological andpharmacological significance NO and MMP-9 interactions. *Biochimica et Biophysica Acta* (2014) 1843, 603–617.

Chapter 8

NITRIC OXIDE CONTRIBUTION TO GENETICS AND AGING

ANNOTATION

Regulation of gene expression of nitric oxide synthases is an essential process that triggers cellular reprogramming toward proliferation, differentiation, development, apoptosis, senescence, carcinogenesis, and aging, in particular. Nitric oxide significantly contributes to the regulation of transcriptional, translational, and post-translational processes. For major human pathologies, including cancer, diabetes, cardiovascular disorders, atherosclerosis, arteriosclerosis, thrombosis, Alzheimer's, Parkinson's, Huntington's disease, and neurodegenerative diseases, aging appeared to be the primary risk. The discovering biological basis of a role of NO in aging, one of the most enigmatic phenomena of Nature, is the greatest remaining challenges for science.

8.1. NITRIC OXIDE IN REGULATION OF TRANSCRIPTIONAL, TRANSLATIONAL, AND POST-TRANSLATIONAL PROCESSES

8.1.1. Introduction

Regulation of nitric oxide synthases (NOSs) at the transcriptional, translational, and post-translational level was well documented by numerous researchers [1–9] and references therein. Figure 8.1 describes some mechanisms responsible for the regulation of eNOS gene expression and translation.

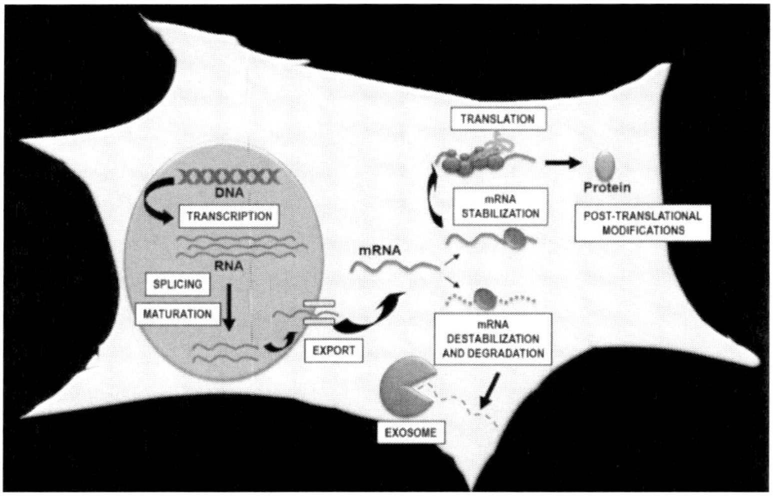

Figure 8.1. Regulation of mRNA stability and translation by RNA-binding proteins (RBPs). They mainly influence the fates of target mRNAs at the post-transcriptional levels. In the cytoplasm, stabilized mRNAs are protected from degradation leading to more protein levels. Destabilized mRNAs are driven to degradation machinery leading to lower protein levels. RBPs can also influence the abundance of mRNAs in the translation machinery [1]. With permission from Intech, 2012.

Various physiological and pathophysiological stimuli, such as lysophosphatidylcholines (LPC, lysoPC), lysophosphatidylcholine, messenger ribonucleoprotein mRNP, messenger ribonucleoprotein complex, and oxidized low-density lipoproteins (LDL), have been shown to modulate eNOS mRNA levels by targeting different stages of gene expression [2].

8.1.2. Transcription

Physiological and pathophysiological stimuli to regulate eNOS expression and their mode of regulation and factors affected on transcriptional and post-transcriptional regulation of the enzyme expression were summarized in [2]. A transcription factor (or sequence-specific DNA-binding factor) is a protein that controls the rate of transcription of genetic information from DNA to messenger RNA, and helps to regulate the expression of genes [1–3]. In normal cells, transcription is a tightly controlled process for cellular proliferation and differentiation. The cytokine-mediated regulation of iNOS gene transcription by the translational factors, interleukin 1 beta (IL-1β) and interferon gamma (IFN-γ) of the iNOS.3 were reported [3]. The pro-inflammatory cytokines 9, a small protein, that is important in cell signaling, IL-1β and (IFN)-γ, decrease functional islet β-cell mass through the increased expression of specific gene of inducible nitric oxide synthase (iNOS). Overproduction of NO, caused by dysregulated iNOS protein accumulation, induces DNA damage, impairs β-cell function, and diminishes cellular viability. Signal transducer and activator of transcription (STAT) proteins are phosphorylated by receptor associated kinases that cause activation and dimerization and finally translocate to the nucleus to work as transcription factors. IL-1β promotes the binding of serine-phosphorylated signal transducer and activator of transcription-1 (STAT1) (Ser727). Phosphorylation at Tyr701 was required for IFN-γ to potentiate the IL-1β response. The role of acetylation, methylation, and phosphorylation in the above-mentioned processes was also discussed [3].

To elucidate the role of •NO in regulation of gene expression in cancers, several mechanisms were suggested. For example, the mechanism of NO action, including regulation of NOS isoform expression and activity, was proposed [4]. As it was found, a combinatorial complex between estrogen receptor (ER)-β and eNOS can repress transcription of prognostic genes. According to [5], NO regulates the activity of the anti-apoptotic transcription factor NF-kappaB, a protein complex that controls transcription of DNA. The inhibition of NF-kappaB was reflected in the level of intracellular S-nitrosothiols. The NO influence on cell death by modulating NF-kappaB activity with the sites of cell type-specific inhibition and NO bioactivity regulating tumor necrosis factor-alpha signaling was also considered. Nitric oxide generated by the inducible isoform of nitric oxide synthase (iNOS) is involved in complex antitumoral mechanisms [6]. Modulation of expression via an activation of the transcription factors NF-kappaB and STAT-1alpha on both the transcriptional and post-transcriptional level is suggested to be the major regulation mechanism for iNOS. Therefore, activation of the iNOS promoter is an essential step for the iNOS induction in human cells.

Transcription factor Ets-1 is a breast cancer oncogene that regulates the expression of genes involved in tumor progression and metastasis. The role of NO signaling in the estrogen receptor-negative (ER-) human breast cancer cells was examined [7]. These effects were studied using either forced NOS2 expression or the use of a chemical NO-donor, diethlylenetriamine NONOate (DETANO). Significant Ets-1 transcriptional activation in ER-breast cancer cells was the result of the NOS2 overexpression and exposure to NO-donors. Studies showed that NO activated Ets-1 transcriptional activity via a mechanism that involved Ras S-nitrosylation. Ras/MEK/ERK is a chain of proteins in the cell that communicates a signal from a receptor on the surface of the cell to the DNA.

The adenovirus E1A oncogene induces an innate immune rejection of mouse and human tumors by NO [8]. Experiments indicated that E1A sensitizes cells to the tumor necrosis factor (TNF). This process can cause cell death apoptosis by repressing its activation of NF-κB-dependent, antiapoptotic defenses. In addition, an E1A-related increase in NO-induced activation of caspasc-2, an initiator of intrinsic apoptosis mediates the oncogene-induced phenotypic conversion of the cellular injury response of cells. NO-induced apoptotic response of E1A-positive cells was eliminated by blocking activation or expression of caspase-2. Some specific aspects of the role of iNOS in transcriptional processes were elicited in [9]. In particular, it was found that NF-kappaB-repressing factor (NRF) interacts with a specific negative regulatory element (NRE) to mediate transcriptional repression of certain NF-kappaB responsive genes. The latter plays a central role in mediating cytokine-stimulated human inducible nitric-oxide synthase (hiNOS) gene transcription. Hydrogen peroxide can also increase eNOS expression.

Effects of S-nitrosylation on gene transcription are [10]: (1) change subcellular localization, which may lead to either import or export out of the nucleus; (2) alteration of DNA-binding activity of certain proteins; and (3) association/dissociation of macromolecular complexes, which may result in dissociation from chromatin. It was shown that S-nitrosylation can affect gene transcription in several ways: (1) upon S-nitrosylation, proteins can change their subcellular localization, which may lead to either import in or export out of the nucleus; (2) S-nitrosylation can alter DNA-binding activity of certain proteins; and (3) SNO can lead to association/dissociation of macromolecular complexes, which may result in dissociation from chromatin.

8.1.3. Translation

A microRNA (miRNA) is a small non-coding RNA molecule that functions in RNA silencing and post-transcriptional regulation of gene expression (Figures 8.1 Regulation of miRNAs by •NO as an important process related epigenetic event was examined [11]. The role of NOS2 in lung cancer driven by the oncogenic *KRAS*^{G12D} allele, *NOS2KO* mice with *Lox-Stop-Lox (LSL)-KRAS*^{G12D} mice, in which the expression of oncogenic KRAS is controlled by a removable transcription termination STOP element, was evaluated. During the inflammatory response, positive correlation between inducible nitric oxide synthase (NOS2) upregulation and sustained NO and lung tumorigenesis was revealed.

Ras is a protein that is involved in transmitting signals within cells. Several studies shed light on key contribution of Ras in oncogenic Ras signaling [12–15]. NOS2 enhances microRNA-21 expression and corresponding proto-oncogene protein (KRAS)-induced lung carcinogenesis, while deletion of NOS2 decreases lung tumor growth and inflammatory responses initiated by oncogenic [12]. These data suggested that both KRAS and NOS2 cooperate in driving lung tumorigenesis and inflammation. Transforming growth factor β1 (TGF-β1), a 25-kDa homodimeric peptide, increased eNOS and mRNA in a time- and dose-dependent manner via augmenting transcription [13].

For continued proliferation, certain tumor cells require presence of activated Ras, an oncogene encoding the guanosine triphosphatase. Protein kinase B (PKB, AKT) which is a target for activated Ras, was proved to be a critical one to sustain tumor growth [14]. Activation of AKT promoted nitrosylation of wild-type Ras with NO, produced by eNOS. If eNOS was depleted with short hairpin RNA, participation of such cells in tumor formation was abolished. Blocking phosphorylation of the AKT substrate, endothelial nitric oxide synthase (eNOS or NOS3), inhibits tumor initiation and maintenance while eNOS enhances the nitrosylation and activation of endogenous wild-type Ras proteins [15].

8.1.4. Post-Translation

Histones are chief protein components of chromatin that package and order the DNA into nucleosomes and play an important role in the post-translational gene regulation. Histone deacetylases are enzymes that remove acetyl groups from an ε-N-acetyl lysine amino acid on a histone. Histone deacetylase 2 (HDAC2) regulates various cellular processes, such a genes cell cycle, senescence, proliferation, differentiation, development, apoptosis, and glucocorticoid function in inhibiting inflammatory response [16]. A mechanism of stimulating NO production and *S*-nitrosylation of HDAC2, that removes acetyl groups from an ε-N-acetyl lysine amino acid on a histone and causes chromatin remodeling, was evaluated [17]. This process is promoted by brain-derived neurotrophic factor. The beneficial effect of deacetylase inhibitors and NO donors in dystrophic muscles allowed to suggest a molecular link among dystrophin, NO signaling, and the histone deacetylases resulted in its by cysteine S-nitrosylation [18].

A major role of NO of the histone post-translational modifications (PTM) in the organization of chromatin structure and subsequent regulation of gene expression was stressed [19]. Nitric oxide regulates gene expression in cancers by controlling histone PTMs directly inhibiting the catalytic activity of JmjC-domain containing histone demethylases. NO exposure promoted oncogenesis via PTMs by changes at fifteen critical lysine residues on the core histones H3 and H4. It was concluded that NO functions, as an epigenetic regulator of gene expression, are mediated by changes in histone PTMs. Nitric oxide can directly inhibit the activity of the demethylase KDM3A by forming a nitrosyl-iron complex in the catalytic pocket, the Jumonji C domain Fe(II) α-ketoglutarate [20]. This NO effect resulted in a significant increase in dimethyl Lys-9 on histone 3 (H3K9me2). The mRNA of several histone demethylases and methyltransferases was also differentially regulated in response to •NO. The authors suggested three following mechanisms of histone methylation: (1) direct inhibition of Jumonji C demethylase activity, (2) reduction in iron cofactor availability, and (3) regulation of expression of methyl-modifying enzymes.

Recent studies that define miRNA's role in maintaining endothelial NO bioavailability in post-transcriptional and transcriptional regulation of endothelial NOS during hypoxia were discussed in review [21]. It was emphasized that miRNAs directly modulate *NOS3* expression or eNOS activity. Not only is eNOS activity influenced by a wide range of protein regulators, but eNOS is itself post-translationally modified by S- nitrosylation in a product feedback relationship that constrains further NO synthesis [22]. Nitric oxide synthases activity can be regulated by a number of post-translational modifications, including fatty acid acylation, substrate, and co-factor availability, degree of phosphorylation, S-nitrosylation, acetylation, and protein–protein interactions [23, 24]. Inducible nitric oxide synthase (iNOS) expression is regulated by both the transcriptional and post-transcriptional level in epithelial cells by tyrosine phosphorylation [25]. The effects of tyrosine phosphorylation on iNOS activity in a human intestinal epithelial cell line stimulated with cytokines was evaluated. It was found that 4-amino-5-(4-chlorophenyl)-7-(*t*-butyl)pyrazolo[3,4-*d*]pyrimidine, a specific inhibitor of Src tyrosine kinases, abolished the pervanadate. This inhibitor induces iNOS tyrosine phosphorylation. These findings indicated a role for iNOS tyrosine phosphorylation in the regulation of iNOS activity and the implication of Src tyrosine kinases in this pathway, processes important for oncogeneses. The ability of aspirin and acetylating analogs to activate eNOS was due to direct acetylation of eNOS protein [26]. Deacetlyation of eNOS, acetylated at lysine 609 is mediated by histone deacetylase 3 (HDAC3).

Hsp90 (heat shock protein 90) is a chaperone protein that assists other proteins to fold properly, stabilizes proteins against heat stress, aids in protein degradation, and stabilizes proteins required for tumor growth [27–32]. In post-translational activation of eNOS, the enzyme regulation can occur through several distinct molecular mechanisms, each of which acts in concert with Ca^{2+}-calmodulin (CaM) [27, 28]. These mechanisms include alterations in protein–protein interactions with caveolin-1, a protein that in humans is encoded by the CAV1 gene, the bradykinin BK B2 receptor, and heat shock protein 90 (Hsp90). Caveolin-1 directly binds to eNOS motif that is located between amino acids 350–358. BK also can stimulate an increase in eNOS activity through phosphorylation of the enzyme at three specific amino acid residues as well as through dephosphorylation at a fourth residue.

As part of their chaperone activities, hsp90 regulates a variety of signal transduction pathways [29]. Hsp90 has been shown to interact with eNOS under resting conditions. Numerous endothelial cell stimuli, including vascular endothelial growth factor (VEGF), histamine, fluid shear stress, and estrogen, promotes increased eNOS activity and NO release [30]. Hsp90 influences eNOS by several mechanisms. The protein binding to eNOS induces a conformational change in eNOS binding to the oxygenase domain of eNOS. This process influences the function of heme or other regulatory proteins such as kinases and phosphatases that can then secondarily influence eNOS function [31]. A novel estrogen signaling mechanism in the vagina suggested that estrogens regulate eNOS post-translationally in the vagina, providing a mechanism to affect NO bioavailability without changes in eNOS protein expression and involving eNOS phosphorylation and eNOS-caveolin-1 interaction [32].

8.2. Nitric Oxide and Aging

8.2.1. Introduction

Aging plays a central role in health problems, morbidity, and mortality in older people and is characterized by a progressive loss of physiological integrity, leading to impaired function and increased vulnerability to death [33–34]. For major human pathologies, including cancer, diabetes, cardiovascular disorders, atherosclerosis, arteriosclerosis, thrombosis, Alzheimer's, Parkinson's, Huntington's disease, and neurodegenerative diseases, aging appeared to be the primary risk. The following hallmarks that represent common denominators of aging in mammalian and other organisms were enumerated: genomic instability, telomere attrition, epigenetic alterations, loss of proteostasis, deregulated nutrient sensing, mitochondrial dysfunction, cellular senescence, stem cell exhaustion, and altered intercellular communication 33–35]. Numerous evidence provided up to date on a reciprocal relationship between nitroxide activity and aforementioned hallmarks.

8.2.2. Genomic Instability and Epigenetics

Genomic instability, one of the primary mechanisms that leads to organismal aging, is an increased tendency of the genome to acquire to a high frequency of mutations when various processes involved in maintaining and replicating the genome are dysfunctional. When genomic maintenance systems, such as nucleotide excision repair, are defective, genomic instability is promoted and causes accelerated aging (progeria) [36–38]. Epigenetics are stable heritable traits (or "phenotypes") that cannot be explained by changes in DNA sequence. The role of DNA damage in vascular aging, and present mechanisms by which genomic instability interferes with regulation of the vascular tone, was considered in detail [32–36]. Genomic instability, which interferes with regulation of the vascular tone, is related to DNA damage (telomeric, non-telomeric, and mitochondrial), which can be caused by reactive oxygen species (ROS) vs. endothelial nitric oxide synthase (eNOS)-cyclic guanosine monophosphate (cGMP) signaling. iNOS inhibitors and NO/RNS-scavengers significantly reduced DNA damages [36–41].

Below, the effects of NO on the genomic instability are illustrated by several examples. Endogenous or exogenous agents can cause genomic instability and telomere attrition lesions in chromosomes by stimulation a variety of DNA. Such lesions can be repaired by a variety of mechanisms. Excessive DNA damage or insufficient DNA repair favors the aging process. Epigenetic alterations in the methylation of DNA or acetylation and methylation of histones and other chromatin-associated proteins can also induce epigenetic changes that contribute to the aging process [36]. Mitochondrial function becomes perturbed by aging-associated DNA mutations, reduced mitochondriogenesis, destabilization of the electron transport chain (ETC) complexes, altered mitochondrial dynamics, or defective quality control by mitophagy. Survival signals to restore cellular homeostasis can be induced by stress signals. Defective mitochondrial function generate ROS below a certain threshold. At higher or continued levels, they can contribute to aging.

In addition, mild mitochondrial damage can induce a hormetic response (mitohormesis) that triggers adaptive compensatory processes [36].

Etiology of vascular aging is based on genomic instability as a causal factor [41]. Progressive vascular aging is a result of unrepaired lesions accumulating during life, which lead to a growing set of pathophysiological changes, either independently or in mutual interaction. Beneficial (increased Nrf2-regulated antioxidants) and detrimental (decreased IGF-1 signaling, pro-inflammatory status) effects may cause the survival response [41].

The effects of stromal oxidative stress, related to NO activity, can be laterally propagated, amplified, and are spread from cell to cell, inducing additional mutation and therefore promoting wide-spread genomic instability [39]. The efficacy of mutagenesis and carcinogenesis depends on the area of the field, the strength of NO/RNS-maintained genomic instability (SGI), and the duration of this field maintenance [40]. It was postulated that reactive nitrogen species produced via iNOS during chronic inflammation may play a key role in stimulation of DNA damages formation and activation of carcinogenesis. Molecular mechanisms of the NO contribution in DNA damage is described in Section7.2.4.

The role of genomic instability on NO-cGMP signaling was described [41]. According to the suggested scheme, genomic instability primarily leads to endothelial eNOS dysfunction in endothelial cells and to increased cGMP metabolism by calcium/calmodulin-dependent 3',5'-cyclic nucleotide phosphodiesterase 1A (PDE1A) and 1C. DNA may affect healthy cells through senescence-associated secretory phenotype/senescence messaging secretome (SASP/SMS), in which plasminogen activator inhibitor-1 (PAI-1) can play a central role. The following relationships were established: (1) genomic instability primarily leads to eNOS dysfunction in endothelial cells and to increased cGMP metabolism by PDE1A, phosphodiesterase 1A, and 1C; (2) cellular senescence caused by unrepaired DNA could affect healthy cells through SASP/SMS pathway; and (3) the affected cells might worsen vascular function through changes in eNOS-cGMP signaling. A putative role of PDE1A and 1C in atherosclerosis, arteriosclerosis, blood flow, and hypertension were discussed. Senescence, imbalanced NO vs. ROS production, inflammation, and changes in insulin signaling was found to be detrimental while autophagy, apoptosis, and stress resistance have a beneficial contribution to vascular aging. IGF-1 can has also a detrimental effect, although this needs further scrutiny [41]. Molecular effectors of genomic instability that contribute to vascular aging, and the potential remedies against, were presented in a diagram.

8.2.3. DNA Damage and Senescence (Biological Aging)

Telomeres are strings of DNA that extend every chromosome beyond its last gene. Terminal telomeres are sacrificed during every mitotic event in human cells (telomere attrition) [41–43]. Cellular oxidative and nitrogen stress accelerate telomere attrition and promotes cellular aging. Telomeres become shorter as individuals age, and may play a role in biological aging [42]. Inflammation and infectious diseases cause telomere erosion, which increases susceptibility to infection. Oxidative stress and related nitric reactive species, can damage telomeres, impair their repair mechanisms, therefore, promote cellular aging. Shortening of telomeres is suggested to be a molecular clock that triggers senescence [43]. Telomere length, its activity during aging of human endothelial cells (ECs) in culture, and the effect of NO on delay in the development of senescence were examined [43]. Experiments indicated that the addition of the NO donor *S*-nitroso-penicillamine significantly reduced EC senescence and delayed age-dependent inhibition of telomerase activity, whereas inhibition of endogenous NO synthesis had an adverse effect. Thus, NO prevents age-related downregulation of telomerase activity and delays EC senescence.

In human umbilical venous endothelial cells (HUVECs), treatment with NO donor (*Z*)-1-[2-(2-aminoethyl)-*N*-(2-aminoethyl)amino]diazen-1-ium-1,2-diolate (DETA-NO) and transfection with endothelial NO synthase (eNOS) into HUVECs decreased the number of SA-β-gal positive cells and increased telomerase activity [44]. The NOS inhibitor N^{G}-nitro-L-arginine methyl ester (L-NAME) abolished the effect of eNOS transfection. Thus, NO can prevent endothelial senescence, thereby contributing to the anti-senile action of estrogen. It was also found that the ingestion of NO-boosting substances, including L-arginine, L-citrulline, and antioxidants, delays endothelial senescence under high glucose.

Epigenetic marks, such as DNA methylation and numerous histone modifications, are emerging as important factors of the overall variation in life expectancy [32, 42–48]. Endothelial cell senescence promotes endothelial dysfunction and may contribute to the pathogenesis of age-associated vascular disorders. NO was revealed as an endogenous epigenetic regulator of gene expression and cell phenotype [46]. Nitric oxide influences key aspects of epigenetic regulation: (1) the histone post-translational modifications, (2) DNA methylation, (3) effects on microRNA levels, (4) regulation of epigenetic protein expression and enzymatic activity resulting in remodeling of the epigenetic landscape to ultimately influence gene expression, and (5) NO signaling epigenetic mechanisms. Molecular mechanisms underlying the biological functions of NO including reaction with heme proteins and regulation of protein activity via modification of thiol residues are also expected by authors. A significant number of transcriptional responses and phenotypes was observed in NO microenvironments [47]. NO signaling provides an explanation for NO-mediated gene expression changes and phenotypes. Epigenetic modifications to chromatin alter the accessibility of DNA to nuclear factors and can involve DNA methylation at cytosine residues in CpG dinucleotides or post-translational modifications to the N-terminal regions of histone tails [48]. These modifications also regulate the expression of NOS genes.

Significant contribution the nitrogen reactive species in the transcriptional, translational and post-translational processes, which directly relates to aging, was discussed in review [47]. Mechanisms of the epigenetic regulation by NO were also summarized.

8.2.4. Loss of Proteostasis

The proteome is the entire set of proteins expressed by a genome, cell, tissue, or organism at a certain time. Proteostasis controls protein synthesis, folding, trafficking, aggregation, disaggregation, and degradation and provides mechanisms for the stabilization of folded proteins. Impaired protein proteostasis can cause some aging-related diseases, such as Alzheimer's, Parkinson's, and Huntington's disease, and, eventually, aging [49, 50].

Post-translational modification of proteome, the entire set of proteins expressed by a genome, cell, tissue, or organism was described in [51]. The expression and the activity of the chaperons, which facilitate folding of polypeptides, are tightly regulated at both the transcriptional and post-translational level at organismal states of increased oxidative and nitrative stresses and, consequently, proteotoxic stress [51]. These processes relate directly or indirectly to various diseases and eventually, to aging. The decisive contribution of ROS and RNS in the regulatory mediators in signaling pathways for chaperons and, eventually, on oxidative and proteotoxic stress was discussed in detail.

8.2.5. Mitochondrial Dysfunction

As cells and organisms age, the efficacy of the respiratory chain tends to diminish, thus, increasing electron leakage and reducing ATP generation. Mitochondrial function has a profound impact on the aging process and mitochondrial dysfunction can accelerate aging in mammals [52, 53]. The mitochondrial free radical theory of aging proposed that the progressive mitochondrial dysfunction that occurs with aging results in increased production of reactive oxygen species. ROS in turn causes further mitochondrial deterioration and global cellular damage [54]. Nevertheless, ROS and RNS can be associated with aging without implying direct damage, but playing a role in mediating a stress response to age-dependent damage [55].

Nitric oxide involved in different mitochondrial signaling pathways that control respiration and apoptosis, and consequently, aging. -15). Changes in the activities of different nitric synthase isoforms lead to the formation of metabolic disorders. Reduced endothelial NOS activity and NO bioavailability are the main factors underlying the mitochondrial dysfunction and formation of metabolic disorders [56]. Decisive contribution of nitric oxide synthases, nitrosative and oxidative stresses in process of mitochondrial dysfunction, and interrelation between nitric oxide and mitochondria in the pathophysiology of metabolic syndrome in the vascular wall, adipocyte, and a hepatocyte were schematically illustrated.

Mitochondrial proteins appear to be especially sensitive to NO-dependent modification [57–59]. In transgenetic mice, elevated levels of peroxynitrite, caused by overexpression of inducible NOS, lead to cardiac enlargement, conduction effects, and, eventually, heart failure. In acute myocardiac disorders, the conversion of tyrosine to 3-nitrotyrosine (3-NT) occurs in reaction with peroxynitrite and other RNS [57]. Superoxide and nitric oxide are generated by mitochondria in skeletal muscle and can be increased by contractile activity [58]. In contrary, these species can be implicated in the loss of muscle mass and function that occurs with aging. Impairment of redox signaling occurred in muscle during aging may contribute to the age-related loss of muscle fibers.

Updated working scheme for sites of ROS/reactive nitrogen species (RNS) generation by skeletal muscle demonstrating the potential role of Nox2 and Nox4 isoforms of NADPH oxidase in generating superoxide in mitochondria and cytosol and acknowledging the lack of evidence for any release of superoxide from mitochondrial during contractile activity was designed [58]. Deregulation of the mitochondrial respiratory mechanism can cause neurodegeneration, oxidative stress, and inflammation [60]. Specifically, binding of NO to cytochrome C oxidase, the terminal acceptor in the mitochondrial electron transport chain, elicits intracellular signaling events, including the diversion of oxygen to nonrespiratory substrates and the generation of reactive oxygen species. In this condition, NO acts as triggers by which mitochondria modulate signal transduction cascades involved in the induction of cellular defense mechanisms and adaptive responses.

Proposed signaling mechanism of angiotensin/Gq and mitochondrial ROS amplification in aging and cardiovascular diseases and mitochondrial-targeted interventions and their therapeutic potential in aging were summarized in schematic diagrams. It was suggested that mitochondrial ROS and mitochondrial DNA damage in Polgm/m mice mediated by NOX are part of a vicious cycle of ROS-induced ROS release [61]. Involving NO in signaling mechanism and mitochondrial ROS amplification in aging and cardiovascular diseases were considered in details [62].

8.2.6. Miscellaneous

Stem cells are cells that have a capacity to self-renew by dividing and to develop into specialized cells [63]. During aging, stem cells lose their ability to divide and the decrease in the renewal of stem cells leads to age-related disorders. The NO activity is important for DNA damage, senescence, and other processes, and therefore this molecule finally contributes to stem cell exhaustion [64–68].

The influence of aging on eNOS uncoupling was investigated [64]. Experiments indicated that shear stress-induced release of NO in vessels of aged rats was significantly reduced and was accompanied by increased production of superoxide. Other consequences of the rats aging are: (1) increasing in the ratio of eNOS monomers to dimers and N(omega)-nitro-l-arginine methyl ester-inhibitable superoxide formation, (2) diminishing of the level of nitrotyrosine in the total protein and precipitated eNOS of aged vessels, and (3) reducing of the level of tetrahydrobiopterin (BH4) in the mesenteric arteries of aged mice as result from the decreased expressions of GTP cyclohydrolase I and sepiapterin reductase, enzymes involved in BH4 biosynthesis. The authors concluded that eNOS uncoupling and increased nitrosylation of eNOS may be important contributors of endothelial dysfunction in aged vessels.

Results of several studies provide strong evidences for role of the NO bioavailability in the aging processes [65–68]. Oxidative stress (OS) reduces the bioavailability of NO, which has been associated with hypertension, arteriosclerosis, and a reduced vasodilatory response [65]. During the aging, high levels of free radicals and the low bioavailability of NO lead to a positive feedback loop of further OS, organelle damage, poor repair, and endothelial dysfunction in aging, associated with impaired vascular function due to endothelial dysfunction and altered redox balance [66]. Regular physical activity improves NO bioavailability, the redox balance, and the plasma lipid profile. In work [67], evidences were presented that during aging, oxidative stress can trigger platelet hyperreactivity by decreasing NO bioavailability. Aging is associated with endothelial dysfunction and, therefore, related to a reduction of NO bioavailability [68]. The redox state of the NO acceptor sGC is one of determinant factor for its bioavailability and is disturbed by ROS known to be increased with age.

REFERENCES

[1] Kotb Abdelmohsen. Modulation of Gene Expression by RNA Binding Proteins: mRNA Stability and Translation. In *Biochemistry, Genetics and Molecular Biology "Binding Protein"* Kotb Abdelmohsen (ed.), Chapter 5, Intech, 2012.

[2] Searles CD. Transcriptional and posttranscriptional regulation of endothelial nitric oxide synthase expression. *Am J Physiol. Cell. Physiol.* (2006) 291, C803–C816.

[3] Burke SJ, Updegraff BL, Bellich RM, Goff MR, Lu D, Minkin SC Jr, Karlstad MD, Collier J. Regulation of iNOS gene transcription by IL-1β and IFN-γ requires a coactivator exchange mechanism. *Mol Endocrinol.* (2013) 27, 1724–1742.

[4] Re A, Aiello A, Nanni S, Grasselli A, Benvenuti V, Pantisano V, Strigari L, Colussi C, Ciccone S, Mazzetti AP, Pierconti F, Pinto F, Bassi P, M. Gallucci M, S. Sentinelli S, F. Trimarchi F, S. Bacchetti S, A. Pontecorvi A, Lo Bello M, Farsetti A. Silencing of GSTP1, a prostate cancer prognostic gene, by the estrogenreceptor-beta and endothelial nitric oxide synthase complex. *Mol. Endocrinol.* (2011) 25, 2003.

[5] Marshall HE, Stamler JS. Nitrosative stress-induced apoptosis through inhibition of NF-kappa B. *The Journal of biological chemistry* (2002) 277, 34223-34228.

[6] Pautz A, Art J, Hahn S, Nowag S, Voss C, Kleinert H. Regulation of the expression of inducible nitric oxide synthase. *Nitric Oxide* (2010) 15 23, 75-93.

[7] Switzer CH, Cheng RY, Ridnour LA, Glynn SA, Ambs S, Wink DA. Ets-1 is a transcriptional mediator of oncogenic nitric oxide signaling in estrogen receptor-negative breast cancer. *Breast Cancer Res.* (2012) 14, R125.

[8] Radke JR, Siddiqui ZK, Miura TA, Routes JM, Cook JL. E1A oncogene enhancement of caspase-2-mediated mitochondrial injury sensitizes cells to macrophage nitric oxide-induced apoptosis. *Journal of Immunology* (2008) 180 8272-8827.

[9] Feng X, Guo Z, Nourbakhsh M, Hauser H, Ganster R, Shao L, Geller DA Identification of a negative response element in the human inducible nitric-oxide synthase (hiNOS) promoter: The role of NF-κB-repressing factor (NRF) in basal repression of the hiNOS gene. *Proc Natl Acad Sci USA* (2002) 99, 14212–14217.

[10] Mengel A, Mounira Chaki M, Shekariesfahlan A, Lindermayr Ch. Effect of nitric oxide on gene transcription – *S*-nitrosylation of nuclear proteins. *Front Plant Sci.* (2013) 4, 293.

[11] Hirokazu Okayama, Motonobu Saito, Naohide Oue, Weiss JM, Stauffer J, Seiichi Takenoshita, Wiltrout RH,[3] S. Perwez Hussain C. Harris C. NOS2 enhances KRAS-induced lung carcinogenesis, inflammation and microRNA-21 expression. *International journal of cancer Journal international du cancer* (2013) 132, 9-18.

[12] Eser S, Schnieke A, Schneider G, Saur D Oncogenic KRAS signalling in pancreatic cancer. *Br J Cancer* (2014) 111, 17-22.

[13] Inoue N, Venema RC, Sayegh HS, Ohara Y, Murphy TJ, and Harrison DG. Molecular regulation of the bovine endothelial cell nitric oxide synthase by transforming growth factor-beta 1. *Arterioscler Thromb Vasc Biol* (1995)15, 1255–1261.

[14] Ray LB. Nitric Oxide-Mediated Oncogenic Loop. *Science, Science Signaling,* AAAS, Washington, DC 20005, USA Science Signaling (2008) *Vol. 1,* ec128.

[15] Lim KH, Ancrile BB, Kashatus DF, Counter CM Tumour maintenance is mediated by eNOS. *Nature* (2008) 452, 646-649.

[16] Hongwei Yao, Irfan Rahman. Role of histone deacetylase 2 in epigenetics and cellular senescence: Implications in lung inflammaging and COPD. *American Journal of Physiology* (2012), 303, L557-L566.

[17] Nott A, Watson PM, Robinson JD, Crepaldi L, Riccio A. S-Nitrosylation of histone deacetylase 2 induces chromatin remodelling in neurons. *Nature* (2008) 455, (7211) 411-415.

[18] Colussi C, Mozzetta C, Gurtner A, Illi B, Rosati J, Straino S, Ragone G, Pescatori M, Zaccagnini G, Antonini A, Minetti G, Martelli F, Piaggio G, Gallinari P,

Steinkuhler C, Clementi E, Dell'Aversana C, Altucci L, Mai A, Capogrossi MC, Puri PL, Gaetano C. HDAC2 blockade by nitric oxide and histone deacetylase inhibitors reveals a common target in Duchenne muscular dystrophy treatment. *Proceed Nat Acad. Sci. USA* (2008) 105, 19183-19187.

[19] Vasudevan D, Hickok JR, Bovee RC, Vy Pham, Mantell LL, Bahroos N, Kanabar P, Cao X-J, Mark Maienschein-Cline M, Benjamin A. Garcia BA Thomas DD. Nitric Oxide Regulates Gene Expression in Cancers by Controlling Histone Posttranslational Modifications. *Cancer Research* (2015) 75, 5299-5308.

[20]. Hickok JR, Vasudevan D, Antholine WE, Thomas DD Nitric oxide modifies global histone methylation by inhibiting Jumonji C domain-containing demethylases. *J Biol Chem* (2013) 288, 16004–16015/

[21] Kalinowski L, Janaszak-Jasiecka A, Siekierzycka A, Bartoszewska S, Woźniak M, Lejnowski D, Collawn JF, Bartoszewski R. Posttranscriptional and transcriptional regulation of endothelial nitric-oxide synthase during hypoxia: the role of microRNAs. *Cell Mol Biol Lett.* (2016) 21, 16.

[22] Dudzinski DM, Igarashi J, Greif D, Michel T. The regulation and pharmacology of endothelial nitric oxide synthase. *Annu. Rev. Pharmacol. Toxicol.* (2006) 46, 235–276.

[23] Qian J, Fulton D. Post-translational regulation of endothelial nitric oxide synthase in vascular endothelium. *Front. Physiol.* 4, PMC3861784.

[24] Sharma NM, Patel KP. Post-translational regulation of neuronal nitric oxide synthase: Implications for sympathoexcitatory states. *Opinion on Therapeutic Targets* (2017), 21, 11-22.

[25] Hausel P, Latado H, Courjault-Gautier F, Felley-BoscoE Src-mediated phosphorylation regulates subcellular distribution and activity of human inducible nitric oxide synthase. *Oncogene* (2006) 25, 198–206.

[26] Taubert D, Berkels R, Grosser N, Schroder H, Grundemann D, Schomig E. Aspirin induces nitricoxide releasefrom vascularendothe-lium: A novel mechanism of action. *Br. J. Pharmacol.* (2004) 143, 159–165.

[27] Venema RC. Post-translational mechanisms of endothelial nitric oxide synthase regulation by bradykinin. *International Immunopharmacology* (2002) 2, 1755–1762.

[28] Smarts EJ, Graf GA, McNiven MA, Sessa WC, Engelman JA, Scherer PE, Okamoto T, Lisanti MP. Caveolins, liquid-ordered domains, and signal transduction. *Mol. Cell. Biol.* (1999) 19, 7289–7304.

[29] Russell KS, Haynes MP, Caulin-Glaser T, Rosneck J, Sessa WC, Bender JR. Estrogen stimulates heat shock protein 90 binding to endothelial nitric oxide synthase in human vascular endothelial cells. Effects on calcium sensitivity and NO release. *J. Biol. Chem.* (2000) 275, 5026–5030.

[30] Pritchard KA, Ackerman AW, Gross ER, Stepp DW, Shi Y, Fontana JT, Baker JE, WC. Heat Shock Protein 90 Mediates the Balance of Nitric Oxide and Superoxide Anion from Endothelial Nitric-oxide Synthase *Biol.Chem.* 276, 17621–17624.

[31] Xu H, Shi Y, Wang J, Jones D, Weilrauch D, Ying R, Wakim B, Pritchard KA Jr. A heat shock protein90 binding domain in endothelial nitric-oxide synthase influence senzymefunction. *J. Biol.Chem.* (2007) 282, 37567–3757.

[32] Musicki B, Liu T, Strong TD, Lagoda GA, Bivalacqua TJ, Burnett AL. Post-translational regulation of endothelial nitric oxide synthase (eNOS) by estrogens in the rat vagina. *J Sex Med.* (2010) 7,1768-1777.

[33] Bautista-Nino PK, Portilla-Fernandez E, Vaughan DE, Danser AH, Jan, Roks AJ. DNA damage: A main determinant of vascular aging. *International Journal of Molecular Sciences* (2016), 17(5), 748/1-748/26.

[34] Lopez-Otin C, Blasco MA, Partridge L, Serrano M, Kroemer G. The hallmarks of aging. *Cell* (2013) 153 1194–1217.

[35] Lopez-Otın C, Blasco MA,2 Partridge L, Serrano M, Guido Kroemer. The Hallmarks of Aging. *Cell* (2013) 153, 1195 -1217.

[36] Wu H, Roks, AJ. Genomic instability and vascular aging: A focus on nucleotide excision repair. *Trends Cardiovasc. Med.* (2014) 24, 61–68.

[37] McAdam E, Haboubi HN, Forrester G, Eltahir Z, Spencer-Harty S, Davies C, Griffiths AP, Baxter JN, Jenkins GJ. Inducible nitric oxide synthase (iNOS) and nitric oxide (NO) are important mediators of reflux-induced cell signalling in esophageal cells. *Carcinogenesis* (2012) 33, 2035–2043.

[38] Whiteside TL. The tumor microenvironment and its role in promoting tumor growth. *Oncogene*. (2008) 27, 5904–5912.

[39] Martinez-Outschoorn UE, Balliet RM, Rivadeneira DB, Chiavarina B, Pavlides S, Wang C, Whitaker-Menezes D, Daumer KM, Lin Z, Witkiewicz AK, Flomenberg N, Howell A, Pes- tell RG, Knudsen ES, Sotgia F, Lisanti MP. Oxidative stress in cancer associated fibroblasts drives tumor-stroma co- evolution: A new paradigm for understanding tumor metabolism, the field effect and genomic instability in cancer cells. *Cell Cycle*. (2010) 9, 3256–327.

[40] Yakovlev VA. Nitric Oxide and Genomic Stability. *Nitric Oxide and Cancer: Pathogenesis and Therapy,* Edition: 2, Chapter: 2, Publisher: Springer, Editors: Benjamin Bonavida, 2015, pp.25-38.

[41] Bautista-Niño PK, Portilla-Fernandez E, Vaughan DE, Danser AH, Roks AJ. DNA damage: A main determinant of vascular aging. *International Journal of Molecular Sciences* (2016), 17(5), 748/1-748/26.

[42] Ilmonen P, Kotrschal A, Penn DJ Masucci MG. Telomere Attrition Due to Infection. *PLoS ONE*. (2008) 3, e2143.

[43] Vasa M, Breitschopf K, Zeiher AM, Dimmeler S. Nitric Oxide Activates Telomerase and Delays Endothelial Cell Senescence. *Circulation Research* (2000) 87, 540-542.

[44] Hayashi T1, Matsui-Hirai H, Miyazaki-Akita A, Fukatsu A, Funami J, Ding QF, Kamalanathan S, Hattori Y, Ignarro LJ, Iguchi A. Endothelial cellular senescence is inhibited by nitric oxide: Implications in atherosclerosis associated with menopause and diabetes. *PNAS* (2006)103, 17018–17023.

[45] Huidobro C, Fernandez AF, Fraga MF. Aging epigenetics: Causes and consequences. *Molecular Aspects of Medicine* (2013) 34, 765–781.

[46] Socco S, Bovee RC, Palczewski MB, Hickok JR, Thomas DD. Epigenetics: The third pillar of nitric oxide signaling. *Pharmacol. Res.* (2017) 121,52-58.

[47] Vasudevan D, Bovee RC, Thomas DD. Nitric oxide, the new architect of epigenetic landscapes. *Nitric Oxide*. (2016) 59, 54-62.

[48] Handy DE, Loscalzo J. Epigenetics and the Regulation of Nitric Oxide. In *Nitrite and Nitrate in Human Health and Disease* Part of the *Nutrition and Health* book series (NH), pp 33-52.

[49] Powers ET, Morimoto RI, Dillin A, Kelly JW, Balch WE. Biological and chemical approaches to diseases of proteostasis deficiency. *Annu. Rev. Biochem.* (2009) 78, 959–991.

[50] Christians ES, Benjamin IJ. Proteostasis and REDOX state in the heart. *Am. J. Physiol. Heart Circ. Physiol.* (2012) 302, H24–H37.

[51] Niforou K, Cheimonidou C, Trougakos IP. Molecular chaperones and proteostasis regulation during redox imbalance. *Redox Biology* (2014) 2 323–332.

[52] Vermulst M, Wanagat J, Kujoth GC, Bielas JH, Rabinovitch PS, Prolla TA, Loeb LA. DNA deletions and clonal mutations drive premature aging in mitochondrial mutator mice. *Nat. Genet.* (2008) 40, 392–394.

[53] Green DR, Galluzzi L, Kroemer G. Mitochondria and the autophagy- inflammation-cell death axis in organismal aging. *Science* (2011) 333, 1109– 1112.

[54] Harman D. *The free radical theory of aging: Effect of age on serum copper levels.* (1965) 20, 151–153.

[55] Hekimi S, Lapointe J, Wen Y. Taking a "good" look at free radicals in the aging process. *Trends Cell Biol.* (2011) 21, 569–576.

[56] Litvinova L. Atochin DN, Fattakhov N, Vasilenko M, Zatolokin P, Kirienkova E. Nitric oxide and mitochondriain metabolic syndrome. *Fronties in Physiology* (2015) 6, 1-10.

[57] Kanski J, Behring A, Pelling J, Schoneich C. Proteomic identification of 3-nitrotyrosine-containing rat cardiac proteins: Effect of biological aging. *Am J Physiol Heart Circ Physiol.* 2004;288:H371–H381.

[58] Jackson MJ. Redox regulation of muscle adaptations to contractile activity and aging. *J Appl. Physiol.* (2015) 119: 163–171.

[59] Litvinova L, Atochin D.N, Fattakhov N, Vasilenko M, Zatolokin P, Kirienkova E. Nitric oxide and mitochondria in metabolic syndrome. *Front Physiol*. 2015; 6: 20.

[60] Erusalimsky JD, Moncada S. Nitric Oxide and Mitochondrial Signaling. From Physiology to Pathophysiology. *Arteriosclerosis, Thrombosis, and Vascular Biology* (2007) 27, 2524-2531.

[61] Durik M. et al., Nucleotide Excision DNA Repair Is Associated with Age-Related Vascular Dysfunction. *Circulation* (2012) 126, 468–478.

[62] Dai D-F, Rabinovitch PS, Ungvari Z. Mitochondria and Cardiovascular Aging. *Circulation Research* (2012) 110, 1109–1124.

[63] Oh J, Lee YD, Wagers AJ. Stem cell aging: Mechanisms, regulators and therapeutic opportunities. *Nat Med*. (2014) 20: 870–880. (2014) 20, 870-880.

[64] Yang YM, Huang A, Kaley G, Sun D. Enos uncoupling and endothelial dysfunction in aged vessels. *Am J Physiol Heart Circ Physiol.* (2009); 297: H1829–H1836. (2009) 297, H1829–H1836.

[65] Diaz M, Degens H, Vanhees L, Austin C, Azzawi M, The effects of resveratrol on aging vessels *Experimental Gerontology* (2016) 85, 41-47.

[66] Gliemann L, Nyberg M, Hellsten Y. Effects of exercise training and resveratrol on vascular health in aging, *Free Radical Biology & Medicine* (2016), 98, 165-176.

[67] Fuentes E, Palomo I. Role of oxidative stress on platelet hyperreactivity during aging. *Life Sciences* (2016), 148, 17-2.3.

[68] Takashi Shimosato, Masashi Tawa, Hirotaka Iwasaki, Takeshi Imamura, Tomio Okamura. Aging does not affect soluble guanylate cyclase redox state in mouse aortas. *Physiol. Rep.* (2016) 4, e12816.

Chapter 9

NITRIC OXIDE AND CANCER

ANNOTATION

Cancer is a group of diseases in lung, stomach, intestines, prostate, a female breast, spleen, kidneys, liver, skin, and other organs. Abnormal cell growth with the potential to invade or spread to other parts of body are typical feature of each member of that family. Nitric oxide produced by NO synthases (NOS) modulates different cancer-related events including angiogenesis, apoptosis, cell cycle, damage of DNA and DNA repair enzymes, transcription, invasion, oncogene expression, and metastasis. The exceptional complexity and diversity of the cancer biochemical processes in combination with a variety of reactions of NO are great challenges for researchers and practitioners in this area so important to mankind.

9.1. INTRODUCTION

Nitric oxide either facilitates cancer-promoting characters or acts as an anti-cancer agent. In some instances, NO· or nitric oxide synthase (NOS) levels correlate with tumor suppression, and in other cases they are related to tumor progression and metastasis. There are wide range of publications on NO in basic biochemical and physiological processes involving cancer [1–13] and references therein. Different cancer-related events modulated by NO, production of NO in tumors, and mechanisms of action of NO on tumor cells are shown in Figure 9.1 and described in detail in the corresponding caption.

Tumor cells often express inducible NO synthase (iNOS) and in some cases endothelial NOS (eNOS) and neuronal NOS (nNOS). The following tumor vascular endothelial cells predominantly express eNOS. The following processes provide the key role of NO in tumor progression [7]: (1) tumor-associated stromal fibroblasts and immune cells express iNOS;

(2) inflammatory cytokines, interferon-γ (IFNγ), interleukin 1β (IL1β), and tumor necrosis factor-α (TNFα) induce iNOS expression through nuclear factor κB (NFκB); (3) hypoxia induces iNOS through hypoxia-inducible factor 1α (HIF1α); (4) angiogenic factors such as vascular endothelial growth factor (VEGF), sphingosine-1-phosphate (S1P), angiopoietins, sex hormones, and shear stress activate eNOS in vascular endothelial cells through adenylate cyclase (AC)–protein kinase A (PKA), phosphoinositide 3-kinase (PI3K)–Akt, phospholipase Cγ (PLCγ)–diacylglycerol (DAG)–protein kinase Cα (PKCα) and PLCγ–cytosolic calcium (Ca2+)–calmodulin (CaM) pathways; and (5) metabolic stress also activates eNOS through AMP kinase (AMPK). NO is produced by all these different sources in tumors [7].

Depending on the dose and duration of NO exposure and on cellular sensitivity to NO, this molecule can induce both tumor progression and metastasis, and tumor regression and inhibition of metastasis [7]. NO can promote tumor progression and metastasis by two ways: direct induction of tumor-cell proliferation, migration and invasion, and indirectly through the expression of angiogenic and lymphangiogenic factors in tumor cells. Tumor regression and inhibition of metastasis can be caused by the cytotoxic effects of NO that are typically induced by high doses promote DNA damage, gene mutation, and tumor-cell death. Gene mutation and/or transformation together with cell death of wild-type cells could contribute to clonal selection of adapted cells and the acquisition of apoptosis resistance, and therefore promote tumor progression [7].

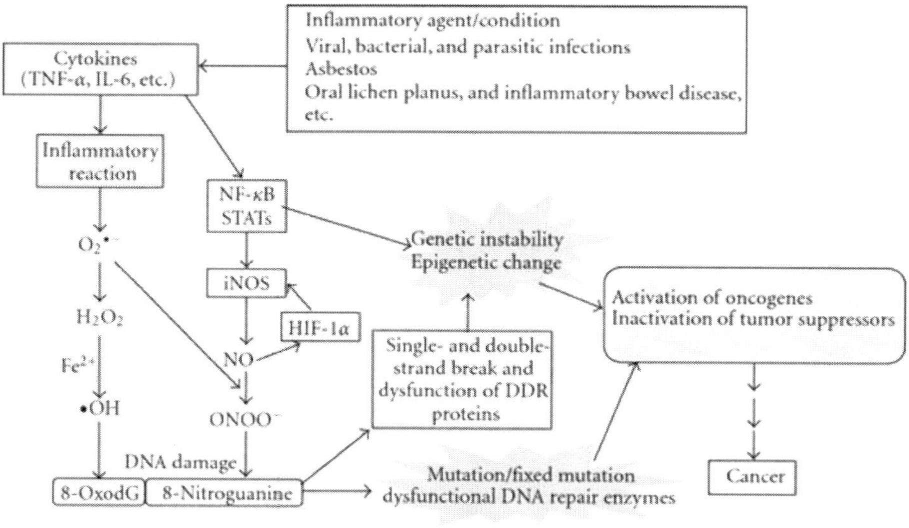

Figure 9.1. Proposed roles of nitrative and oxidative DNA damage in inflammation-related carcinogenesis. NF-κB is the transcription factor, STATs regulate the expression of a variety of genes, HIF-1 is the transcriptional regulator hypoxia-inducible factor 1 DDR proteins belong to discoid in domain receptor family [see details in [35]. With permission from Hindawi, 2012.

Precise cellular responses are differentially regulated by specific NO concentration [3, 11]. At lower concentrations, NO promotes in angiogenesis, which stimulates tumor progression, for example, by giving blood flow access to the tumor and subsequently results in cell proliferation. In contrary, higher levels of NO tend to be cytotoxic to cancer cells. Another significant aspect of NO effects in organs in real condition is temporal and space distribution.

Nitric oxide can directly affect metalloenzymes functional activity by binding NO to haem prosthetic group, which blocks its chemical reaction. Indirect influence of NO on processes in organs infected by a cancer relates to formation of products of reactions of NO with oxygen and superoxide. In the modulation of carcinogenesis, peroxinitrite, nitric dioxide and other radicals attack proteins, enzymes, nucleic acids, membranes, etc. (chapter 4). Nitrosation of thiol and tyrosine protein groups has been proved to also be an important factor of cell signaling.

9.2. Nitric Oxide and Apoptosis Cancer

Apoptosis is process of programmed cell death that occurs in multicellular organisms. A defect in apoptosis can cause cancer or autoimmunity, while enhanced apoptosis may cause degenerative diseases. Thus, the apoptotic signals contribute to safeguarding the genomic integrity, while defective apoptosis may promote carcinogenesis [12]. Apoptosis is closely associated with numerous fundamental processes, which in turn can be exposed to NO action [10–13].

High levels of extracellular NO can induce apoptosis by direct membrane damage, inhibition of ribonucleotide reductase, and inhibition of cellular ATP generation by mitochondrial electron transport enzymes aconitase and mitochondrial glyceraldehydes-3-phosphate dehydrogenase [12]. Apoptosis can be induced by NO through *S*-nitrosylation of NF-κB, glyceraldehydes-3-phosphate dehydrogenase, Fas receptor, and Bcl-2 family of regulator proteins (Bcl-2). The signals of carcinogenesis modulate the central control points of the apoptotic pathways, including inhibitor of apoptosis (IAP) proteins and FLICE-inhibitory protein (c-FLIP). The noncancerous cells have a DNA repair machinery, while defects of a DNA normal repair machinery can trigger cell suicide.

Nitric oxide affects cellular decisions of life and death either by proapoptotic or antiapoptotic pathways [13]. Proapoptotic pathways of NO involves the tumor suppressor p53 as a target for NO [9]. Resistance of tumor cells to apoptosis is provided by the expression of antiapoptotic proteins such as (Bcl-2) or by the downregulation or mutation of proapoptotic proteins such as Bcl-2-associated X protein (BAX). In the cells, p53 is trapped in the nucleus, is phosphorylated, ubiquitinated, and transcriptionally active. Mutated p53 causes enhanced iNOS expression, and therefore activation of the reactive nitrogen oxide species system.

Antiapoptotic actions of NO are numerous, ranging from an immediate interference with proapoptotic signaling cascades to long-lasting effects based on expression of cell protective proteins [14]. Reactive nitrogen intermediates (RNI) may promote reversible S-nitrosylation/-nitrosation (-SNO) of pro caspases with denitrosation predominating under cell. Special interest is the ability of NO-redox species to block a caspase-3-like group (caspase-3 and -7), caspase-2, and caspase-9, central components in a cascade triggered by NO by thiol-nitrosation Apoptosis can be also induced by NO through *S* -nitrosylation of NF-κB, glyceraldehydes-3-phosphate dehydrogenase, Fas receptor, and Bcl-2'. Intramolecular oxidation (-S-S-) or mixed disulfide formation (-S-SG), cytochrome c, released from injured mitochondria, may contribute in the process. NO can also inhibit apoptosis via cell death protective protein expression, radical–radical interferences, and S-nitrosation.

Caspases and mitogen-associated protein kinases may modulate cytochrome c release through their effects on the Bcl-2 family of proteins [13]. The cytochrome c release from mitochondria is an important mechanism for the activation of caspase-3 and the initiation of cell death.

RNI may promote reversible S-nitrosylation/-nitrosation (-SNO) of pro caspases with denitrosation predominating under cell, that is, Fas-activation. RNI may indirectly block processing of caspases at the DISC or the apoptosome, which attenuates the procaspases – active caspase conversion. Once an active enzyme is produced, it is subjected to (poly)-S-nitrosation, intramolecular oxidation (-S-S-), or mixed disulfide formation (-S-SG), all of which may attenuate caspase activity and thereby block substrate cleavage [14].

Photodynamic therapy (PDT), involving light and a photosensitizing chemical substance in conjunction with molecular oxygen in exited singlet, is popularly used to elicit cell death. PDT-mediated apoptosis in carcinoma cells may be modulated by nitric oxide (NO) in addition to oxidative stress [15]. To investigate the generation of nitric oxide (NO) and its role on cancer cell death in human hepatocellular carcinoma (HepG2), induced by PDT human hepatocellular carcinoma (HepG2) cells, Hypocrellin B (HB), a natural perylenequinone pigment, as a photosensitizer was employed [16]. Exposure of the cells to HB/light resulted in inducible nitric oxide synthase (iNOS) activation and significant increase in NO generation. HB/light-induced caspase-3, -9 activation, and apoptosis were markedly enhanced under action of NO donor monomethyl-L-arginine (L-NMMA) and an NO scavenger 2-(4-carboxyphenyl)-4, 4, 5, 5-tetramethyl-imidazoline-1-oxyl-3-oxide (cPTIO). Incubation of the cells with an exogenous NO donor sodium nitroprusside rescued cells from HB/light-induced apoptosis. The NO-releasing ability of a new series of NO-donating derivatives of evodiamine was detected in BGC-823, Bel-7402, and L-02 cells [17]. The percentage of apoptotic cells after 72 h treatment, were found to be 19%–52% at different cells, reagents, and concentrations.

The induction of iNOS in target cells by immune cytokines and resulting in the sensitization of resistant tumor cells to death ligands-induced apoptosis was evaluated [18].

Experiments showed endogenous/exogenous mediation of immune sensitizing effect, inhibition of NF-κB activity, and downstream of its anti-apoptotic gene targets. In tumor cells, a dysregulated pro-survival/anti-apoptotic loop consisting of NF-κB/Snail/YY1/RKIP/PTEN was identified. Modification of these species by NO was responsible: (1) for the reversal of chemo and immune resistance, (2) sensitization to apoptotic mechanisms by cytotoxic agents, (3) the inhibition of NF-κB activity via S-nitrosylation, and (4)NF-κB target gene. Transcription repression, epithelial-mesenchymal transition, induction of RKIP and PTEN, and modification of each gene product modified by NO in the loop, involved in chemo-immunosensitization, were also revealed in the experiments.

9.3. Nitric Oxide and Angiogenesis

Angiogenesis, the physiological process through which new blood vessels form from pre-existing vessels, is a key mediator for tumor progression. Tumors induce angiogenesis by secreting various growth factors. Unlike normal blood vessels, tumor blood vessels are dilated with an irregular shape [6, 7, 19–21]. Highlighted roles of iNOS and eNOS enzymes in the development of angiogenesis in cancer have been evaluated [22].

Understanding of the role of NO in tumor progression, especially in relation to angiogenesis and vascular functions, was summarized in [7]. Nitric oxide that is predominantly synthesized by eNOS in vascular endothelial cells promotes angiogenesis directly and functions both upstream and downstream of angiogenic stimuli [7]. NO can promote or inhibit angiogenesis, depending on concentration and distribution of NO, which acts as a downstream mediator of multiple angiogenic effectors involving multiple pathways [3, 23]. Angiogenic factors such as vascular endothelial growth factor (VEGF), sphingosine-1-phosphate (S1P), angiopoietins, sex hormones, and shear stress activate eNOS in vascular endothelial cells through adenylate cyclase (AC)–protein kinase A (PKA), phosphoinositide 3-kinase (PI3K)–Akt, phospholipase C (PLC)– diacylglycerol (DAG)–protein kinase C (PKC), and PLC–cytosolic calcium (Ca^{2+})–calmodulin (CaM) pathways [7]. Metabolic stress also activates eNOS through AMP kinase (AMPK). NO is produced by all these different sources in tumors [7, 23]. Angiogenic stimuli induce NO production by eNOS in endothelial cells, in which endogenous and/or exogenous NO triggers multiple signaling pathways through S-nitrosylation, and/or cyclic guanylyl mono phosphate cGMP S-nitrosothiol (SNO) can be formed by NO [24]. The processes can run through three separate pathways: (1) a direct reaction that is followed by electron abstraction, (2) through auto-oxidation of NO (NOx), or (3) through catalysis at metal centers (Me-NO) [24].

Using NOS2 inhibitors, experiments *in vivo* indicated that NO effect on cancer biochemical processes including angiogenesis [7]. For example, employing an inhibitor of iNOS (N^6-(1-iminoethyl)-L-lysine-dihydrochloride (L-nil), it was found that this enzyme and NO enhance carcinogenesis and tumor progression in human melanoma *in vivo,* stimulate angiogenesis, support tumor growth, promote metastasis, and inhibit T cell–dependent immune responses [25]. The tumor vascular effects of inhibition of non-small-cell lung, prostate, and cervical cancers by the nitric oxide synthase inhibitor and N-nitro-L-arginine (L-NNA) have been reported [26]. Thus, NO has been implicated in tumor angiogenesis and in the maintaining of vasodilator tone of tumor blood vessels.

9.4. The Soluble Guanylate Cyclase–Cyclic Guanylate Monophosphate Pathway in Cancer

The enzyme soluble guanylate cyclase (sGC) is a heterodimer composed of one alpha and one heme-binding beta subunits (chapter 6). Binding of NO to the sixth coordinating position of the haem iron of sGC induces a conformational change that results in 200-fold activation of the enzyme and increased synthesis of cGMP3. The (sGC)–cGMP pathway has an important role in NO-mediated key processes directly or nondirectly related to cancer [7, 9, 27, 28].

The biochemical role of sGC stimulated by NO is illustrated in Figures 9.2 and 9.3 (See also chapter 6).

NO-cGMP pathway exhibits a diverse role in cancer [29]. A differential expression of nitric oxide signaling components in ES cells an enhanced differentiation of cells into myocardial cells with NO donors and sGC activators was well established. A differential expression of the sGC subunits, NOS-1 and PKG mRNA, and protein levels was also evaluated in various human cancer models. Specifically, robust levels of sGC β_1 were observed in OVCAR-3 (ovarian) and MDA-MB-468 (breast) cancer cells, which correlated well with the sGC activity. In addition, a marked increase in cGMP levels upon exposure to the combination of a NO donor and a sGC activator was described. The authors concluded that the effects of activators/inhibitors of NO-sGC-cGMP in tumor cell proliferation is mediated by both cGMP-dependent and independent mechanisms.

The Notch signaling pathway, which is a signaling system of notch receptors, each of them is a single-pass transmembrane receptor protein, plays an oncogenic role in ovarian cancer [30] Three ovarian cancer cell lines express a higher level of GUCY1B3 (the β subunit of sGC) compared to non-cancerous immortalized ovarian surface epithelial (IOSE) cell lines. In IOSE cells, forced activation of Notch3 increases the expression of guanylate cyclase soluble subunit beta-1 (GUCY1B3), NO-induced cGMP production, and the expression of cGMP-dependent protein kinase (PKG). Enhancing NO^- and cGMP leads to phosphorylation of vasodilator-stimulated phosphoprotein, a direct PKG substrate protein, while inhibition of sGC decreases growth of ovarian cancer cells Thus, Notch is a positive regulator of NO/sGC signaling in IOSE and ovarian cancer cells.

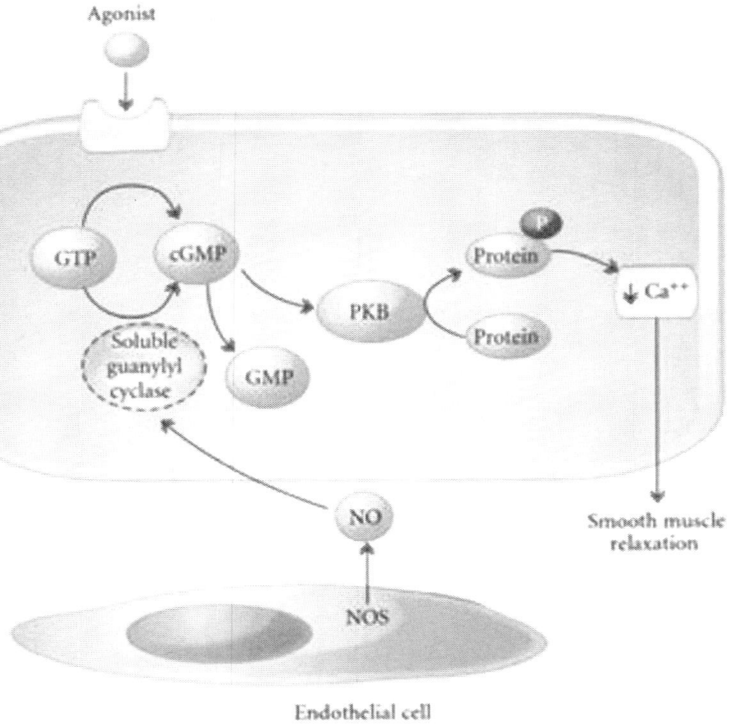

Figure 9.2. Simplified role of NO (nitric oxide) stimulating soluble guanylyl cyclase smooth muscle relaxation. PKB (protein kinase B), NOS (nitric oxide synthase) And guanosine 3′,5′-monophosphate (cGMP) derived from guanosine triphosphate (GTP) [9]. With permission from Hindawi, 2012.

9.5. DNA Damage in Cancer

DNA damage occurs as a break in a strand of DNA, a base missing from the backbone of DNA, or a chemically changed base. Reactive oxygen species, reactive nitrogen species, reactive carbonyl species, lipid peroxidation products and alkylating agents, and hydrolysis cleaving chemical bonds in DNA can be involved in the process [31–35].

Figure 9.1 is an illustration of the scheme and significance of the DNA repair processes and proposed roles of nitrative and oxidative DNA damage in inflammation-related carcinogenesis, respectively [36]. Extensive oxidative damage is observed in cellular DNA in activated macrophages. In this condition, N_2O_3 or NO^+ caused deamination products such as xanthine and oxanosine as result of nitrosylation of DNA bases. DNA damage induced by oxidizing agents, such as NO, is repaired through the base excision repair pathway.

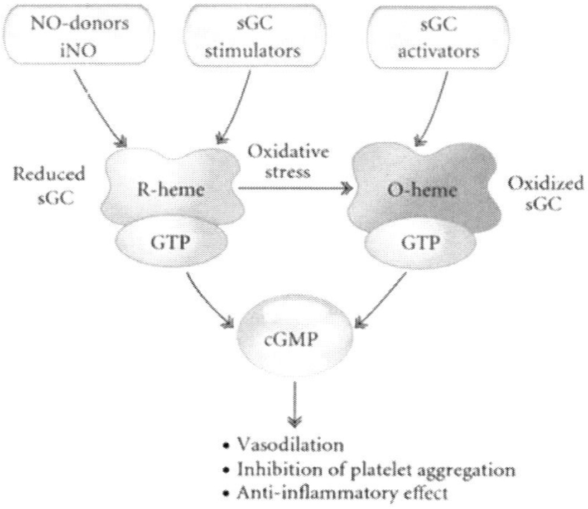

Figure 9.3. Role of NO (nitric oxide), inhaled NO, and soluble guanylyl cyclase stimulators in stimulating the reduced heme of sGC and the role of sGC activators in stimulated oxidized sGC to stimulate cGMP leading to vasodilation, inhibition of platelet aggregation, and an anti-inflammatory effect [9]. With permission from Hindawi, 2012.

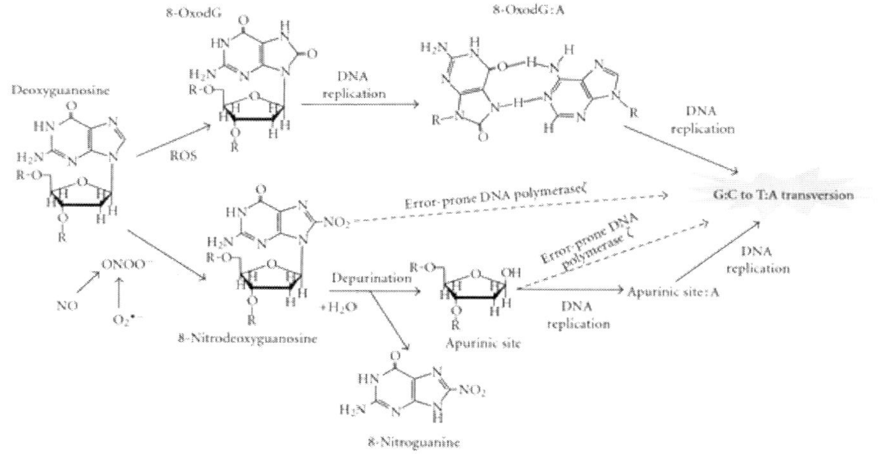

Figure 9.4. Proposed mechanism of point mutation induced by 8-nitroguanine and 8-oxodG through induction of the G : C → T : A transversion [31]. With permission from Hindawi, 2012.

The effects of high concentrations of NO generated by decomposition of dipropylenetriamine-NONOate (DPTA-NONOate) in cultured mammalian fibroblasts were examined [37]. The effects include: (1) the steady-state levels of oxidative DNA base modifications; (2) the susceptibility of the cells to the induction of additional DNA damage and micronuclei by H_2O_2; and (3) the repair kinetics of various types of DNA modifications [37]. In the iNOS-transfected fibroblasts, the endogenous generation of NO was associated with a protection from DNA single-strand break formation and micronuclei induction by H_2O_2. It was also shown that NO can protect from DNA damage by H_2O_2 and selectively inhibits the repair of oxidative DNA base modifications. Nitric oxide from pro-inflammatory cytokines induces DNA damage and ultimately β-cell death and also activates a functional repair process [38]. The mechanisms activated by NO that facilitate the repair of damaged β-cell DNA were evaluated. The following finding were presented: (1) c-Jun N-terminal kinase (JNK) attenuates the repair of nitric oxide-induced DNA damage; (2) expression of growth arrest and DNA damage (GADD) 45α, a protein that is encoded by the GADD45A, DNA repair gene, was stimulated by NO in a JNK-dependent manner; (3) the process is regulated by tumor suppressor p53-dependent and p53-independent pathways; (4) and blocking the GADD45 α expression attenuates the repair of nitric oxide-induced β-cell DNA damage. Thus, β-cells have the ability to repair nitric oxide-damaged DNA and JNK and GADD45α mediate the p53-independent repair of DNA damage.

The DNA damage, by guanine (G) and 8-oxoguanine (8-oxoG) nucleobases, via oxidation reactions induced by N_2O_3 and ONOO can occur through direct strand breaks or base modification [39] and by inhibition of DNA repair enzymes, thereby enhancing the damaging actions of nitric oxide [40]. For example, aim of work [40] was to determine whether NO directly inhibits activity of human 8-oxoguanine glycosylase (hOgg1) in repairing excision of 8-oxoguanine, a ubiquitous oxidative DNA lesion. NO-generating enzyme inducible NO synthase, NO donor *S*-nitroso-*N*-acetyl-D-L-pencillamine, and peroxynitrite completely inhibited activity of hOgg1 from KMBC and in a cell-free system. The inhibition of hOgg1 protein was characterized by formation of *S*-nitrosothiol adducts and loss/ejection of zinc ions. It was suggested that the DNA bases nitrosation can occur via a generation of radical intermediates (for example guanidinylidene radical), which recombine with the $^{\cdot}NO_2$ formed during peroxynitrite degradation [41].

Mechanisms of nitrative and oxidative DNA damage in inflammation-related carcinogenesis was examined using asbestos-fiber-induced iNOS-dependent formation of 8-nitroguanine and 8-oxo-7, 8-dihydro-2′-deoxyguanosine (8-oxodG) in cancer tissues [31]. Proposed mechanisms of point mutation induced by 8-nitroguanine and 8-oxodG through induction of the $G:C \rightarrow T:A$ transversion and roles of nitrative and oxidative DNA damage in inflammation-related carcinogenesis are presented in Figure 9.4.

Overproduction of NO causes the generation of peroxynitrite ($ONOO^-$), which leads to the formation of 8-nitroguanine, an indicator of nitrative DNA damage. 8-Nitroguanine undergoes spontaneous depurination in DNA, resulting in the formation of an apurinic site. The formation of 8-nitroguanine and 8-oxodG at sites of carcinogenesis in various clinical specimens and animal models in relation to inflammation-related carcinogenesis was also investigated. Obtained results suggested that in inflammation-related cancer tissues DNA base damage leads to double-stranded breaks. Nitric oxide synthases, which cause damage to the cellular DNA, were shown to be involved in cancer progression [42]. Twist is a basic helix-loop-helix transcription factor family normally expressed during embryonic development and activated in variety of tumors. To determine the involvement of iNOS and its correlation with Twist expression in breast cancer lesions in 85 breast biopsies, which include 19 non-cancer and 66 cancerous lesions, mRNA levels of iNOS were measured in different stages of breast cancer lesions. Increased expression of iNOS was observed in higher stage of human breast carcinomas and was found to positively correlate with the Twist expression.

The effect of low NO level (10 μM) on runt-related transcription factor (2RUNX2), associated with osteoblast differentiation and B-cell lymphoma (2Bcl-2) expression, was examined in prostate cancer cells *in vitro* using a number of methods, including gene and protein expression analyses, nitrite quantitation, protein-DNA interaction assays, viability assays, and LNCaP xenografts [43]. The experiments suggested that RUNX2 overexpression in LNCaP tumors *in vivo* decreased the time to tumor presentation and increased tumor growth, while these tumors exhibited improved tumor angiogenesis and oxygenation.

9.6. Nitric Oxide and Metastasis

Metastasis is the spread of a cancer or other disease from one organ or part of the body to another without being directly connected with it [44–46]. Over past decades, NO has emerged as a molecule of interest in carcinogenesis and tumor growth progression, specifically in the metastasis process [46]. The typical progression of tumor cell metastasis and role of NO, as a signaling molecule, in the regulation of cancer formation, progression, and metastasis, were considered in [47]. The action of NO on cancer relies on multiple factors, including cell type, metastasis stage, and organs involved.

Metastatic cell behavior can be controlled through direct or indirect action of NO on proteins. Direct actions of NO involve nitrogen dioxide (NO_2) peroxynitrite ($ONOO^-$), nitrosonium cation, and NO^- (nitroxyl anion) leading to oxidation, nitrosation (addition of NO to an amine, thiol or hydroxyl aromatic group), and nitration (addition of NO_2 to aromatic groups) The production of reactive oxygen species and/or protective effects via $ONOO^-$ can also occur. As regards the "indirect" mechanism, NO can affect through sGC involving cascade processes: (1) conversion guanosine triphosphate (GTP) to the secondary messenger, cyclic guanosine monophosphate (cGMP); and (2) activation of cGMP-dependent protein kinase (PKG), PKG phosphorylation a variety of proteins. The emerging downstream effect is S-nitrosylation of cysteine residues of proteins, analogous to phosphorylation.

The metastatic cascade includes a series of basic cellular processes such as proliferation, apoptosis, adhesion, secretion, migration, invasion, and angiogenesis. In determining the precise role of NO in metastatic diseases, the following factors have to be taken into consideration [48]: (1) specificity of NOS(s) associated with given cancers; (2) NO concentration; (3) stage in the metastatic process; (4) type of cancer/cell and cellular origin; (5) NO role in various components of the metastatic cascade; (6) multiple mutations accumulating during proliferative activity; (7) changes in levels of gene expression (epigenetics); (8) disruption of epigenetic regulatory mechanisms, such as DNA methylation and histone modification; (9) changes in adhesive properties; (10) proteolysis; and (11) the metastatic cascade, which is the production of new blood vessels to enable the secondary tumors to continue growing. The process of cell translocation across the extracellular matrix comprises three main steps: local proteolysis, cell process extension, migration. All in all, tumor cells escape from the primary site into circulation blood and/or lymph.

Nitric oxide production enhances cell migration, decreases cell adhesion in some cell types, and, conversely, reduces cell migration and increases cell adhesion in other cell types [49], and references therein. Downregulation of the expression of NOS causes increasing aggressiveness of breast tumor cells associating with metastatis NOS. Treatment with NOS inhibitors increased motility in MDA-MB-231 and T47D cells, and exogenous NO resulted in increased adhesion in the MCF-7 and MDA-MB-231 cell lines. These results are consistent with the suggestion that loss of NOS expression may be associated with the progression of breast cancers via increase in motility and loss of adhesion.

In the metastasis including cancer progression, adhesion detachment is a key process. The adherence properties of many different cell types can change during a variety of activities, including cancer progression. For example, if a cell has very strong adhesion to its neighbors and/or the substrate, it may prevent moving from the primary tumor and spread to distant sites. From another side, strong adhesiveness can aid such a cell to attach to distant sites. Nitric oxide has been shown to inhibit adhesion to the extracellular matrix (ECM) components of many cell types, including, eosinophils [62] and neutrophils. Cell adhesion molecules [50–52]. Endothelial NO limits platelet activation, adhesion, and aggregation proliferation of vascular smooth muscle cell relaxes both vascular and nonvascular smooth muscles and inhibits adhesion of leukocyte to the endothelium [53]. For example, application of N-nitro-L-arginine (a general NOS inhibitor) and aminoguanidine (a specific iNOS inhibitor) increased neutrophil adhesion to the endothelium. Adhesion of eosinophils from rats to fibronectin administered with the NOS inhibitor N-nitro-L-arginine methyl ester hydrochloride (L-NAME) was inhibited. Nitric oxide may also prevent cancer metastasis from its direct vasodilation and inhibition of cell adhesion molecules (CAMs) [53]. NO donor S-nitrosocaptopril (CAP-NO) produces direct vasorelaxation that can be antagonized by typical NO scavenger hemoglobin and guanylate cyclase inhibitor and inhibits expression of the stimulated CAMs, which were reversed by hemoglobin. It was concluded that CAP-NO interfered with the hetero-adhesion between cancer cells and human umbilical vein endothelial cells by downregulation of CAMs induced by cytokines.

The activation of the metalloproteinases MMPs by modulating MMP expression with NO has an important function in tumor cell migration and invasion [54]. The effects of non-toxic concentrations of NO on human non-small cell lung cancer (NSCLC) cells, including the expression of integrins—the transmembrane receptor that facilitates cell-extracellular matrix adhesion and the migration of the cell—was assessed [52]. The migration capacities of these cells were evaluated by wound healing and transwell migration assays. It was found that NO treatment resulted in a significant increase in the expression of integrin αv and β1 in three NSCLC-derived cell lines tested. A role of NO in the regulation of integrin expression and the migratory capacity of NSCLC was discussed.

9.7. Nitric Oxide and Pancreatic and Prostate Cancer

Pancreatic cancer is hallmarked by aggressive biology and extreme lethality and very high mortality rates. Inflammation, autophagy, and obesity are common features in the pathogenesis of pancreatitis and pancreatic cancer. Pancreatic cancer arises when cells in the pancreas begin to multiply out of control and form a mass (https://en.wikipedia.org/wiki/Pancreatic_cancer). Pancreatic cancer is characterized by aggressive development and extreme lethality and very high mortality rates [54–58]. Strong association between NOS2 and the disease aggressiveness was provided [58–60]. Overproduction of NO imposes an adverse effect on the tumor microenvironment, which causes genetic and epigenetic changes in tumor and tumor stromal cells. This effect promotes survival and growth advantage of malignant cells [58]. From another hand, NOS2-deficiency: (1) enhanced survival and reduced tumor severity in the genetically engineered mouse model of pancreatic cancer (KPC mouse); (2) inhibited proliferation, migration, and invasion of primary tumor cells and enhanced apoptosis in tumors from NKPC mice; (3) reduced tumor macrophage infiltration, chemokine ligand-2/Mast cell protease 1MCP1, and a mammalian microRNAmir-21 expression in KPC mice; and (4) inhibited extracellular signal-regulated kinases, ERK-signaling, and phosphorylation of FOXO3, a human protein. In addition, genetic deletion of NOS2 enhanced survival in mice with autochthonous pancreatic ductal adenocarcinoma, PDAC, and targeting NOS3/eNOS reduced the abundance of precursor lesions in mice.

Nitric oxide, which plays a role in the post-translational modification of proteins, exhibits tumoricidal activity. The regulation of insulin receptor substrate (IRS)-1 protein expression and insulin/insulin-like growth factor (IGF) signaling by NO, which is involved in the proliferation and invasion of pancreatic cancer cells, was investigated [68]. NO donor inhibited insulin/IGF-I-stimulated phosphorylation of insulin receptor/IGF-I receptor, IRS-1, Akt/PKB, and glycogen synthase kinase-3beta, decreased expression of IRS-1 protein in MIAPaCa-2 cells, and enhanced the phosphorylation of extracellular signal-regulated kinase-1/2. Thus, NO inhibits the proliferation and invasion of pancreatic cancer cells, through upregulation of IRS-1 protein degradation and resultant downregulation of the insulin/IGF-I-Akt pathway [55].

Prostate cancer (PC) is the development of cancer in the prostate, a gland in the male reproductive system (https://en.wikipedia.org/ wiki/Prostate_cancer). The cancer cells may spread from the prostate to other parts of the body, particularly the bones and lymph nodes. Involving NO in various steps of the prostate cancer has been well documented [61–66].

The results of a meta-analysis [61] suggested that the eNOS gene 894G > T polymorphism might be a risk factor in the onset of PCa. An association between eNOS gene 894G > T polymorphism and prostate cancer (PCa) risk was established. Several variants within gene-encoding endothelial isoform of nitric oxide synthase have been reported to confer prostate cancer susceptibility and/or progression and support the involvement of NOS3 variants in molecular pathogenesis of PCa [62]. The meta-analysis also showed the evidence that NOS3 rs1799983 polymorphism rs1799983 polymorphism was associated with a risk of prostate cancer development in overall populations [63].

Androgens, any natural or synthetic compound, usually a steroid hormone, that stimulates or controls the development and maintenance of male characteristics in vertebrates by binding to androgen receptors, are essential for the development of the prostate and prostate cancer [64, 65]. Findings presented in [65] indicate that increased NO production by acquired increased expression of activated eNOS could contribute to the antiandrogen-resistant growth of prostate cancer cells. This process occurs via a mechanism of NO-mediated suppression of androgen receptor (AR) activity. Development of antiandrogen-resistance in advanced prostate cancer involves multiple AR-dependent and -independent pathways. An overexpression of eNOS pattern in hormone-refractory prostate cancer and several models of advanced hormone-resistant prostate cancer were reported. The effects of testosterone (T) and hormone prolactin (PRL) on plasma-membrane carboxypeptidase-D (CPD) expression, and the role(s) of CPD in NO production and survival of prostate cancer cells, were examined [66]. It was found that T and PRL upregulate CPD and NO levels in PCa cells, while CPD increases NO production to promote PCa cell survival.

Soluble guanylyl cyclase α1 (sGCα1) and p53 cytoplasmic are involved in sequestration and downregulation in prostate cancer [67]. Tumor protein p53, also known as cellular tumor antigen p53, is any isoform of a protein encoded by homologous genes in various organisms. sGCα1 was identified as a novel androgen-regulated gene essential for prostate cancer cell proliferation. The sGCα1 expression is highly elevated in prostate tumors, contrasting with the low expression of sGCβ1, when sGCα1 dimerizes to mediate NO signaling. Results of work [69] suggested that anti-cancer effects of citrus peel flavonoids in human prostate xenograft tumors were accompanied by mechanistic downregulation of the protein levels of inflammatory enzymes (inducible nitric oxide synthase, iNOS and cyclooxygenase-2, COX-2), metastasis (matrix metallopeptidase-2, MMP-2 and MMP-9), angiogenesis (vascular endothelial growth factor, VEGF), and proliferative molecules, the induction of apoptosis in prostate tumors. COX-2, detected in the muscle fibers of the hyperplastic stroma of some control prostates, and nitric oxide synthase-2 were expressed in the lesions of prostatic intraepithelial neoplasia (PIN) in control prostates from 12 patients [70]. The both enzymes are important in angiogenesis.

The expression of iNOS was evaluated in prostate cancer in 198 PC patients [71]. The results demonstrate an association between strong iNOS expression and rapid cancer cell proliferation rate, dedifferentiation, and advanced stage cancer indicating that the enzyme has stimulative and suppressive effects on cancer cell growth. In pancreatic cancer cells, NOS plays a critical role in PI3K-Akt signal transduction pathway [PI3K (phosphatidylinositol 3-kinase)] and Akt (protein kinase B)] and guanosine triphosphate (GTPase) signaling pathway [72]. For example, irradiation of PANC-1 cells promoted invasion and production of NO, which activated the PI3K-AKT signaling pathway. NOS in conjunction with inhibition of PI3K, Rho-associated kinase, and serine protease can suppress the radiation-enhanced invasion of PANC-1 cells, suggesting as possible targets for the management.

References

[1] Choudhari SK, Chaudhary M, Bagde S, Gadbail AR, Joshi V. Nitric oxide and cancer: A review. *World J Surg Oncol.* (2013) 11, 118.

[2] Hussain SP, He P, Subleski J, Hofseth LJ, Trivers GE, Mechanic L, Hofseth AB, Bernard M, Schwank J, Nguyen G, Mathe E, Djurickovic D, Haines D, Weiss J, Back T, Gruys E, Laubach VE, Wiltrout RH, Harris CC. Nitric oxide is a key component in inflammation-accelerated tumorigenesis. *Cancer Res.* (2008) 68, 7130–7136.

[3] Huzefa Vahora, Munawwar Ali Khan, Usama Alalami and Arif Hussain. The Potential Role of Nitric Oxide in Halting Cancer Progression Through Chemoprevention. *J. Cancer Prev.* (2016) 21, 1–12.

[4] El-Sehemy A, Postovit L-M, Yang Xin Fu. Nitric oxide signaling in human ovarian cancer: A potential therapeutic target. *Nitric Oxide* 1 (2016) 54, 30–37.

[5] Vannini F, Kashfi K, Nath N. iNOS in cancer. *Redox Biology* (2015) 6, 334–343.

[6] Burke AJ, Francis J. Sullivan FJ, Giles FJ, Glynn SA. The yin and yang of nitric oxide in cancer progression. *Carcinogenesis* (2013) 34 (3): 503-512.

[7] Fukumura D, Kashiwagi S, Jain RK. The role of nitric oxide in tumour progression. *Nature Reviews Cancer* (2006) 6, 521-534.

[8] Hickok JR, Thomas DD. Nitric Oxide and Cancer Therapy: The Emperor has NO Clothes. *Curr. Pharm. Des.* (2010) 16 381391.

[9] Nossaman B, Pankey E, Kadowitz P. Stimulators and Activators of Soluble Guanylate Cyclase: Review and Potential Therapeutic Indications. *Critical Care Research and Practice* Volume (2012), 2012, Article ID 290805, 12 pages.

[10] Xu W, Liu L Z, Loizidou, AM. The role of nitric oxide in cancer. *Cell Research* (2002) 311-320.

[11] Vahora H, Khan MA, Alalami U, Hussain A. The Potential Role of Nitric Oxide in Halting Cancer Progression Through Chemoprevention. *J Cancer Prev.* (2016) 21, 1-12.

[12] Hassan M, Watari H, AbuAlmaaty A, Ohba Y, Sakuragi N. Apoptosis and Molecular Targeting Therapy in Cancer. *BioMed Research InternationalVolume* (2014) 2014, Article ID 150845, 23 pages.

[13] Boyd CS, Cadenas E. Nitric oxide and cell signaling pathways in mitochondrial-dependent apoptosis. *Biol. Chem.* (2002) 383, 411-423.

[14] Melino G, Brüne B Nitric oxide: NO apoptosis or turning it ON? *Cell Death and Differentiation* (2003) 10, 864-869.

[15] Reeves KJ, Reed MVR, Brown NJ. Is nitric oxide important in photodynamic therapy? *J. Photochem. Photobiol. B Biol.* (2009) 95, 141–14.

[16] Yuan Yuan Ji, Yan Jun Ma, JianWen Wang. Cytoprotective role of nitric oxide in HepG2 cell apoptosis induced by hypocrellin B photodynamic treatment. *J. Photochem. Photobiol. B Biol.* (2016) 163, 366–373.

[17] Nan Zhao, Kang-tao Tian, Ke-guang Cheng, Tong Han, Xu Hua, Da-hong Li, Zhan-lin Li, Hui-ming Hua. Antiproliferative activity and apoptosis inducing effects of nitric oxide donating derivatives of evodiamine. *Bioorganic & Medicinal Chemistry* (2016) 24, 2971–2978.

[18] Bonavida B, Garban H. Nitric oxide mediated sensitization of resistant tumor cells to apoptosis by chemo-immunotherapeutics. *RedoxBiology* (2015) 6, 486–494.

[19] Gonzalez-Perez RR, Rueda, Bo R. 2013. *Tumor angiogenesis regulators* Boca Raton: Taylor & Francis.

[20] Sessa WC. eNOS at a glance. *J. Cell Sci.* 117, (2004) 2, 427–2429.

[21] Jones MK, Tsugawa K, Tarnawski AS, Baatar D. Dual actions of nitric oxide on angiogenesis: Possible roles of PKC, ERK, and AP-1. *Biochem. Biophys.* (2004) *Res. Comm.* 318, 520–528.

[22] Halder AK, Mukherjee A, Adhikari N, Saha A, Jha T. Nitric Oxide Synthase (NOS) Inhibitors in Cancer Angiogenesis. *Current Enzyme Inhibition (*2016), 12, 49-66.

[23] Cooke JP. NO and angiogenesis *Atheroscler. Suppl* (2003) 4, 5360.

[24] Gow AJ, Farkouh CR, Munson DA, Posencheg MA, Ischiropoulos H. Biological significance of nitric oxide-mediated protein modifications. *Am. J. Physiol. Lung Cell. Mol. Physiol.* (2004) 287, L262–L268.

[25] Sikora AG, Gelbard A, Davies MA, Sano D, Ekmekcioglu S, Kwon J, H ailemichael Y, Jayaraman P, Myers JN, Grimm EA, Overwijk WW. Targeted inhibition of

inducible nitric oxide synthase inhibits growth of human melanoma *in vivo* and synergizes with chemotherapy. *Clin. Cancer Res.* (2010) 16, 1834-1844.

[26] Ng QS, Goh V, Milner J, Stratford MR, Folkes LK, Tozer GM, Saunders MI, Hoskin PJ. Effect of nitric-oxide synthesis on tumour blood volume and vascular activity: A phase I study. *Lancet Oncol.* (2007) 8, 111-118.

[27] Bellamy TC, Wood J, Garthwaite J. On the activation of soluble guanylylcyclase by nitric oxide. *PNAS* (2002) 99, 507–510.

[28] Friebe A, Koesling D. Regulation of nitric oxidesensitive guanylyl cyclase. *Circ. Res.* 93, (2003) 96–105.

[29]. Mujoo K, Sharin VG, Martin E, Byung-Kwon Choi, Courtney Sloan, Nikonoff LE, Kots AY, Murad F. Role of soluble guanylyl cyclase-cyclic GMP signaling in tumor cell proliferation. *Nitric Oxide* (2010) 22, 43–50.

[30] El-Sehemy A, Chang AC, Azad AK, Gupta N, Xu Z, Steed H, Karsan A, Fu Y. Notch activation augments nitric oxide/soluble guanylyl cyclase signaling in immortalized ovarian surface epithelial cells and ovarian cancer cells. *Cell Signal.* (2013) 25, 2780-2787.

[31] Murata M, Thanan R, Ma N, Kawanishi S. Role of Nitrative and Oxidative DNA Damage in Inflammation-Related Carcinogenesis. *J Biomed Biotech*, Volume 2012(2012), Article ID 623019, 11 pages.

[32] De Bont R, van Larebeke N. Endogenous DNA damage in humans: A review of quantitative data. *Mutagenesis* 19, (2004) 169-185.

[33] Hoeijmakers JH. DNA damage, aging, and cancer. *N Engl J Med.* (2009) 361, 1475-1485.

[34] Clancy S. DNA damage & repair: Mechanisms for maintaining DNA integrity. *Nature Education* (2008) 1, 103.

[35] Mariko Murata, Raynoo Thanan, Ning Ma, Shosuke Kawanishi. Role of Nitrative and Oxidative DNA Damage in Inflammation-Related Carcinogenesis. *J Biomed Biotech Volume 2012* (2012), Article ID 623019, 11 pages.

[36] deRojas-Walker T, Tamir S, Ji H, Wishnok JS, Tannenbaum SR. Nitric oxide induces oxidative DNA damage in addition to deamination in macrophage DNA. *Chem. Res. Toxicol.* (1995) 8, 47–477.

Phoa N, Epe B.Influence of nitric oxide on the generation and repair of oxidative DNA damage in mammalian cells. *Carcinogenesis* (2002) 23, 469-475.

[38] Katherine J. Hughes KJ, Gordon P. Meares GP, Kari T. Chambers KT, Corbett JA. Repair of Nitric Oxide-damaged DNA in b-Cells Requires JNK-dependent GADD45a Expression. *J. Biol. Chem.* (2009) 284, 27402–27408.

[39] Niles JC, Wishnok JS, Tannenbaum SR. Peroxynitrite-induced oxidation and nitration products of guanine and 8-oxoguanine: Structures and mechanisms of product formation. *Nitric Oxide* (2006) 14, 109–121.

[40] Jaiswal M, LaRusso NF, Nishioka N, Nakabeppu Y, Gores GJ. Human Ogg1, a Protein Involved in the Repair o 8-Oxoguanine, Is Inhibited by Nitric Oxide. (2001) *Cancer Res.* 61, 6388–6393.

[41] Schormann N, Ricciardi R, Chattopadhyay D. Uracil-DNA glycosylases—Structural and functional perspectives on an essential family of DNA repair enzymes. *Protein Sci.* (2014) 23, 1667–1685.

[42] Santhalakshmi Ranganathan, Arunkumar Krishnan, Niranjali Devaraj Sivasithambaram. Significance of twist and iNOS expression in human breast Carcinoma. *Mol Cell Biochem* (2016) 412:41–47.

[43] Nesbitt H, Browne G, O'Donovan KM, Byrne NM, Worthington J, McKeown SR, McKenna DJ. Nitric Oxide Up-Regulates RUNX2 in LNCaP Prostate Tumours Implications for Tumour Growth In Vitro and In Vivo *J. Cell Physiol.* (2016) 231,473-482.

[44] Klein CA (September 2008). "Cancer. The metastasis cascade". *Science.* 321, (5897): 1785–1787.

[45] Halin Bergstroem S, Haeggloef C, Thysell E, Bergh A, Wikstroem P, Lundholm M. Extracellular Vesicles from Metastatic Rat Prostate Tumors Prime the Normal Prostate Tissue to Facilitate Tumor Growth. *Scientific Reports* (2016), 6, 31805.

[46] Sheetal Korde Choudhari, Minal Chaudhary, Sachin Bagde, Amol R Gadbail, and Vaishali Joshi. Nitric oxide and cancer: A review. *World J Surg Oncol.* (2013) 11, 118.

[47] Cheng H, Wang L, Mollica M, Re AT, Wu S, Zuo L. Nitric oxide in cancer metastasis. *Cancer Lett.* (2014) 353, 1-7.

[48] Williams EL Djamgoz MBA. Nitric oxide and metastatic cell behavior. *BioEssays* (2005) 27, 1228–1238.

[49] Lahiri M, Martin JHJ. Nitric oxide decreases motility and increases adhesion in human breast cancer cells. *Onclogy Reports* (2009) 21, 275-281.

[50] Dal Secco D, Moreira AP, Freitas A, Silva JS, Rossi MA, Ferreira SH, Cunha FQ. Neutrophil migration in inflammation: Nitric oxide inhibits rolling, adhesion and induces apoptosis. *Nitric Oxide* (2003) 9,153–164.

[51] Ferreira HH, Costa RA, Jacheta JM, Martins AR, Medeiros MV, Macedo-Soares MF, De Luca IM, Antunes E, De Nucci G. Modulation of eosinophil migration from bone marrow to lungs of allergic rats by nitric oxide. *Biochem Pharmacol* (2004) 68, 631-639.

[52] Saisongkorh V, Maiuthed A, Chanvorachote P. Saisongkorh, Vhudhipong; Maiuthed, Arnatchai; Chanvorachote, Pithi Nitric oxide increases the migratory activity of non-small cell lung cancer cells via AKT-mediated integrin αv and β1 upregulation. *Cellular Oncology* (2016) 39, 449-462.

[53] Lu Y, Yu T, Liang H, Wang J, Xie J, Shao J, Gao Y, Yu S, Chen S, Wang L, Jia L.. Nitric Oxide Inhibits Hetero-adhesion of Cancer Cells to Endothelial Cells: Initiating Metastatic Cascade. *Sci Rep.* (2014) 4, 4344.

[54] Jodele S, Blavier L, Yoon JM, DeClerck YA, Modifying the soil to affect the seed: Role of stromal-derived matrix metalloproteinases in cancer progression. *Cancer Metastasis Rev* (2006) 25, 35-43.

[55] Gukovskaya A, Karin M. Inflammation, autophagy, and obesity: Common features in the pathogenesis of pancreatitis and pancreatic cancer. *Gastroenterology* (2013)144, 1199–1209.

[56] Sugita H, Furuhashi S. Baba H. Nitric Oxide Regulates Growth Factor. In Srivastava Sanjay K. (Ed) *Signaling in Pancreatic Cancer Cells, Medicine, Oncology, "Pancreatic Cancer - Molecular Mechanism and Targets,"* 2012, CC BY 3.0 license.

[57] Wang J, Hussain SP. NO• and Pancreatic Cancer: A Complex Interaction with Therapeutic Potential. *Antioxidants & Redox Signaling*. (2017) 26, 1000-1008.

[58] Wang L, Xie K. Nitric oxide and pancreatic cancer pathogenesis, prevention, and treatment. *Curr Pharm Des.* (2010) 16, 421-427.

[59] Wang J, Yang S, He P, Schetter AJ, Gaedcke J, Ghadimi BM3, Ried T, Yfantis HG, Lee DH, Gaida MM, Hanna N, Alexander HR, Hussain SP. Endothelial Nitric Oxide Synthase Traffic Inducer (NOSTRIN) is a Negative Regulator of Disease Aggressiveness in Pancreatic Cancer. *Clinical Cancer Reserch* (2016) 22, 5992-6001.

[60] Wang J, He P, Gaida M, Yang S, Schetter AJ, Gaedcke J, Ghadimi BM, Ried T, Yfantis H, Lee D, Weiss JM, Stauffer J, Hanna N, Alexander HR, Hussain SP. Inducible nitric oxide synthase enhances disease aggressiveness in pancreatic cancer. *Oncotarget* (2016) 7, 52993-53004.

[61] Zhao C, Yan W, Zu X, Chen M, Liu L, Zhao S, Liu H, Hu X, Luo R, Xia Y, Qi L. Association between endothelial nitric oxide synthase 894G>T polymorphism and prostate cancer risk: A meta-analysis of literature studies. *Tumour Biol.* (2014) 35, 11727.

[62] Nikolić ZZ, Pavićević DLj, Romac SP, Brajušković GN. Genetic variants within endothelial nitric oxide synthase gene and prostate cancer: A meta-analysis. *Clin Transl Sci.* (2015) 8, 23–31.

[63] Wu JH, Yang K, Ma HS, Xu Y. Association of endothelia nitric oxide synthase gene rs1799983 polymorphism with susceptibility to prostate cancer: A meta-analysis. *Tumour Biol.* (2014) 35, 7057–7062.

[64] Vaarala MH, Hirvikoski P, Kauppila S, Paavonen TK. Identification of androgen-regulated genes in human prostate. *Mol Med Rep.* (2012) 6, 466-472.

[65] Yu S, Jia L, Zhang Y, Wu D, Xu Z, Ng CF, et al. To KK, Huang Y, Chan FL. Increased expression of activated endothelial nitric oxide synthase contributes to

antiandrogen resistance in prostate cancer cells by suppressing androgen receptor transactivation. *Cancer Lett.* (2013) 328, 83–94.

[66] Thomas LN, Morehouse TJ, Too CK. Testosterone and prolactin increase carboxypeptidase-D and nitric oxide levels to promote survival of prostate cancer cells. *Prostate.* (2012) 72, 450–460.

[67] Cai C, Hsieh CL, Gao S, Kannan A, Bhansali M, Govardhan K, Dutta R, Shemshedini L. Soluble guanylyl cyclase α1 and p53 cytoplasmic sequestration and downregulation in prostate cancer. *Mol Endocrinol.* (2012) 26, 292–307.

[68] Sugita H, Kaneki M, Furuhashi S, Hirota M, Takamori H, Baba H. Nitric oxide inhibits the proliferation and invasion of pancreatic cancer cells through degradation of insulin receptor substrate-1 protein. *Mol Cancer Res.* (2010) 8, 1152-1163.

[69] Lai CS, Li S, Miyauchi Y, Suzawa M, Ho CT, Pan MH. Potent anti-cancer effects of citrus peel flavonoids in human prostate xenograft tumors. *Food Funct.* (2013) 4, 944–949.

[70] Uotila P, Valve T, Martikainen P, Nevalainen M, Nurmi M, Härkönen P. Increased expression of cyclooxygenase-2 and nitric oxide synthase-2 in human prostate cancer. *Urological Research* (2001) 29, 25–28.

[71] Aaltoma SH, Lipponen PK, Kosma VM. Inducible nitric oxide synthase (iNOS) expression and its prognostic value in prostate cancer. *Anticancer Research* (2001) 21, 3101-3106.

[72] Fujita M, Imadome K, Endo S, Shoji Y, Yamada S, Imai T. Nitric oxide increases the invasion of pancreatic cancer cells via activation of the PI3K-AKT, and RhoA signaling pathways after carbon ion irradiation. *FEBS Lett.* (2014) 588, 3240-3250.

Chapter 10

NITRIC OXIDE IN ARTERIAL AND BONE DISEASES

ANNOTATION

Cardiovascular diseases (CVD) involve the heart or blood vessels. Nitric oxide (NO), produced by both endothelial cells and macrophages, significantly affects the processes occurring in vascular physiology and pathophysiological development of CVD. Nitric oxide has a number of intracellular effects in cardiovascular diseases, including atherosclerosis, intimal hyperplasia, diabetic vascular disease, thrombosis problems, and aneurysmal disease that lead to vasorelaxation, endothelial regeneration, inhibition of leukocyte chemotaxis, vascular lesion, and platelet adhesion.

10.1. CARDIOVASCULAR DISEASES

10.1.1. Introduction

The cardiovascular system incorporates the heart blood vessels and lymphatic vessels. Vascular function and dysfunction of NO is critical to the pathophysiology of vascular disease and to endothelial dysfunction, which is defined as impairment of physiologic endothelium-dependent relaxation [1–6]. The cardioprotective roles of NO also include regulation of blood pressure and vascular tone, and prevention of smooth muscle cell proliferation [1–11]. The vascular effects of NO on biochemical and physiological processes related to cardiovascular diseases are summarized in [1].

Roles of different cells from the arterial vessel wall in atherosclerosis were discussed in [4]. Endothelial cells, smooth muscle cells (SMCs), fibroblasts, and adipocytes from the tunica intima, media, adventitia, and perivascular adipose tissue and their related cytokines participate in the inflammatory response of atherosclerosis. Endothelial dysfunction, smooth muscle cell migration and proliferation, the transformation of fibroblasts into myofibroblasts, and adipokines produced by perivascular adipose tissue are implicated in the pathological process of atherosclerosis.

Arteriosclerosis is the thickening, hardening, and loss of elasticity of the walls of arteries that restricts the blood flow. Invasion and accumulation of white blood cells and proliferation of intimal-smooth-muscle cell creating can cause atherosclerosis. NO, produced by endothelial cells and macrophages, significantly affects the processes occurring in vascular physiology and pathophysiological development of atherosclerosis. Atherosclerosis, as an inflammatory disease, can be a subject of the NO effects [1–12]. Atherosclerosis involves a cascade of pathologic processes such as endothelial dysfunction, tissue perfusion, limb loss, and heart attack occurring when blood flow stops to a part of the heart causing damage to the heart muscle (Figure 10.1). During atherosclerosis, the following processes occur [4]: (1) oxidation of low-density lipoprotein (oxLDL) in the arterial intima leading to endothelial cell dysfunction; (2) inflammation of the endothelium secretes inflammatory cytokines, chemokines, and adhesion molecules that recruit circulating immune cells into the artery wall; (3) monocytes transmigration into the subendothelial space take up oxLDL and differentiation into lipid-laden macrophages; (4) immune components, including dendritic cells, T cells, and B cell antibodies that are specific for oxLDL, accumulate as the lesion expands, having both pro- and anti-inflammatory effects; (5) a subset of vascular smooth muscle cells enters a proliferative state and migrates to form the fibrous cap, whereas other SMCs phenotypically switch into macrophage-like cells within lesions; (6) pro- and anti-inflammatory immune cell subsets are present within the adventitia and perivascular adipose tissue (PVAT) at homeostasis; and (7) these cells increase in number in response to atherogenic stimuli and organize to form arterial tertiary lymphoid organs (ATLO) with distinct B cell and T cell zones [4].

Factors affecting the efficiency of NO in the process of atherosclerosis were formulated as following [6]: (1) impairment of membrane receptors in the arterial wall; (2) reducing concentrations of both of inducible and endothelial NO synthase and l-arginine; (3) impaired release of NO from the atherosclerotic damaged endothelium including NO diffusion from the endothelium to vascular smooth muscle cells; (4) degradation of NO by the reactive oxygen species; (5) limitation of cyclic GMP production as a result of interaction impaired interaction of NO with guanylate cyclase; and (6) eNOS gene polymorphism, which might be an additional risk factor that may contribute to cardiovascular events. Endothelial NO production inhibits TNF-stimulated vascular cell adhesion molecule-1 (VCAM-1) expression in endothelial cells. Metabolic intervention with antioxidants reduces arterial oxidation-speciWc epitopes and systemic oxidative stress.

Depending on the source and amount of production, NO can exert both atherosclerotic and protective effects. A reduction in NO production or activity is one of major mechanism of endothelial dysfunction and a contributor to atherosclerosis [7]. In atherosclerotic blood vessels, impaired NO production could contribute to a number of processes, such as enhanced vasoconstriction, increased adhesion of platelets and monocytes, and migration and proliferation of vascular SMCs. Nitric oxide can also inhibit the following processes: monocyte and platelet adhesion, leukocyte chemotaxis, expression of adhesion-molecules, oxidation of LDL, SMC proliferation and migration, and also promote endothelial regrowth. In atherosclerosis, the pathophysiology of vascular disease and to endothelial dysfunction, which is defined as impairment of physiologic endothelium-dependent relaxation, is critically caused by vascular function and dysfunction of NO [8]. Nitric oxide maintains vascular homeostasis and thereby prevents vascular atherosclerotic changes [9]. Endothelium-derived NO has also been shown to modulate angiogenesis *in vitro* and *in vivo* [9].

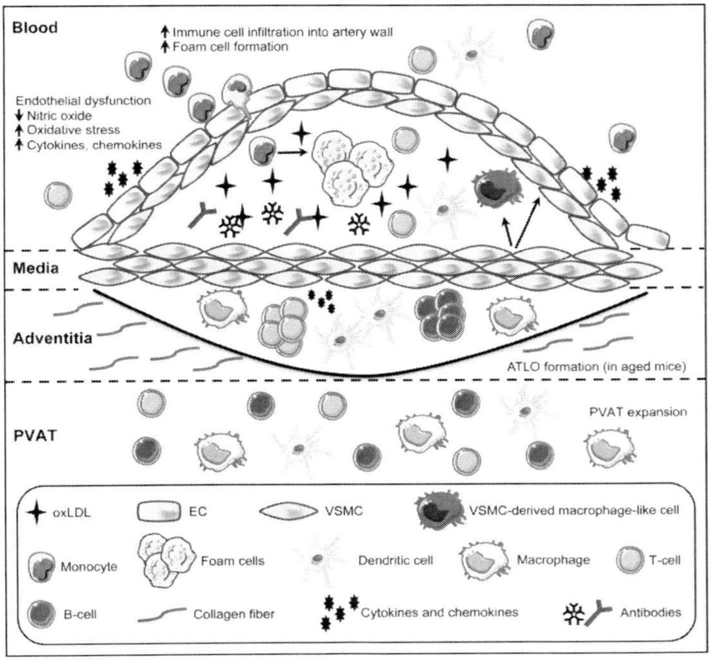

Figure 10.1. Cell types and structure of the vessel wall in atherosclerosis. During atherosclerosis, buildup of oxidized low-density lipoprotein (oxLDL) in the arterial intima leads to endothelial cell dysfunction. The resulting inflamed endothelium secretes inflammatory cytokines, chemokines, and presents adhesion molecules that recruit circulating immune cells into the artery wall. Monocytes that transmigrate into the subendothelial space take up oxLDL and differentiate into lipid-laden macrophages (foam cells). Immune components, including dendritic cells, T cells, and B cell antibodies that are specific for oxLDL, accumulate as the lesion expands, having both pro- and anti-inflammatory effects, depending on cell type. A subset of vascular smooth muscle cells (VSMCs) enters a proliferative state and migrates to form the fibrous cap, whereas other SMCs phenotypically switch into macrophage-like cells within lesions. Pro- and anti-inflammatory immune cell subsets are present within the adventitia and perivascular adipose tissue (PVAT) at homeostasis. These cells increase in number in response to atherogenic stimuli and organize to form arterial tertiary lymphoid organs (ATLO) with distinct B cell and T cell zones that are similar to secondary lymphoid organs in aged mice [4]. With permission from Mary Ann Liebert, Inc., 2017.

Figure 10.2 illustrates various aspects of the biochemistry of atheroprotective estrogen actions in endothelial cells.

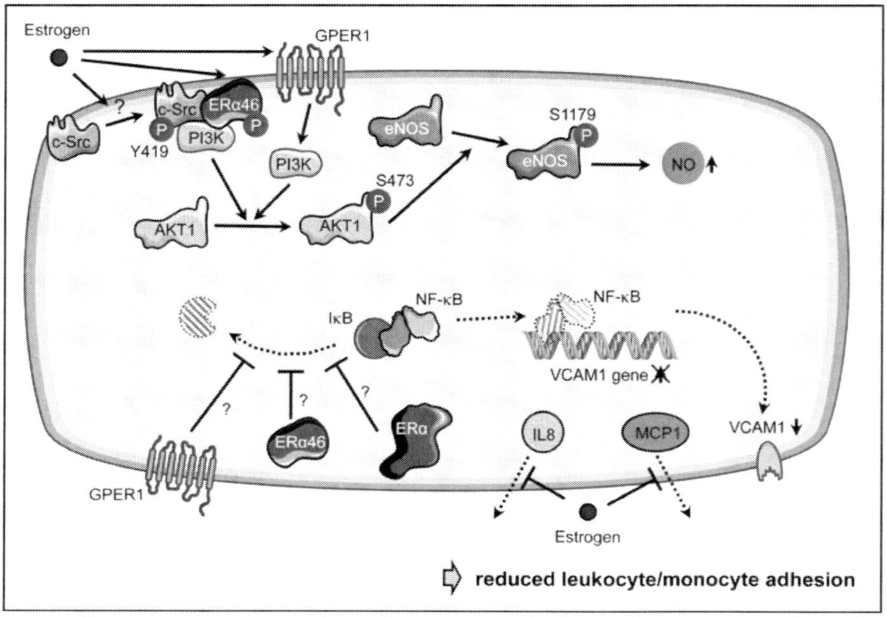

Figure 10.2. Atheroprotective estrogen actions in endothelial cells. Estrogen can prevent endothelial dysfunction in multiple ways. On one hand, it induces phosphorylation of membrane-bound s-Src on tyrosine 419, which then phosphorylates a46 kDa variant of estrogen receptor a (ERa46), favoring the assembly of a complex containing c-Src, ERa46, and PI3-kinase(PI3K). The latter can phosphorylate AKT1, which, in turn, activates eNOS by phosphorylation on serine 1179, leading to increased NO-production. However, also the G-protein-coupled estrogen receptor 1 (GPER1) was shown to increase eNOS phosphorylation in a PI3K-dependent manner. Furthermore, estrogen can reduce the NF-jB-dependent upregulation of VCAM1 by interfering with the degradation of its inhibitor nuclear factor kappa B inhibitor a (IjBa) and thus, nuclear translocation of NF-jB. However, it is currently not known as to which estrogen receptor is responsible for this effect and how it is mediated. Finally, estrogen blocks the secretion of interleukin 8 (IL8) and monocyte chemo-attractant protein-1(MCP1) without affecting their expression; this, together with the diminished VCAM1 expression, prevents adhesion of leukocytes and monocytes [4]. With permission from Mary Ann Liebert, Inc., 2017.

Highlights of the role of NO and endothelial nitric oxide synthase (eNOS) and their roles in vascular physiology and pathophysiological development of atherosclerosis were summarized in review [10]. It was concluded that the NO/$\bullet O_2^-$ equilibrium is tightly regulated under normal physiological conditions, and that this balance can be disturbed by endothelial dysfunction. The impact of NO synthases in the initiation and development of atherosclerosis were summarized in review [11]. In particular, it was stressed that oxidation of tetrahydrobiopterin may cause eNOS uncoupling and consequent oxidative stress and reduction of eNOS-derived NO, which is a protective principle in the vasculature.

Cardiovascular risk factors activate reactive oxygen species (ROS)-producing enzyme systems and/or inhibit ROS-detoxifying systems and eventually lead to uncoupling of eNOS, which in turn potentiates the pre-existing oxidative stress by producing superoxide [11]. Cardiovascular risk factors also inhibit eNOS activity, and the reduced bioavailability of endothelial NO manifests as endothelial dysfunction. Under physiological conditions, tissue levels of BH4 are optimal for eNOS catalytic activity, and activation of eNOS generates NO and L-citrulline [12]. Tissue levels of BH4 are reduced with hypercholesterolemia and atherosclerosis, when oxidative stress is increased. In the presence of suboptimal levels of BH4, activation of eNOS leads to uncoupling of NOS with subsequent generation of superoxide rather than NO. Eventually, NO generated by eNOS serves as an anti-atherogenic molecule.

10.2. Nitric Oxide and Cardiac Function

10.2.1. General

Nitric oxide acting on the heart is produced by vascular and endocardial endothelial NOS, as well as neuronal and inducible synthases and plays a significant role in regulating cardiac contractility and heart rate participating in the control of contractility and heart rate, limits cardiac remodeling after an infarction, and contributes to the protective effect of ischemic pre- and postconditioning [13]. Low concentrations of NO, with production of small amounts of cGMP, induces the following processes: (1) inhibiting phosphodiesterase III, thus preventing the hydrolysis of cAMP; (2) subsequent activation of a protein-kinase A; (3) opening of sarcolemmal voltage-operated and sarcoplasmic ryanodin receptor $Ca(2^{+})$ channels; and (4) increasing myocardial contractility. High concentrations of NO cause the production of larger amounts of cGMP followed by a cardiodepression in response to an activation of protein kinase G (PKG) with a blockade of sarcolemmal $Ca(2^{+})$ channels. NO is also involved in reduced contractile response to adrenergic stimulation in heart failure, and NO-synthase (NOS) inhibition leads to reduction of heart rate. It was suggested that NO can limit the deleterious effects of cardiac remodeling after heart stroke possibly via the cGMP pathway, and the protective effect of NO is mediated by the guanylyl cyclase-cGMP pathway. The last process results in activation of PKG with opening of mitochondrial ATP-sensitive potassium channels and inhibition of the mitochondrial permeability transition pores. The authors concluded that cell persistence in the high concentration range (micromolar) may turn to the pathological way with the production of peroxynitrite and other reactive oxygen and nitrogen species.

Cardiac myocytes, which are essential for cardiac excitation-contraction coupling, express both eNOS and nNOS in muscle tissue [14]. Several mechanisms can account for endothelial dysfunction including: (1) changes in eNOS mRNA or protein levels; (2)

decreased substrate availability; (3) decreased cofactor availability; (4) improper subcellular localization; (5) abnormal phosphorylation; and (6) scavenging of NO by superoxide (O^{-2}) to form peroxynitrite anion ($ONOO^{-}$).

10.2.2. Ischemic Stroke

Ischemic stroke occurs when an artery to the brain is blocked, causing a shortage of oxygen, when plaque builds up inside blood vessels and restricts the normal flow of blood [15–23]. Ischemic stroke continues to hold the leading position among causes of morbidity, mortality, and disability of the population in the world. Cerebral ischemia initiates a cascade of detrimental events, including glutamate-associated excitotoxicity, intracellular calcium accumulation, formation of reactive oxygen species (ROS), membrane lipid degradation, DNA detrimental events, and acute inflammation [16]. Nitric oxide can act both as a signaling molecule and a neurotoxin, and involved in the mechanisms of cerebral ischemia. The role of cerebral ischemia-stroke reperfusion injury is significant because of NO's ability to modulate both oxidative stress and the inflammatory response [17]. The paradox of beneficial and damaging effects of NO was discussed in review [18].

Nitric oxide donors are able to modulate both oxidative stress and the inflammatory response, and therefore can serve as neuroprotective agents in the pathophysiology of cerebral ischemia-reperfusion injury [16]. Different pharmacokinetic and dynamic profiles of the nitric oxide donors (NODs) that determine the type and extent of their biological effects of the inhibitors were considered in details [16] The main determinant of these effects is the manner in which NO is released, the amount of NO generated, and the time during which it is released. eNOS-derived NO scavenges ROS and inhibits the expression of cellular adhesion molecules, platelet aggregation, and leukocyte adhesion. Cerebral ischemia-reperfusion activates two major signaling pathways, which exert an effect on NODs in experimental cerebral ischemia used in the work (Figure 10.3). Nitric oxide donors are able to modulate both oxidative stress and the inflammatory response, and therefore can serve as neuroprotective agents in the pathophysiology of cerebral ischemia-reperfusion injury [16].

During ischemia in mice, NO concentration decreases because of oxygen deficiency immediately after reperfusion *in vivo*. Biosynthesis of this molecule is triggered mainly by overactivation of NOS [20]. Data suggest that NO production in the striatum after reperfusion is closely related to activities of both nNOS and eNOS.

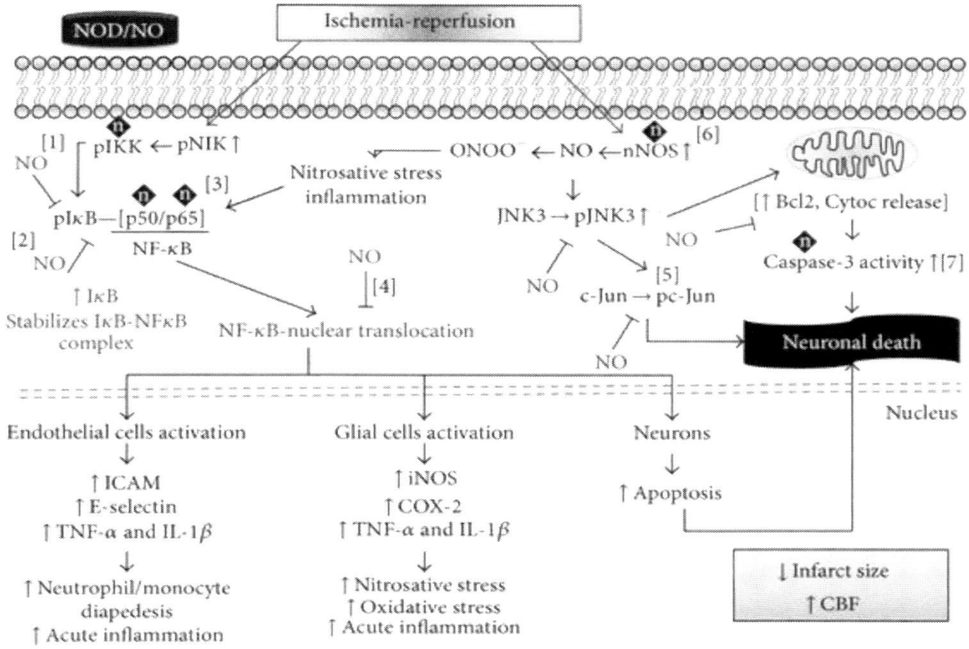

Figure 10.3. Cerebral ischemia-reperfusion activates two major signaling pathways, which exert an effect on NODNF-κB pathway involving oxidative stress and inflammatory stimuli phosphorylate NIK. In the frame of this pathway, the nuclear translocation of NF-κB, therefore, the expression of their target genes is prevented.Cerebral ischemia-reperfusion increases nNOS activity. NO can react with free radicals to produce ONOO−and activates the JNK-3: (c-Jun N-terminal kinase-pathway) (See details in [16]. With permission from Hindawi, 2013.

Nitric oxide donors' ability of organic nitrates (e.g., nitroglycerin, isosorbide-5-mononitrate, nicorandil, pentaerythritol tetranitrate); S-nitrosothiols (e.g., S-nitroso-N-acetylpenicillamine and S-nitroso-glutathione); sydnonimines (e.g., molsidomine, SIN-1); NONOates (JS-K, SPERMINE-NONOate, and PROLI-NONOate), sodium nitroprusside, and others to release NO or an NO-related species, such as the nitrosonium ion (NO^+) or the nitroxyl anion (NO^-), in the brain was examined [21]. Various new classes of NO donors, such as S-nitrosothiols, diazeniumdiolates, furoxans, zeolites, and so on, were also reported [22]. Cerebral NO donor agents, for example sodium nitrite, appeared to replicate the effects of eNOS-derived NO, and therefore have neuroprotective properties [23]. In order to deliver NO, organic nitrates, the most common NODs utilized in coronary artery disease, require enzymatic bioactivation [24]. It is surprising that NO produced by the neuronal or inducible isoform of nitric oxide synthase (nNOS, iNOS) in acute ischemic stroke (IS) is detrimental, whereas that derived from the endothelial isoform is beneficial [25].

Infarction is tissue death (necrosis) due to inadequate blood supply to the affected area. NOS inhibitors, selective to nNOS or iNOS, reduce infarct volume in IS. During stroke recovery, intertissue restorative events occur in the adult ischemic brain, which include angiogenesis, neurogenesis, oligodendrogenesis, astrogliosis, and neurite outgrowth [26].

Experimental results suggested that elevated cyclic guanosine 3',5'-monophosphate (cGMP) levels induced by NO donors and phosphodiesterase 5 (PDE5) inhibitors act on cerebral endothelial cells, neural stem cells, and oligodendrocyte progenitor cells. The kinetics of NO induced *in vivo* by either eNOS or neuronal NO synthase (nNOS) after transient global forebrain ischemia and ischemic changes to hippocampal CA1 neurons in eNOS knockout (-/-) mice and nNOS (-/-) mice during cerebral ischemia and reperfusion was measured [27]. The data indicated that NO production in the striatum after reperfusion is closely related to activities of both nNOS and eNOS, and to nNOS following reperfusion.

Oxidative and nitrosative stresses play a key role in ischemic-reperfusion situations [28]. In cerebral ischemia, an increase in the production of NO catalyzed by enzymes expressed in cells occurs in a pro-oxidant environment. For example, the increase of NO production in a pro-oxidant environment contributes to brain damage and the involvement of $ONOO^-$ in ischemia-mediated damage $ONOO^-$ decomposition and in brain damage reduction has been correlated [28]. The magnetic resonance imaging (MRI), taken just after estimated stroke onset disclosed brain lesions, provides evidences for fluctuation of serum NO_x at the onset of spontaneous stroke [29]. This process is accompanied by the appearance of peroxynitrite in brain lesions in malignant stroke-prone spontaneously hypertensive rats (M-SHRSP) with developed hypertension [29].

In a rat model of focal cerebral ischemia, administration of NO modulator, S-nitrosoglutathione (GSNO) provided neuroprotection and after the onset of ischemia reduced infarction and improved cerebral blood flow [31]. Specifically, treatment of the ischemic brain with GSNO induced a cascade of processes. The treatment: (1) reduced the expression of tumor necrosis factor-a, interleukin-1b, and iNOS; (2) inhibited the activation of microglia/macrophage; (3) downregulated the expression of leukocyte function-associated antigen-1 and intercellular adhesion molecule-1; (4) decreased the number of apoptotic cells and the activity of caspase-3; and (5) inhibited cytokine-mediated expression of iNOS and activation of NF-jB. Thus, GSNO protects the brain against ischemia/reperfusion injury by modulating NO systems, resulting in a reduction in inflammation and neuronal cell death. In the ischemic brain, during permanent focal cerebral ischemia a wide range of stimuli including nNOS and iNOS may trigger activation of NF-κB [32]. In male Wistar rats, NO acts as a mediator of ischemia-associated neuronal damage, as inhibitors of NO synthesis ameliorate neuronal and can directly inhibit the DNA binding activity of NF-κB family proteins. It was concluded that NO of neuronal and inducible origin promotes NF-κB activation via IkB-alpha modulation and mediates ischemic-related damage in the brain.

In mice subjected to chronic mild stress model (CMSmice ice), hippocampal NO production and their regulation by protein kinase C (PKC) in the memory impairment was investigated [33]. After PKC activity blockade, a correlation of chronic stress induced a diminished NO production by nNOS with an increment in gamma and zeta PKC isoenzymes, and partial restoration of nNOS activity was established. Interestingly, ROS formation was higher in the presence of nNOS inhibitor in both control and CMS mice. Loss of hippocampal nNOS contributes to the stress-related deficit in learning and memory and suppresses neuronal apoptotic cell death [34].

In adult female rats with ischemia, the effects of exogenous administration of GSNO, an endogenous redox modulating anti-neuroinflammatory agent, which hastens functional recovery in a cauda equina (CE) compression (CEC) in adult female rats with ischemia, was examined [35]. It was found that administration of GSNO after CEC decreased inflammation, hyperalgesia, increased sensitivity to pain, and cell death leading to improved locomotor function of CEC rats. This effect is related with S-nitrosylation of the Cys-179 of IKKß (mutant version of κB kinase), which causes a decrease in the enzyme ability to phosphorylate IκB, a nuclear factor of kappa light polypeptide gene enhancer in B-cells inhibitor. The latter inhibits the transcription factor, NF-κB [36]. The identification of IKKβ, as a target for S-nitrosylation, provides insights into the mechanisms of inhibition of NF-κB by NO.

10.3. Nitric Oxide and Bones

10.3.1. Role of Nitric Oxide in Bone Function and Remodeling

Bone tissue is made up of osteoblasts and osteocytes involved in the formation and mineralization of bone; osteoclasts are involved in the resorption of bone tissue. Nitric oxide (NO) has important effects on bone cell function [37–40]. In bone, the expressed endothelial isoform of nitric oxide synthase (eNOS) is essential for normal osteoblast function on a constitutive basis, whereas inducible NOS is only expressed in response to inflammatory stimuli and acts as a mediator of cytokine effects.

Nitric oxide and related compounds have a variety of effects on bone: (1) decrease in NO production may cause cardiovascular events, sexual dysfunction, and osteoporosis; (2) at medium doses, NO suppressed osteoclastic bone resorption and promoted growth of osteoblasts; and excess local production of NO aggravates bone destruction in inflammatory arthropathies; and (3) nitroglycerin, glyceryl trinitrate, and nitrates have beneficial effects on bone resorption (release calcium, decreased osteoclastic activity), and have an anabolic action on bone formation (i.e., enhanced osteoblastic activity). In a fracture healing, a marked iNOS expression is observed. Increased expression of eNOS also occurs during the healing of fractures [41]. Significant elevation of calcium-dependent NOS activity occurs in the cortical blood vessels and in osteocytes in the early phase of fracture repair. Increased eNOS concentration in bone blood vessels mediates the increased blood flow during fracture healing. Interactions of cytokines, estrogen, growth factors, mechanical stresses, and stimulating eNOS and iNOS in bone homeostasis and bone metabolism were investigated in [38].

Bone remodeling is a process where mature bone tissue is removed from the skeleton and new bone tissue is formed. eNOS mediates the effects of mechanical loading on the skeleton, promotes bone formation, and suppresses bone resorption in a process of remolding [42]. In bone cells, pro-inflammatory cytokines IL-1 and TNF cause activation of the iNOS pathway in bone cells, and NO derived from this pathway provokes cytokine and inflammation that induced bone loss [42]. Interferon gamma is found to be a stimulator of NO production when combined with other cytokines and produces high concentrations of NO. As evidenced by the experiments on eNOS knockout mice having osteoporosis, the eNOS isoform regulates osteoblast activity and bone formation, and acts as a mediator of the effects of oestrogen in bone. Osteoclasts differentiate from haemopoetic precursors in the monocyte lineage in response to activation of receptor activator of nuclear factor κ B (RANK), by its ligand receptor activator of nuclear factor kappa-B ligand (RANKL), expressed on stromal cells. This interaction is blocked by osteoprotegerin OPG, a decoy receptor for RANKL. Osteoblasts differentiate from mesenchymal precursors in bone marrow in response to activation of the osteoblast-specific transcription factor Cbfa1 [42]. The authors suggested that increased NO production and cytokine activation are relevant to the pathogenesis of osteoporosis in inflammatory diseases such as rheumatoid arthritis. Bone adaptation to mechanical loading is promoted by activated osteocytes, which produce signaling of NO molecules and modulate the activity of the bone-forming osteoblasts and the bone-resorbing osteoclasts [43].

Sodium nitroprusside or 1-hydroxy-2-oxo-3,3-bis(3-aminoethyl)-1-triazene, affects cyclic guanosine monophosphate (cGMP), and thereby modifies bone resorption by osteoclasts [44]. Studies of cGMP regulation in avian osteoclasts, and the roles of NO and natriuretic peptides, which induces the excretion of sodium by the kidneys, revealed the following events: (1) in response to the NO donors, osteoclasts produce cGMP and express NO-activated guanylate cyclase and cGMP-dependent protein kinase (G-kinase); (2) the process reduces membrane HCI transport activity and phosphorylates a 60-kD osteoclast membrane protein; and (3) bone degradation was reduced by activity of the NO generators and hydrolysis-resistant cGMP analogues and the osteoclast centers on HCI secretion, that dissolves bone mineral. In contrast, cGMP antagonists increased activity. It was concluded that cGMP is produced in response to NO made by other cells and acts as a negative regulator of osteoclast activity. Administration of the NOS inhibitor N(G)-monomethyl-L-arginine (L-NMMA) L-NMMA also markedly reduced the extent of bone loss and the percentage of MMP-1-synthesizing osteoblasts stimulated by lipopolysaccharide [45]. NO produced by osteoblasts stimulates osteoblast proliferation and increases bone mass in homeostasis and under hormonal stress [46].

10.3.2. Nitric Oxide and Arthritis

Arthritis is a disorder that affects joints involving chondrocytes, cartilage, and other joint tissues and is an inflammatory process. There are two common forms of arthritis: osteoarthritis (OA), a degenerative disease, affects the fingers, knees, and hips; and rheumatoid arthritis (RA), which is an autoimmune disorder, affects the hands and feet [47]. Nitric oxide and its redox derivatives are associated the development and progression of OA and may also play protective roles in in inflammation and pain perception the joint OA [47–50]. Thus, NO may exhibit a complex mixture of positive and negative effects in osteoarthritis. In a vascular pain model of rats, pain was reduced by infusing L-NAME, an inhibitor of NOS. The pain may be restored by infusing the NO donor sodium nitroprusside. Low level production of NO increases vasodilatation, improves circulation, and reduces nerve irritation and inflammation that lead to the relief of pain. In contrary, increased NO production after activation of iNOS by inflammatory cytokines can increase pain perception.

Experiments demonstrated that NO involves in the catabolic development of OA, mediates the inflammatory response, promotes degradation of matrix metalloproteinases, inhibits the synthesis of collagen and proteoglycans, and helps to mediate apoptosis [48]. For example, in cultured chondrocytes, NO prevents the nuclear localization of the transcription factor nuclear factor-κB that resulted in the inhibition of pro-inflammatory activation. In addition, NO can stimulate collagen synthesis in cultured rat fibroblasts and human tendon cells. In contrast, peroxynitrite enhances the inflammatory response by sustaining the nuclear localization of nuclear factor-κB. According to [51], osteoarthritis associated with increased levels of reactive nitrogen and oxygen species and pro-inflammatory cytokines, such as interleukin-1 (IL-1). NO can mediate corresponding catabolic effects. IL-1 induces DNA damage through both strand breaks and base modifications by increasing NO or superoxide. A NOS2 inhibitor and superoxide dismutase (SOD) reduce IL-1-mediated DNA damage.

In rheumatoid arthritis (RA), the disease of the joints of hands, feet, shoulder, and knee, NO is implicated in inflammation, angiogenesis, and tissue destruction [52, 53]. During RA in the synovial joints affected by RA, the localized overproduction of NO is produced by iNOS, expressed in the synovial macrophages and the fibroblast-like synoviocytes [52]. The pro- and anti-inflammatory cytokine signaling network is involved in the regulation of iNOS expression. A proposed model explained the impact of interferon and interleukin-10 pathways on signaling G protein Rac2-iNOS interaction. This interaction leads to overproduction of NO and causes chronic inflammation in the RA synovium. Consequences of increased nitric oxide production in NO-dependent tissue of RA and a variety of rheumatic diseases were discussed in detail in review [53]. For instance, NO regulates T cell functions under physiological conditions. Overproduction of NO may contribute to T lymphocyte dysfunction: T cells from RA patients produce 2.5 times more

NO than healthy donor T cells. Suppression of arthritis in streptococcal cell wall (SCW) fragment-treated rats by inhibitors of NOS was established [54]. Experiments showed that administration of NG-monomethyl-L-arginine, an inhibitor of NOS, reduced NO production, elevated in the inflamed joints of rats, and corresponding synovial inflammation and tissue damage [54].

Increased levels of NO in the serum of patients with RA was reported [55]. Oxidative stress generated by RNS within inflammatory joints of RA patients induced autoimmune phenomena and joint destruction. Radical species with oxidative activity act as mediators of inflammation and cartilage damage. Concentration of NO correlates with disease activity, inflammatory markers, and radiological joint status.

Finding that NO production is elevated in the inflamed joints of SCW-treated rats was confirmed in [56]. Administration of NG-monomethyl-L-arginine profoundly reduced the synovial inflammation and tissue damage. These studies implicate the NO pathway in the pathogenesis of an inflammatory arthritis and demonstrate the ability of a NOS inhibitor to modulate the disease.

References

[1] Barbato JE, Edith Tzeng E. Nitric oxide and arterial disease. *Journal of Vascular Surgery* (2004) 40, 187–193.

[2] Wang D, Wang Z, Zhang L, Wang Y. Di Wang, Zhiyan Wang, Lili Zhang, Yi Wang. Roles of Cells from the Arterial Vessel Wall in Atherosclerosis. *Mediators of Inflammation*. Volume 2017 (2017), Article ID 8135934, 9 pages.

[3] Naseem KM. The role of nitric oxide in cardiovascular diseases. *Molecular Aspects of Medicine* (2005) 26, 33–65.

[4] Kohlgrueber S, Upadhye A, Dyballa-Rukes N, McNamara, CA., Altschmied J. Regulation of Transcription Factors by Reactive Oxygen Species and Nitric Oxide in Vascular Physiology and Pathology. *Antioxidants & Redox Signaling* (2017), 26), 679-699.

[5] Terrence Pong, Huang PL. Effects of Nitric Oxide on Atherosclerosis. In Hong Wang Patterson C (Eds) *Atherosclerosis: Risks, Mechanisms, and Therapies.* John Wiley & Sons, 2015.

[6] Napoli C, de Nigris F, Ignarro WS, Pignalosa O, Sica V, Ignarro LJ, Nitric oxide and atherosclerosis: An update. *Nitric Oxide* (2006) 15, 265-279.

[7] Mukadder Yasa, Türkseven S. Vasoprotective Effects of Nitric Oxide in Atherosclerosis. *J. Pharm. Sci.,* (2005) 30, 41-53.

[8] Grange R, Isotani E, Lau K, Kamm K, Huang P, Stull J. Nitric oxide contributes to vascular smooth muscle relaxation in contracting fast-twitch muscles. *Physiol Genomics.* (2001) 5, 35–44.

[9] Murohara T, Asahara T. Nitric oxide and angiogenesis in cardiovascular disease. *Antioxid Redox Signal*. (2002) 4, 825-831.

[10] Wang H, Patterson C, Pong T, Huang PL. Effects of Nitric Oxide on Atherosclerosis. In: *Atherosclerosis: Risks, Mechanisms, and Therapies*. Wiley, 2015.

[11] Wang H, Patterson C, Pong T, Huang PL., Sarti P. *Nitric Oxide in Human Health and Disease*. In: eLS. John Wiley & Sons Ltd, Chichester, 2013 http://www.els.net [doi 10.1002/9780470015902. a0003390.pub2]

[12] Huige Li, Horke S, Förstermann U. Vascular oxidative stress, nitric oxide and atherosclerosis. *Atherosclerosis* (2014) 237, 208–219.

[13] Seinosuke Kawashima S, Mitsuhiro Yokoyama M. Dysfunction of Endothelial Nitric Oxide Synthase and Atherosclerosis. *Arteriosclerosis, Thrombosis, and Vascular Biology* (2004) 24, 998-1005.

[14] Rastaldo R, Pagliaro P, Cappello S, Penna C, Mancardi D, Westerhof N, Losano G. Nitric oxide and cardiac function. *Life Sci*. (2007) 81,779-793.

[15] Liu VW, Huang PL. Cardiovascular roles of nitric oxide: A review of insights from nitric oxide synthase gene disrupted mice. *Cardiovascular Research* (2008) 77, 19–29.

[16] Donnan GA, Fisher M, Macleod M, Davis SM. "Stroke." *Lancet* (2008) 371, 1612–1623.

[17] Godinez-Rubi M, Rojas-Mayorquin AE, Ortuno-Sahagun D. Nitric oxide donors as neuroprotective agents after an ischemic stroke-related inflammatory reaction. Oxidative *Medicine and Cellular Longevity* (2013) 297357, 16 pp.

[18] Claire L. Gibson CL, Teresa C. Coughlan TC, Sean P. Murphy SP Glial nitric oxide and ischemia. *GLIA (*2005) 50, 417–426.

[19] Guix FX, Uribesalgo I, Coma M, Muñoz FJ. The physiology and pathophysiology of nitric oxide in the brain. *Progress in Neurobiology* (2005) 76, 126–152.

[20] Chen Z-q, Mou Ru-t, FengD-x, Wang Z, Chen G. The role of nitric oxide in stroke. *Med Gas Res*. (2017);7:194–203

[21] Araki N. Nitric oxide production during cerebral ischemia and reperfusion in eNOS- and nNOS-knockout mice. *Curr Neurovasc Res*. (2010)7, 23-31.

[22] Ignarro LJ, Napoli C, Loscalzo J. Nitric Oxide Donors and Cardiovascular Agents Modulating the Bioactivity of Nitric Oxide: An Overview. *Circ Res*. 2002. 90:21-28.

[23] Scatena R, Bottoni PA, Pontoglio A, Giardina B Pharmacological modulation of nitric oxide release: new pharmacological perspectives, potential benefits and risks. *Current Medicinal Chemistry* (2010) 17, 61–73.

[24] Terpolilli NA, Moskowitz MA, Plesnila N. Nitric oxide: considerations for the treatment of ischemic stroke. *J Cereb Blood Flow Metab*. (2012) 32: 1332–1346.

[25] Thatcher GRJ, Nicolescu AC, Bennett BM, Toader V. Nitrates and no release: Contemporary aspects in biological and medicinal chemistry. *Free Radical Biology & Medicine* (2004) 37, 1122–1143.

[26] Willmot M, Gibson Cl, Gray L, Murphy S, Bath P. Nitric oxide synthase inhibitors in experimental ischemic stroke and their effects on infarct size and cerebral blood flow: A systematic review. *Free Radical Biology & Medicine* (2005) 39, 412-425.

[27] Zhang RL, Zhang ZG, Chopp M. Targeting nitric oxide in the subacute restorative treatment of ischemic stroke. *Expert Opinion on Investigational Drugs* (2013) 22, 843-851.

[28] Ito Y, Ohkubo T, Asano Y, Hattori K, Shimazu T, Yamazato M, Nagoya H, Kato Y, Araki N. Nitric oxide production during cerebral ischemia and reperfusion in eNOS- and nNOS-knockout mice. *Current Neurovascular Research* (2010) 7, 23–31.

[29] Moro MA, Cardenas A, Hurtado O, Leza JC, Lizasoain I. Role of nitric oxide after brain ischaemia. *Cell Calcium* (2004) 36, 265–275.

[30] Tabuchi M, Umegaki K, Ito T, Suzuki M, Ikeda M, Tomita T. Fluctuation of NOx concentration at stroke onset in a rat spontaneous stroke model (M-SHRSP). Peroxynitrite formation in brain lesions. *Brain Res.* (2002) 949, 147–156.

[31] Guix FX, Uribesalgo I, Coma M, Muñoz FJ. The physiology and pathophysiology of nitric oxide in the brain. *Progress in Neurobiology* (2005) 76, 126–152.

[32] Khan M, Sekhon B, Giri S, Jatana M, Gilg AG, Ayasolla K, Elango C, Singh AK, Singh I. S-Nitrosoglutathione reduces inflammation and protects brain against focal cerebral ischemia in a rat model of experimental stroke. *Journal of Cerebral Blood Flow and Metabolism* (2005) 25, 177–192.

[33] Greco R, Mangione AS, Amantea D, Bagetta G, Nappi G, Tassorelli C. IkappaB-alpha expression following transient focal cerebral ischemia is modulated by nitric oxide. *Brain Research* (2011) 1372, 145–151.

[34] Palumbo ML, Fosser NS, Rios H, Zorrilla Zubilete MA, Guelman LR, Cremaschi GA, Genaro AM. Loss of hippocampal neuronal nitric oxide synthase contributes to the stress-related deficit in learning and memory. *Journal of Neurochemistry* (2007) 102, 261–274.

[35] Contestabile A, Ciani E. Role of nitric oxide in the regulation of neuronal proliferation, survival and differentiation. *Neurochemistry International* (2004) 45, 903–914.

[36] D'Acquisto F, Maiuri MC, de Cristofaro F, Carnuccio R. Snuclear factor-κB and nuclear factor-interleukin-6 activation. *Naunyn-Schmiedeberg's Archives of Pharmacology* (2001) 364, 157–165.

[37] Reynaert NL, Ckless K, Korn SH, Vos N, Guala AS, Wouters EF, van der Vliet A, Janssen-Heininger YM. Nitric oxide represses inhibitory κB kinase through S-nitrosylation. *Proc. Natl Ac. Sci. USA* (2004) 101, 8945–8950.

[38] van't Hof RJ, MacPhee J, Libouban H, Helfrich MH, Ralston SH. Regulation of Bone Mass and Bone Turnover by Neuronal Nitric Oxide Synthase. *Endocrinology* (2004) 145 5068-5074.

[39] Wimalawansa SJ. Nitric oxide and bone. *Ann. N Y Acad. Sci.* (2010) 1192, 391-403.

[40] Klein-Nulen DJ, van Oers RFM, Bakker AD, Bacabac RG. Nitric oxide signaling in mechanical adaptation of bone. *Osteoporosis International* (2014) 25, 1427–1437.

[41] Wimalawansa SJ. Nitric oxide: Novel therapy for osteoporosis. *Exp. Opin. Pharmacother.* (2008) 9, 3025–3044.

[42] Corbett SA, Hukkanen M, Batten J, McCarthy ID, Polak JM, Hughes SP. Nitric oxide in fracture repair: Differential localisation, expression and activity of nitric oxide synthases. *J. Bone Joint Surg. Br.* (1999) 81, 531–537.

[43] van't Hof RJ, Ralston SH. Nitric oxide and bone. *Immunology* (2001) 103, 255-261.

[44] Klein-Nulen DJ, van Oers RFM, Bakker AD, Bacabac RG. Nitric oxide signaling in mechanical adaptation of bone. *Osteoporosis International* (2014) 25, 1427–1437.

[45] Dong SS, Williams JP, Jordan SE, Cornwell T, Blair HC. Nitric oxide regulation of cGMP production in osteoclasts. (1999) *J. Cell Biochem.* 73: 478–487.

[46] Lin SK, Kok SH, Kuo MY, Lee MS, Wang CC, Lan WH, Hsiao M, Goldring SR, Hong CY. Nitric oxide promotes infectious bone resorption by enhancing cytokine-stimulated interstitial collagenase synthesis in osteoblasts. *J. Bone Miner. Res.* (2003) 18, 39–46.

[47] Grover M, Nagamani S, Erez A, Lee B. Dissecting the role of osteoblast derived nitric oxide in bone remodeling. *Bone Abstracts* (2013) **2** OP15 DOI:10. 1530/boneabs.2.OP1.

[48] Mackenzie IS, Rutherford D, MacDonald TM. Nitric oxide and cardiovascular effects: New insights in the role of nitric oxide for the management of osteoarthritis. *Arthritis Research & Therapy* (2008) 10(Suppl 2): S3 pp 1-12.

[49] Abramson SB. Nitric oxide in inflammation and pain associated with osteoarthritis. *Arthritis Res. Ther.* (2008) 10(Suppl 2): S2.

[50] Bonnet CS, Walsh DA. Osteoarthritis, angiogenesis and inflammation. *Rheumatology* (Oxford). (2005) 44, 7–16.

[51] Hancock CM, Riegger-Krugh C. *Modulation of pain in osteoarthritis: the role of nitric oxide. Clin. J. Pain.* (2008) 24, 353-365.

[52] Davies CM, Guilak F, Weinberg JB, Fermor B. Reactive nitrogen and oxygen species in interleukin-1-mediated DNA damage associated with osteoarthritis. *Osteoarthritis Cartilage* (2008) 16,624–630.

[53] McInnes IB, Schett G. Cytokines in the pathogenesis of rheumatoid arthritis. *Nat Rev Immunol.* (2007) 7, 429–442.

[54] Nagy G, Koncz A, Telarico T, Fernandez D, Érsek B, Buzás E, Perl A. Central role of nitric oxide in the pathogenesis of rheumatoid arthritis and sysemic lupus erythematosus. *Arthritis Res. Ther*. (2010) 12, 210.

[55] McCartney-francis N, Allen BJ, Mizel DE. Suppression of arthritis by an inhibitor of nitrice oxide synthase. *J Exp Med.* (1993)178, 749–754.

[56] Ali AM, Habeeb RA, El-Azizi NO, Khattab DA, Abo-Shady RA, Elkabarity RH. Higher nitric oxide levels are associated with disease activity in Egyptian rheumatoid arthritis patients. *Rev. Bras. Reumatol.* (2014). 54, 446-451.

[57] Gonzalez-Gay MA, Llorca J, Sanchez E, Lopez-Nevot MA, Amoli MM, Garcia-Porrua C, Ollier WE, Martin J. Inducible but not endothelial nitric oxide synthase polymorphism is associated with susceptibiy to rheumatoid arthritis in northwest Spain. *Rheumatology* (Oxford) (2004) 43, 1182–1185.

Chapter 11

NITRIC OXIDE AND THE CENTRAL NERVOUS SYSTEM

ANNOTATION

The central nervous system (CNS) is the part of the nervous system consisting of the brain and spinal cord. Nitric.oxide, as the signaling molecule, neurotransmitter, as well as initiator of redox chemical reactions, strongly implicate in the CNS key biochemical and physiological processes. The level of NO in cells of organisms is also significantly affecting the sclerotic phenomena.

11.1. INTRODUCTION

Three isoforms of NOS involving the CNS have been identified as NOS1 or neuronal NOS (nNOS), NOS2 or inducible NOS (iNOS), and NOS3 or endothelial NOS (eNOS) [1–5]. These three isoforms differ in their location: (1) nNOS localizes to synaptic spines, astrocytes, in neuronal cell bodies, especially in the cortex, hippocampus, hypothalamus, olfactory bulb, claustrum, amygdala, and thalamus, and the loose connective tissue surrounding blood vessels in the brain; (2) iNOS is expressed by glial cells, in macrophages, infiltrating neutrophils, and, to some extent, neurons; and (3) eNOS is present in cerebral vascular endothelial cells, motor neurons, and choroid plexus. The physiological roles of NO, as a Janus-faced molecule, in the CNS depend on its local concentrations, as well as its availability and the nature of downstream target molecules.

11.2. Nitric Oxide as a Signaling Molecule

In the CNS, nitric oxide is an important regulator of biochemical physiological processes [6–10]. NO is a beneficial physiological agent utilized for essential functions, such as (1) differentiation or neurotransmission promoting optimal cerebral blood flow, (2) consolidating memory processes, (3) facilitating long-term potentiation, (4) maintaining sleep-wake cycles, (5) regulating of synaptic plasticity, hormone secretion, cell viability, and (6) assisting in normal olfaction. At pathological levels, NO adversely affects brain function, such as (1) producing nitroxidative stress, (2) causing or exacerbating central nervous system disease and injury, and (3) promoting development of neurodegenerative diseases including Alzheimer's disease, Parkinson's disease, and other disorders of the CNS (see also chapter 12). A variety of factors, such as the cellular environment and the rate of NO flux, dictate whether NO is helpful or harmful. The availability of second-messenger cascades for utilization by NO for beneficial or toxic cell signaling are also factors of great importance.

The nitric oxide signaling pathway in brain includes the following steps: (1) NO is synthetized by two Ca^{2+}-dependent or one independent Ca^{2+}-mediated processes, namely, (i) NOS1 or neuronal NOS (nNOS)-catalyzed reaction converts L-arginine into L-citrulline in the presence of O_2, nicotinamide adenine dinucleotide phosphate (NADPH) and tertiary-butyl hydroperoxide (TBH) after the activation of the NMDA receptor by Ca^{2+} and (ii) intracellular Ca^{2+} activates eNOS to release NO from brain microvessels; (2) NO binds to sGC receptors, which trigger a cGMP-dependent pathway and interacts with its downstream effectors (cGKI, CNG, PKG, PDE), the ultimate mediators of the NO's physiological response; (3) NO initiates the synthesis of ONOO– when O^{2-} is present, which results in a dysfunctional uncouple variety of NOS that produces O^{2-} rather than NO; and (4) NO is synthesized following the transcriptional expression of a Ca^{2+}-independent iNOS isoform in glial cells, astrocytes, and microglia after cytokine exposure [5].

Nitric oxide bears a variety of cellular functions in the brain [4]. Under pathological conditions, free radicals may deplete NO produced by eNOS through the formation of ONOO–, thus decreasing the vascular bioavailability of NO, which results in the blood-brain barrier (BBB) dysfunction. This process causes endothelial damage, edema development, and hypoxia. The NO produced by iNOS in glial cells or by nNOS under excitotoxic process can form O^{2-}, $ONOO^-$, •OH, and NO_2, thereby producing several deleterious effects on tissue, such as through tyrosine nitration and cysteine oxidation in various proteins.

Different steps in the NO signaling cascade under physiological/pathological conditions in the brain were described [4]: (1) NOS1 or neuronal NOS (nNOS) catalyze the NO synthesis after the activation of the NMDA receptor by Ca^{2+}; (2) under excitotoxic conditions, excessive Ca^{2+} leads to nNOS hyperactivity, whereas excessive NO production can combine with superoxide to form peroxynitrite, which is responsible for tissue damage due to several biological effects, including blockage of the eNOS pathway and BBB impairment; (3)NO is synthesized following the transcriptional expression of a Ca^{2+}-independent NOS2 or iNOS isoform in glial cells after cytokine exposure, thereby contributing to neuroinflammation and tissue damage in the brain; (4) intracellular Ca^{2+} activates NOS3 or eNOS to release NO from brain microvessels; (5) NO binds to soluble guanylyl-cyclase (sGC) receptors, which triggers a cGMP-dependent pathway and interacts with its downstream mediators of the physiological regulation of vasodilation and vascular resistance, platelet adhesion and aggregation, leukocyte-endothelial interaction, and BBB integrity maintenance [4].

11.3. Nitric Oxide as Neurotransmitter

Synaptic transmission is the process by which signaling molecules, neurotransmitters, are released by the presynaptic neuron and bind to and activate the receptors of the postsynaptic neuron [3, 8, 11–17]. Neurotransmission is essential for the process of communication between two neurons. In the nervous system, transduction is an event wherein a physical stimulus is converted into an action potential, which is transmitted along axons toward the central nervous system. Nitric oxide is a key messenger that regulates synaptic transmission in the cerebral cortex [8]. NO binds to guanylyl-cyclase and, through cGMP-mediated signaling, acts as a neurotransmitter serving either as a post- or a presynaptic messenger. NO may activate the cGMP-dependent protein kinase G (PKG) pathway that phosphorylates synaptophysin (a major synaptic vesicle protein p38). The phosphorylation leads to potentiating and facilitating neurotransmission, and also acts on inhibitory gamma-aminobutyric acid (GABA)-ergic synaptic transmission via cGMP-dependent pathways as well as on ion channels and channels and exchangers.

Nitric oxide, as a transmitter molecule, can also alter neural activity without direct synaptic connections [15]. GABA is found to be the chief inhibitory neurotransmitter. The whole-cell recordings under voltage clamp was used to investigate the effect of NO on spontaneous GABAergic synaptic transmission in the mechanically isolated rat auditory cortical A1 neurons preserving functional presynaptic nerve terminals. The NO donor, S-nitroso-*N*-acetylpenicillamine (SNAP), reduced the GABAergic induced pluripotent stem cell (IPSC) frequency and inhibits spontaneous GABA release by activation of cGMP-dependent signaling and inhibition of presynaptic Ca^{2+} channels in the presynaptic nerve terminals of A1 neurons. It was also revealed that the application of S-nitroso-N-acetyl-dl-penicillamine (SNAP, 100 μM), an NO donor, suppressed uIPSC amplitudes in 31% of the connections, whereas 39% of the connections showed IPSC facilitation. The NO scavenger, 2-phenyl-4,4,5,5-tetramethylimidazolineoxyl-1-oxyl-3-oxide (PTIO), or the inhibitor of guanylate cyclase, 1H-[1,2,4]oxadiazolo[4,3-a]quinoxalin-1-one (ODQ), abolished the SNAP-induced uIPSC modulation.

In NO signaling pathways at a neuronal synapse, glutamate release activates postsynaptic NMDA and AMPA receptors (NMDAR, AMPAR) leading to Ca^{2+}-induced nNOS activation. NO diffuses to activate sGC to produce cGMP, which affects presynaptic neurotransmitter release and target several ion channels. nNOS also associates with CAPON to activate a downstream MAP kinase cascade [4]. The increase in synaptic puncta (interacting synaptic proteins that colocalize) involving signaling through the NO-cGMP-cGK pathway and the possible roles of two classes of molecules that regulate the actin cytoskeleton (Ena/VASP proteins and Rho GTPases) were examined [13]. It was found that NO, cGMP, cGK, actin, and Rho GTPases including RhoA act directly on potentiation in both the presynaptic and postsynaptic neurons. As a neurotransmitter, NO may activate the cGMP-dependent protein kinase G (PKG) pathway that phosphorylates synaptophysin, which is critical for fusion of presynaptic vesicles, thereby potentiating and facilitating neurotransmission.

In experiments with mouse diaphragm muscle, electrophysiological and fluorescence techniques were employed to investigate the effects of a nitric oxide donor, SNAP, and an NO-synthase blocker, NG-nitro-L-arginine methyl ester (LNAME), on transmitter release and processes of exo- and endocytosis of synaptic vesicles in the motor nerve ending [16]. Results suggested that exogenous and endogenous NO in the mouse neuromuscular synapse caused the depression of neurotransmitter release. Due to a decrease in endocytosis or/and mobilization of synaptic vesicles from a recycling pool to the exocytosis sites, the suppression of synaptic-vesicle recycling occurs.

Nitric oxide, produced enzymatically in postsynaptic structures in response to the activation of the excitatory amino acid receptor, influences neurotransmitter release and synaptogenetic processes and modulates the synaptic plasticity [17]. NO plays significant roles in the developing and mature brain. The synaptic modulation, the learning and memory, the neurotoxicity, and the neuron death are processes of the NO influence. NO acts as an unconventional neurotransmitter that diffuses from one neuron to another without binding to any receptors. Activation of NOS can result in formation of peroxynitrite, which can activate the enzyme poly ADP ribosyl synthase, eventually leading to the depletion of ATP and initiate oxidative stress.

At high levels, NO is involved in the following events [18]: (1) rapid inhibition of mitochondrial respiration; (2) slow inhibition of glycolysis; (3) induction of mitochondrial permeability transition, and/or (iv) activation of poly-ADP-ribose polymerase, which induces energy depletion-induced necrosis. NO can also: (1) induce apoptosis, via oxidant activation of: p53, p38 MAPK pathway or endoplasmic reticulum stress; (2) at low levels block cell death via cGMP-mediated: vasodilation, activation, or block of mitochondrial permeability transition; (3) at high levels protect by killing pathogens, activating NF-kappaB or S-nitro(sy)lation of caspases and the NMDA receptor. GAPDH, Drp1, mitochondrial complex I, matrix metalloprotease-9, Parkin, XIAP, and protein-disulphide isomerase can also be S-nitrosylated. Neurons are sensitive to NO-induced excitotoxicity because NO rapidly induces both depolarization and glutamate release, which together activate the *N*-methyl-D-aspartate NMDA receptor [18]. The nNOS activation may contribute to excitotoxicity, most probably via peroxynitrite activation of poly-ADP-ribose polymerase and/or mitochondrial permeability transition.

The role of extracellular calcium in the interaction between intracellular cAMP and nitric oxide (NO)/cGMP on the contractility of rat diaphragm pretreated with aminophylline, competitive nonselective phosphodiesterase inhibitor, which raises intracellular cAMP, was investigated [19]. Experiments showed that in a Ca^{2+}-free medium, L-NAME depresses aminophylline-induced potentiation of tension developed (Td). Verapamil and nicardipine, significantly antagonized the potentiating effect of L-NAME on Td in a calcium-containing medium. In the optic nerve of a rat, NO released from blood vessels depolarizes axons [20]. Detected changes in the axonal membrane potential suggested that the tonic NO production is maintained by phosphorylation of eNOS and that PI3 kinase-mediated eNOS phosphorylation is partially responsible for the process. NO plays key roles as a neurotransmitter in blood vessels of autonomic efferent nerves [21]. The physiological roles of this nerve in the control of smooth muscle tone of the artery, vein, and corpus cavernosum, and pharmacological and pathological implications of neurogenic NO have been reviewed. Decreased formation of NO from endothelial cells, autonomic nitrergic nerves, or brain neurons and increase production of reactive oxygen species, which impairs the cerebral blood flow, was reported [22].

Glial cells, are non-neuronal cells that maintain homeostasis, form myelin, and provide support and protection for neurons in the central and peripheral nervous systems. Various signaling cascades converge to activate several transcription factors that control the transcription of iNOS in glial cells. For example, the regulation of iNOS in astroglia, a group of star-shaped glial cells in the brain and spinal cord, prevents from the regulation of this gene in other immune cells in periphery such as macrophage. Neuroinflammation represents the coordinated cellular response to tissue damage and is characterized by the microglial release of pro-inflammatory factors, such as cytokines, proteases, and toxic free radicals [5]. Neuroinflammation-induced cell death is caused by the increase of reactive oxygen and nitrogen species (RONS). The species play a major role in eliciting apoptotic cell death through irreversible oxidative or nitrosative injury to neuronal elements [23].

General overview of MAP-kinase pathways (MKP), transcription factors, and iNOS transactivation in glial cells was reported [23]. It was shown that activation of cytokine receptors induces a downstream signal, which is mediated by MKPs, and several transcription factors act downstream of a specific MKP, which then translocate to the nucleus to transactivate iNOS. The central role of p38 MKP in iNOS induction was stressed (see details in [23]). In review [24], a large body of evidence indicates that NO plays an important role in the processing of persistent inflammatory and neuropathic pain in the spinal cord. As an example, multiple downstream signaling mechanisms of NO was proved by finding that spinal delivery of NO donors caused dual pronociceptive and antinociceptive effects. Nociception is the sensory nervous system's response to certain harmful or potentially harmful stimuli.

11.4. Nitric Oxide in Multiple and Amyotrophic Lateral Sclerosis

Multiple sclerosis (MS) is an inflammatory disease in which the myelin of the SNC has damaged insulating covers of nerve cells in the brain and spinal cord and results in physical, mental, and psychiatric problems [25] Amyotrophic lateral sclerosis (ALS) is a disease that causes the death of motor neurons in the CNS, which control muscles and glands, and results in weakness, paralysis, atrophy of the muscles, and death [26, 27].

Biochemical and physiological studies of the cerebrospinal fluid (CSF), blood, and urine of patients, and the pathological study of MS lesions themselves, provided unambiguous evidences that the production of NO is significantly raised within MS lesions [28–30]. There are numerous evidences that NO is involved in multiple sclerosis, including disruption of the BBB, oligodendrocyte injury and demyelination, axonal degeneration, and that it contributes to the loss of function by impairment of axonal conduction [31]. For example, the concentrations of nitrate and nitrite are raised in the CSF, blood, and urine of patients with MS. The net effect of NO production in MS can also has several beneficial immunomodulatory effects. In inflammation of multiple sclerosis, immune T-cells bypassing the BBB affect oligodendrocyte structure and activate glial response through NF-κB and AP-1. Reactive oxygen species (ROS) and NO secretion by activated microglia and astrocytes further contribute to myelin damage, axon degradation, and ultimate neuronal death (see details in text and [5]).

The genetic basis of three NOS genes, NOS1, NOS2A, and NOS3, with susceptibility to MS, was examined [34]. The results suggested that: (1) two NOS3 markers were associated with susceptibility to MS and early disease development; (2) haplotypes obtained from NOS2A and NOS3 showed increased susceptibility to MS; and (3) NOS1 indicates no significant association with MS. Thus, this study evidenced for the association between selected NOS2 and NOS3 markers and MS susceptibility. During neuroinflammation of MS, the lesions from nerve terminals, the induction of iNOS, or the release of neurotransmitters affected by NO occur [10]. The inflammatory process is associated with a disturbance of the BBB and the cerebral vessels. NO has two major effects of NO on cerebral vessels, namely, vasodilation—the widening of blood vessels—and a disturbance of the BBB. Findings obtained in work [35] suggested that the NOS inhibitor NG-monomethyl-L-arginine (L-NMMA) and noncompetitive inhibitor of the action of nitric oxide-sensitive guanylyl cyclase H-[1,2,4]Oxadiazolo[4,3-a]quinoxalin-1-one (ODQ) inhibits the expression of vascular endothelial growth factor (VEGF), that may contribute to disruption of the BBB. NO and its reactive derivative peroxynitrite are implicated in the pathogenesis of MS [27]. For instance, molecular imaging and pharmacological experiments showed that macrophage-derived RONS can trigger mitochondrial pathology and initiate "focal axonal degeneration" [27].

Sporadic amyotrophic lateral sclerosis (sALS) is associated with both excess NO metabolites and decreased protective superoxide dismutase (SOD) activity in the CSF [32]. Experiments revealed stable NO metabolite levels to be significantly higher and SOD activity lower in the CSF of sALS patients. These results provide evidence *in vivo* suggesting that NO products and SOD activity significantly contribute to the oxidant/antioxidant imbalance in sporadic ALS. Oxidative damage is a common and early feature of ALS. In ALS, physiological levels of NO promote survival of motor neurons, but the same concentrations can stimulate motor neuron apoptosis and glial cell activation under pathological conditions providing a complex mechanism involving multiple cell types in the pathogenesis of ALS [33]. In response to injury or chronic oxidative stress, damaged motor neurons upregulate expression of critical genes involved in their survival/death including nNOS. Activation of upregulated receptors to amplify damage in affected motor neurons, leading to apoptosis, is caused by expressive release of NO and soluble proapoptotic factors (FasL and NGF).

Investigation of glial-induced neuroinflammation and neurotoxicity in ALS revealed that [5]: (1) reactive astrocytes contribute to the degenerative process by influencing the activity of microglial and immune cells; (2) an upregulation of filament glial fibrillary acidic protein (GFAP) occurs; (3) astrocytes increase the release of pro-inflammatory markers including NO and ROS; (4) when mutated SOD1 accumulates within microglia, and the later generates substances potentially harmful to other cells; (5) demyelinization and progressive loss of cholesterol is also observed after oligodendrocyte damage; and (6) a reduction in the monocarboxylate transporter 1 (MCT1), in which in turn the energy supplies to the neuron is difficult [5].

The release of pro-inflammatory markers, including NO and ROS, causing DNA damage by oxidation, as well as demyelinization, was also reported. Peroxynitrite-mediated tyrosine nitration has been suggested as a key for triggering neuronal degeneration in ALS [36]. In particular, the appearance of both free and protein-linked 3-nitro-l-tyrosine (nitrotyrosine), increased free nitrotyrosine levels in the spinal cord of transgenic mice expressing ALS-linked superoxide dismutase mutants. There are evidences that free nitrotyrosine may play a role in the induction of motor neuron apoptosis in ALS and ROS and NO secretion by activated microglia and astrocytes contribute to myelin damage, and also to axon degradation, and ultimate neuronal death.

REFERENCES

[1] Steinert JR, Chernova T, Forsythe ID. Nitric oxide signaling in brain function, dysfunction, and dementia. *Neuroscientist* (2010) 16, 435-452.

[2] Tardivo V, Crobeddu E, Pilloni G, Fontanella M, Spena G, Panciani PP, Berjano P, Ajello M, Bozzaro M, Agnoletti A, Altieri R, Fiumefreddo A, Zenga F, Ducati A, Garbossa D.

[3] Say "no" to spinal cord injury: is nitric oxide an option for therapeutic strategies? *International Journal of Neuroscience* (2015), 125, 81-90.

[4] Olivia May, *Nitric Oxide Contribution in the CNS: A NO brainer.* 2010-10-01 (https://www.caymanchem.com/Article/2159).

[5] Reis PA, Cassiano Felippe Gonçalves de Albuquerque CF, Maron-Gutierrez T, Silva AR, Hugo Caire de Castro Faria Neto. Role of Nitric Oxide Synthase in the Function of the Central Nervous System under Normal and Infectious Conditions. In *Biochemistry, Genetics and Molecular Biology, Nitric Oxide Synthase - Simple Enzyme-Complex Roles,* Seyed Soheil, Saeedi Saravi (eds), Chapter 4, 2017, pp 55-70.

[6] Yuste JE, Tarragon E, Campuzano CM, Ros-Bernal F. Implications of glial nitric oxide in neurodegenerative neurodegenerative diseases. *Front. Cell. Neurosci.* (2015) 9, 322.

[7] Dzoljic E, Grbatinic I, Kostic V. Why is nitric oxide important for our brain? *Functional Neurology*. (2015) 30, 159-163.

[8] Duncan AJ, Heales SJ. Nitric oxide and neurological disorders. *Molecular Aspects of Medicine* (2005) 26, 67-96.

[9] Yamamoto K, Takei H, Koyanagi Y, Koshikawa N, Kobayashi M. Presynaptic cell typedependent regulation of GABAergic synaptic transmission by nitric oxide in rat insular cortex. *Neuroscience* (2015) 284, 65-77.

[10] Akyol O, Zoroglu SS, Armutcu F, Sahin S, Gurel A. Nitric oxide as a physiopathological factor in neuropsychiatric disorders. *In Vivo*. (2004) 18, 377-390.

[11] Garry PS, Ezra M, Rowland MJ, Westbrook J, Pattinson KT. The role of the nitric oxide pathway in brain injury and its treatment—from bench to bedside. *Experimental Neurology*. 2015;263:235-243.

[12] Piedrafita B, Cauli O, Montoliu C, Felipo V. The function of the glutamate-nitric oxide-cGMP pathway in brain in vivo and learning ability decrease in parallel in mature compared with young rats. *Learn Mem*. (2007) 14, 254-258.

[13] Yassin L, Radtke-Schuller S, Asraf H, Grothe B, Hershfinkel M, al. Forsythe ID, Kopp-Scheinpflug C. Nitric oxide signaling modulates synaptic inhibition in the superior paraolivary nucleus (SPN) via cGMPdependent suppression of KCC2. *Frontiers in Neural Circuits*. 2014) 8, 65.

[14] Wang HG, Lu FM, Jin I, Udo H, Kandel ER, et al. Kandel ER, de Vente J, Walter U, Lohmann SM, Hawkins RD, Antonova I. Presynaptic and postsynaptic roles of NO, cGK, and RhoA in long-lasting potentiation and aggregation of synaptic proteins. *Neuron.* (2005) 45, 389-403.

[15] Förstermann UF, SessaWC. Nitric oxide synthases: regulation and function. *Eur Heart J.* (2012) 33: 829–837

[16] Garry PS, Ezra M, Rowland MJ, Westbrook J, Pattinson KT. The role of the nitric oxide pathway in brain injury and its treatment—from bench to bedside. *Experimental Neurology.* (2015) 263, 235-243.

[17] Lee JJ. Nitric oxide modulation of GABAergic synaptic transmission in mechanically isolated rat auditory cortical neurons. *Korean J Physiol Pharmacol* (2009) 13,461-467.

[18] Yakovleva OV, Shafigullin MU, Sitdikova GF. The role of nitric oxide in the regulation of neurotransmitter release and processes of exo- and endocytosis of synaptic vesicles in mouse motor nerve endings. *Neurochemical Journal* (2013), 7(2), 103-110.

[19] Bao-Lu Zhao. Nitric oxide in neurodegenerative diseases. *Frontiers in Bioscience* (2005) 10, 454-46.

[20] Brown GC. Nitric oxide and neuronal death. *Nitric Oxide* (2010) 23, 153-1651.

[21] Stojanovic R, Todorovic Z, Nesic Z, Vuckovic S, Cerovac-Cosic N, Prostran M. NG-nitro-L-argininemethyl ester-induced potentiaton of the effect of aminophylline on rat diaphragm: The role of extracellular calcium. *Journal of Pharmacological Sciences.* (2004) 96, 493-498.

[22] Bartus K. *Nitric oxide-mediated cGMP signal transduction in the central nervous system.* (2010) Doctoral thesis, UCL (University College London.

[23] Toda N, Okamura TS. The pharmacology of nitric oxide in the peripheral nervous system of blood vessels. *Pharmacological Reviews* (2003) 55, 271-324.

[24] Toda N, Ayajiki K, Okamura T. Cerebral blood flow regulation by nitric oxide in neurological disorders. *Canadian Journal of Physiology and Pharmacology* (2009) 87, 581-594.

[25] Saha RN, Pahan K. Regulation of inducible nitric oxide synthase gene in glial cells. *Antioxid. Redox Signal* (2006) 8, 929-947.

[26] Schmidtko, Achim. Nitric Oxide-Mediated Pain Processing in the Spinal Cord. *Handbook of Experimental Pharmacology.* (2015), 227(Pain Control), 103-117.

[27] Milo R, Kahana E Multiple sclerosis: Geoepidemiology, genetics and the environment. *Autoimmun. Rev.* (2010) 9, A387-94.

[28] Long Kv, Nguyễn LT. Roles of vitamin D in amyotrophic lateral sclerosis: Possible genetic and cellular signaling mechanisms. *Mol Brain.* (2013) 6, 16.

[29] Nikić I, Merkler D, Sorbara C, Brinkoetter M, Kreutzfeldt M, Bareyre FM, Brück W, Bishop D, Misgeld T, Kerschensteiner. A reversible form of axon damage in

experimental autoimmune encephalomyelitis and multiple sclerosis. *Nat. Med.* (2011) 17, 495-499.

[30] AlFadhli S, Mohammed EM, Al Shubaili A. Association analysis of nitric oxide synthases: NOS1, NOS2A and NOS3 genes, with multiple sclerosis. *Ann Hum Biol* (2013) 40, 368-375.

[31] *Multiple sclerosis: psychosomatic origins and the role of nitric oxide:* https://www.researchgate.net/publication/253441999_multiple_sclerosis_psychosomatic_origins_and_the_role_of_nitric_oxide [accessed Jul 23, 2017].

[32] Encinas JN, Manganas L, Enikolopov G. Nitric oxide and multiple sclerosis. *Current Neurology and Neuroscience Reports* (2005) 5: 232–238.

[33] Smith KJ, Lassmann H. The role of nitric oxide in multiple sclerosis. *The Lancet Neurology* (2002) 1, 232-241.

[34] Boll MC, Alcaraz-Zubeldia M, Montes S, Murillo-Bonilla L, Rios C. Raised nitrate concentration and low SOD activity in the CSF of sporadic ALS patients. *Neurochem Res.* (2003) 28, 699-703.

[35] Drechsel DA, Estévez AG, Barbeito L, Beckman JS. Nitric oxide-mediated oxidative damage and the progressive demise of motor neurons in ALS. *Neurotox Res.* 2012,22, 251-264.

[36] AlFadhli S, Mohammed EM, Al Shubaili A. Association analysis of nitric oxide synthases: NOS1, NOS2A and NOS3 genes, with multiple sclerosis. *Ann Hum Biol.* (2013) 40, 368-375.

[37] Mayhan WG. VEGF increases permeability of the blood-brain barrier via a nitric oxide synthase/cGMP-dependent pathway. *Am. J. Physiol.* (1999) 276, C1148-C1153.

[38] Peluffo H, Shacka JJ, Ricart K, Bisig CG, Martìnez-Palma L, Pritsch O, Kamaid A, Eiserich JP, Crow JP, Barbeito L, Estèvez AG. Induction of motor neuron apoptosis by free 3-nitro-L-tyrosine. *J Neurochem.* (2004) 89, 602-612.

Chapter 12

NEURODEGENERATIVE DISEASE

ANNOTATION

Neurodegenerative diseases including Alzheimer's, Parkinson's, and Huntington's diseases, depression, and migraine are caused by neurodegeneration, which is the progressive loss of structure or function of neurons. Nitric oxide (NO) plays a central role in the cascade of biochemical and physiological events leading to the pathology of these diseases. This chapter is dedicated to the common characteristics of the neurodegenerative diseases and peculiarities of each of them from the point of view of the participation of NO and its derivatives.

12.1. INTRODUCTION

Alzheimer's disease (AD), is a chronic disease accompanying loss of short-term memory and motivation, problems with language and orientation, not managing self-care, and behavioral issues [1, 2]. Parkinson's disease (PD) is a long-term degenerative disorder of the central nervous system [1, 2]. Shaking, rigidity, slowness of movement, and difficulty with walking are the most common symptoms. Huntington's disease (HD), is a disorder with problems of mood, mental abilities, lack of coordination, and an unsteady gait, which results in death of brain cells [1, 2].Regarding to Huntington's disease, its biochemistry, physiology, symptoms, and reactivity to NO are similar to that for the Parkinson's diseases. The key difference between the both diseases is that Parkinson's disease is a disorder with rigidity, tremors, slowing of movements, postural instability, and gait disturbances usually occurring in old age due to degeneration of the substantia nigra of the midbrain, while Huntington's disease is a neurodegenerative disorder usually occurring in a younger population, characterized by emotional problems, loss of cognition, repetitive, rapid movement, and involuntary movements.

12.2. Common Characteristics of Neurodegenerative Disease

Figure 12.1 presents schematically NO signaling pathways in the nervous system physiological and pathological states [3]. Under physiological conditions, basal levels of NO from nNOS, stimulated by normal synaptic activity of N-methyl-D-aspartate-type glutamate receptors (NMDARs), provide neuroprotection via sGC/cGMP pathways as well as S-nitrosylation of NMDARs and caspases. Under pathophysiological conditions, overactivation of NMDARs results in increased production of NO. Excessive toxic NO, generated by iNOS, can lead to S-nitrosylation of dynamin-related protein (Drp1), which contributes to neuronal synaptic injury via excessive mitochondrial fission and bioenergetic impairment. Additionally, S-nitrosylation of proteins such as PDI, Parkin, XIAP, and GAPDH can contribute to neurodegenerative disorders (see details in [3]).

Mitochondrial dynamics related to neurodegenerative disorders are schematically illustrated in Figures 12.1 and 12.2.

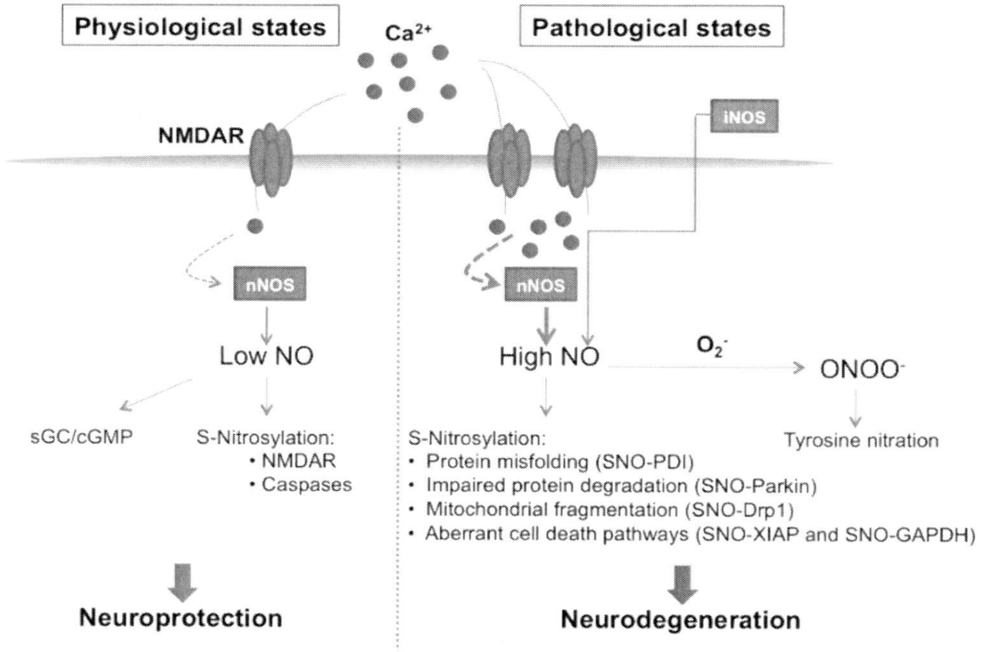

Figure 12.1. NO signaling pathways in the nervous system can contribute to neurodegenerative disorders (see details in text and in [3], with permission from SAGE Ltd., 2013.

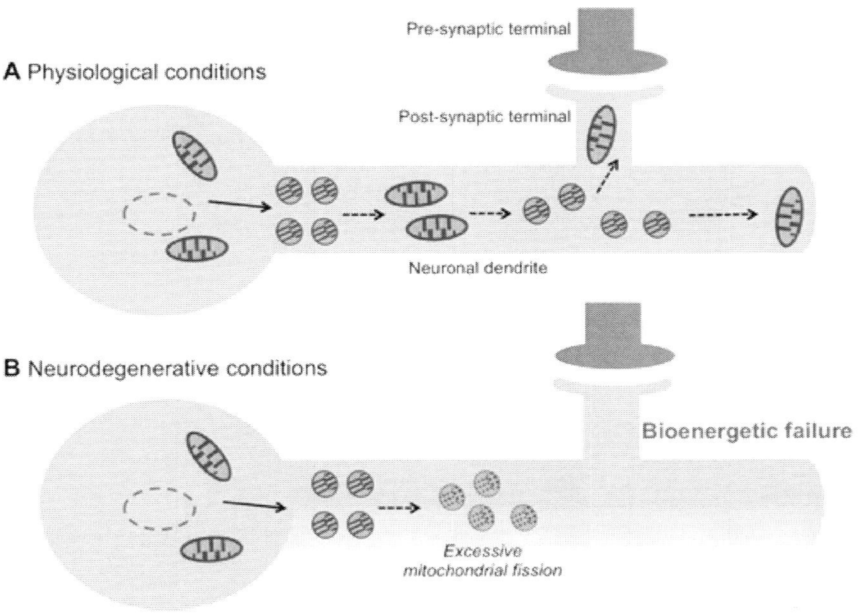

Figure 12.2. Proposed model of excessive mitochondrial fission contributing to neurodegenerative disorders. Balanced mitochondrial dynamics (mitochondrial fission and fusion events) facilitate proper distribution of neuronal mitochondria within neuronal dendrites and into synaptic termini. In addition, the precise control of mitochondrial fission and fusion events in axons is needed to support normal neuronal function (e.g., via production of ATP and buffering of Ca^{2+}). Malfunction of the fission or fusion machinery can result in excessive mitochondrial fragmentation under neurodegenerative conditions, leading to bioenergetic failure and subsequent synaptic damage [3], with permission from SAGE Ltd., 2013.

Emerging evidences suggest that mitochondrial dysfunction is a major causative factor in AD, PD, and HD [3–5]. A mechanism reactive nitrogen species (RNS) can affect mitochondrial function and thus neuronal survival occurs via S-nitrosylation of the mitochondrial fission protein Drp1, forming S-nitrosothiol (SNO)-Drp1. Subsequently, the formation of SNO-Drp1 leads to synaptic damage and neuronal death. With aging or environmental toxins that generate excessive NO, aberrant S-nitrosylation reactions can occur and affect protein misfolding, mitochondrial fragmentation, synaptic function, apoptosis or autophagy, the natural, regulated, destructive mechanism of the cell. The abnormal mitochondrial morphology, mediated by RNS, was observed in many neurodegenerative disorders, including AD, PD, and HD [3].

S-nitrosylation plays a dynamic role in both normal and aberrant neuronal signal transduction pathways [5]. Initially, activation of soluble guanylate cyclase, with consequent increase in production of cGMP, was identified as an NO-mediated signal transduction pathway. Additionally, peroxynitrite (ONOO–) can mediate neurotoxicity in part via a protein post-translational modification of tyrosine residues termed nitration. Under physiological conditions, NO is produced in neurons predominantly by nNOS, which is activated by calcium influx through NMDAR-associated ion channels. Excessive

NO generation can lead to aberrant protein S-nitrosylation in disease states (Figure 12.2). S-nitrosylation is a determinant of CNS disease progression. Physiological levels of NO can mediate neuroprotective effects by S-nitrosylating caspase and histone deacetylase inhibitor 2 (HDAC2). Additionally, during periods of moderate stress, NO can facilitate protection of neurons, for instance, via S-nitrosylation of NMDARs to downregulate excessive activity, representing a negative feedback mechanism [5]. A crucial role of S-nitrosylated proteins in the pathogenesis of neurodegenerative diseases, including Alzheimer's and Parkinson's diseases, was demonstrated in [5].

Aberrant protein S-nitrosylation plays a pathological role in many neurodegenerative conditions: (1) production of excessive NO can S-nitrosylate parkin, PDI, Drp1, Cdk5, Prx2, MMP-9, COX-2, GAPDH, PTEN/Akt, JNK/IKKβ, MAP1B, and XIAP; (2) these S-nitrosylation reactions trigger neurotoxic signaling pathways leading to ER stress, protein misfolding, mitochondrial fragmentation, bioenergetic compromise, and consequent synaptic/neuronal damage; and (3) the processes can contribute to the pathogenesis of PD, AD, HD, ALS, stroke, and potentially other neurodegenerative disorders (see details in [5]).

Neurodegenerative signaling pathways triggered by aberrant S-nitrosylation are associated with neurodegenerative diseases and occur via several processes [5]: (1) S-nitrosylation of specific proteins can cause loss-of-function by impairing E3 ligase ubiquitin ligase activity (parkin and XIAP) or isomerase/molecular chaperone activity (PDI); (2) this stage can contribute to accumulation of neurotoxic proteins and activation of apoptotic pathways; (3) oligomeric Ab peptide can result in increased generation of NO and S-nitrosylation of dynamin-1-like protein, a GTPase that regulates mitochondrial fission (Drp1); (4) this process leads to excessive mitochondrial fragmentation, bioenergetic compromise, and consequent synaptic damage; (5) Cdk5 is activated when Ab increases calpain activity to cleave the Cdk5 regulatory subunit p35 to p25; (6) The resulting neurotoxic kinase activity of Cdk5 is further enhanced by S-nitrosylation; and (7) formation of SNO-Cdk5 may also contribute to Ab-induced spine loss by transnitrosylating Drp1, with the resultant SNO-Drp1 participating in mitochondrial fragmentation and synaptic loss [5].

12.3. Nitric Oxide and Alzheimer Disease

Chronic disorder syndrome of AD, first described by Alois Alzheimer in 1905, evolves with loss of neurons and their synapses and with deficits of memory, intellectual and emotional dysfunctions [6]. Age and cardiovascular diseases are main factors affected on Alzheimer state of patients, which in turn are strongly depended on NO status in cells. Nitric oxide plays a central role in the cascade of events leading to the AD. Impaired bioavailability of NO, synthesized by eNOS and nNOS, leads to cerebral hypoperfusion and cognitive decline and neurodegeneration in AD [7]. nNOS and eNOS are involved in a multitude of inter- and intra-cellular signaling pathways being expressed constitutively in the normal brain, while iNOS was widespread in the central nervous system [8].

Nitric oxide/sGC/cGMP signaling is important for modulating synaptic transmission and plasticity in the hippocampus and cerebral cortex [10]. Physiological concentrations of NO are critical for learning and memory and also elicit anti-apoptotic/prosurvival effects against various neurotoxic challenges and brain insults. Depression of the NO/sGC pathway is a feature of AD attributed to amyloid-β neuropathology, and altered expression and activity of NOS, sGC, and a phosphodiesterase (PDE) enzyme. Nitric oxide significantly contributes in the causative relationships of endothelium-mediated mechanisms of vascular dysfunction in AD pathogenesis.

The mitochondria plays a significant role in Alzheimer's disease, and mitochondrial dysfunction is related to neurotoxic β-amyloid (Aβ), a key mediator that induces neuronal injury in the pathogenesis of this disease [9–13]. In addition, it may be imported into human brain mitochondria, where it inhibits key enzymes of respiratory metabolism [12]. Nitric oxide is produced in response to Aβ induces S-nitrosylation of the mitochondrial division protein, dynamin-related protein 1, which leads to excessive mitochondrial fission, synaptic loss, and neuronal damage.

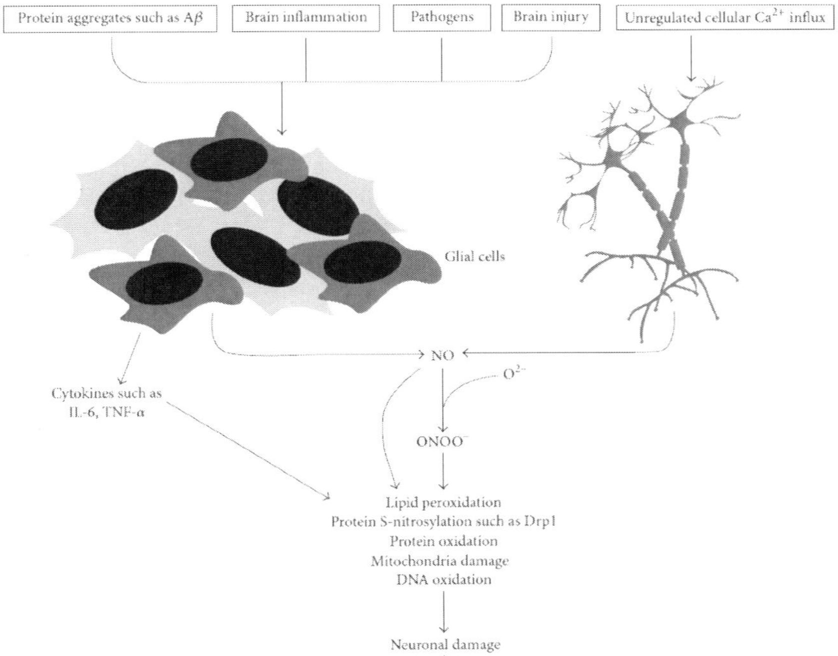

Figure 12.3. Hypothetical illustration of nitric oxide (NO) and how it is linked to Alzheimer's disease (see details in text and [13]), with permission from Springer Nature, 2016.

Activated immune cells such as the glia and astrocytes secrete induced nitric oxide synthase (iNOS), leading to the generation of NO (Figure 12.3). NO also can directly nitrosylate macromolecules and inactivates respiratory enzymes leading to a reduction in ATP production, hence disrupting bioenergetics. In parallel, glutamate acts on N-methyl-D-aspartate receptors and triggers inflow of calcium ions. In AD, ryanodine receptor expression is increased resulting in upregulated calcium ions influx, activating nNOS neurons to express nNOS, and leading to the synthesis of NO. Sustained calcium inflow results in increased NO synthesis to induce t oxidative stress/nitrosative stress.

In Alzheimer disease, endothelial nitric oxide modulates expression and processing of amyloid precursor protein (APP) whose proteolysis generates Aβ [14]. It was found that inhibition of eNOS, with the specific NOS inhibitor L-NAME, leads to increased APP and BACE1 (β-site APP-cleaving enzyme1) protein levels, as well as to increased secretion of Aβ. In turn, NOS operates as central mediators of amyloid beta-peptide (Aβ) action, giving rise to elevated levels of NO that contributes to the maintenance, self-perpetuation, and progression of the disease [15]. Redox active heam—heme−Cu−Aβ peptide complex is derived from the iron accumulated in the amyloid plaques originating from amyloid β (Aβ) peptides.The complex mediates the formation of partially reduced oxygen species (PROS) and corresponding oxidative stress [16]. The structure of the complex, which involves three amino acid residues of Aβ and effect NO on the heam release, and reaction of Heme-Fe(III)–Cu(I)–Aβ with Noa,are shown in Figures 12.3 and 12.4, respectively. Nitric oxide enforces the detrimental effects of heme−Aβ and Cu bound heme−Aβ complexes by removing heme from the heme−Aβ complex (12.4).

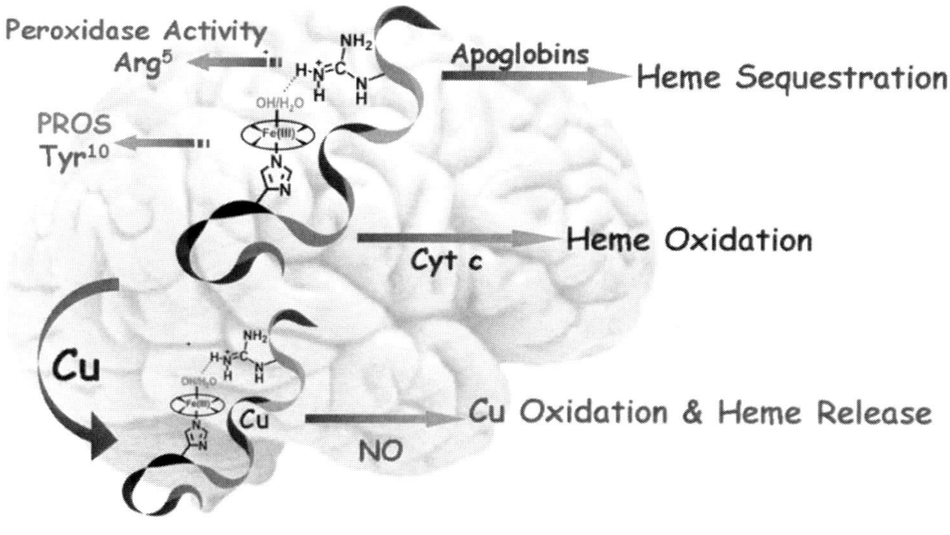

Figure 12.4. Structure of heme−Cu−Aβ peptide complex and effect NO on the heam release [16]. Copyrights 2015 American Chemical Society.

Aβ induces S-nitrosylation of dynamin-related protein (Drp1), which is required for mitochondrial division in mammalian cells [17]. This process leads to excessive mitochondrial fission, synaptic loss, and neuronal damaging brains of patients with Alzheimer's disease which contain high amounts of S-nitrosylated Drp-1. A critical cysteine residue (Cys644) in the GTPase effector-domain of Drp-1 was thought to be responsible for increase of the enzyme activity, as well as for higher order assembly and mitochondrial division activity [17]. Nitrosylation of Drp-1 at this residue increased dimer formation and GTPase activity and caused fragmentation of mitochondria, which can trigger synaptic damage. In addition, NO-induced mitochondrial fragmentation in neurons was abrogated the expression of the C644A Drp-1 variant defective for S-nitrosylation [18]. The oxidative stress, caused by NO in the brain, plays a critical role in the pathogenesis of AD related to proteolytic dysfunction of an integral membrane protein (APP) [19]. Inhibition of constitutive NO production in the human brain microvessel endothelial cell (HBMEC) cultures increases endothelial permeability during inflammation [19]. In contrast, treatment with the NO donors sodium nitroprusside (SNP) and DETA NONOate or the cGMP agonist 8-Br-cGMP significantly increased monolayer resistance.

The main risk factor for sporadic AD is aging because aging cells are affected by increasing oxidative stress and perturbed energy homeostasis, which causes progressive mitochondrial damage [20]. NOS activation in neurons confers oxidative stress-resistance on neurons, while generation of ROS by Aβ is a major action in promoting NO degradation [21]. Obesity is also one of the most risk factors for AD. Obesity induces endothelial dysfunction because cerebral hypoperfusion enhances the production of β- amyloid that in turn impairs endothelial function [22]. In addition, obesity also decreases cerebral blood flow that is attributed with decreased synthesis and actions of NO derived from the endothelium. Pathways by which NO is liberated from endothelial cells to vascular smooth muscle cells affected by obesity and promotes the pathogenic changes leading to AD was discussed in [22].

12.4. Nitric Oxide and Parkinson's Disease

Parkinson's disease is geriatric degenerative disorder of the CNS that affects the motor system [23–27]. The disease is characterized by several abnormalities including inflammation, mitochondrial dysfunction, iron accumulation, and oxidative stress. Biochemical and physicochemical mechanisms of these processes are discussed in corresponding sections.

Review [24] summarized some important implications in the pathogenesis of PD: protein–protein interactions, NO mediated apoptosis, and NO-mediated mitochondrial dysfunction. Nitric oxide having both, neuroprotective or neurotoxic actions and depending on its redox state, is involved in many of the processes leading to PD [25–27]. NO• in small doses serves as a neuroprotective agent in the brain, while NO• from nNOS contributes to neurodegeneration in PD, and NO• produced by inducible NOS from proliferating microglia as inflammatory responses to neuronal insults, mediates the disease progression [27]. The significant contribution of NO in excitotoxicity, inflammation, oxidative stress, mitochondrial function impairment, DNA damage, and S-nitrosylation of diverse proteins was confirmed in studies in humans and in experimental models of parkinsonism [25, 27]. For example, inhibitors of NO• synthase protect against dopamine-ergic neuronal death, and NOS-deficient mice resistant to PD-producing neurotoxins.

The sequence of events in progressive neurodegeneration in PD is [27]: (1) normal human substantia nigra pars compacta (SNpc) shows a balanced antioxidant defense mechanism present in the astrocytes and the neuronal population, with very few activated microglia; (2) due to aging or toxin exposure, oxidative stress-mediated dopaminergic cell death occurs; (3) this leads to inflammation; and (4) increased inflammation causes microglial proliferation, leading to more nitrosative stress. Parkinsonian neurotoxins impair mitochondrial integrity and lead to the generation of reactive oxygen species. Due to aging or toxin exposure, oxidative stress-mediated dopaminergic cell death occurs as result of inflammation and corresponding nitrosative stress [27]. Microglia contribute to the production of NO, which diffuses across to dopaminergic neurons to complete the breakdown of the cell's defense mechanism. The basal ganglia (nuclei) is a group of subcortical nuclei in the brains of vertebrates, including humans. In the basal ganglia signaling, NO is involved in the following cascade [27]:·(1) the glutamate-mediated phosphorylation of dopamine (DA) and cAMP-regulated protein phosphatase; (2) activation of neuronal nitric oxide synthase; and (3) catalyzing generation of NO. In addition, NO causes an increase in soluble cyclic GMP levels and, in turn, activates protein kinase G (PKG) phosphorylation. During PD, balance in the circuitry is lost due to the depletion of DA levels, and increase of nNOS expression can create a further imbalance. Upon activation, microglia cells cause the upregulation of inducible NO synthase [26].

The NO–sGC–cGMP signaling pathway in the striatum and one of the nuclei in the subcortical basal ganglia of the forebrain is found to be a target for the treatment of PD [28, 29]. In the intact striatum, transient elevations in intracellular cGMP act to increase neuronal excitability and to facilitate glutamatergic corticostriatal transmission. The following aspects of the NO involving in biochemical processes in this system were in detail considered in [29]: (1) NO signaling in the striatum; (2) modulation of striatal NOS activity by dopamine D1 and D2 receptor activation; (3) impact of dopamine–glutamate interactions on neuronal NOS activity; (4) role of NOS interneurons in the regulation of corticostriatal transmission; and (5) striatal pathophysiology in PD: involvement of NO–sGC–cGMP signaling pathways.

In PD, significant changes in behavioral, electrophysiological, and molecular responses to pharmacological manipulations of DA and glutamate can be caused by disruption of striatal NO–sGC–cGMP signaling cascades [30]. Abnormal striatal NO–

sGC–cGMP signaling is followed by dopamine depletion, which contributes to pathophysiological changes observed in basal ganglia circuits. It was concluded that in the parkinsonian striatum, cGMP synthesis and catabolism, as well as the temporal and spatial patterning of NO–sGC–cGMP signaling may be perturbed in medium spiny neurons (MSNs). Two functionally distinct groups of MSNs that form the striatonigral MSNs and the striatopallidal MSNs pathways were reported [31]. Nitric oxide, synthesized when GABA (*gamma*-aminobutyric acid, the chief inhibitory neurotransmitter) binds to receptors on the surface of the plasma membrane diffuses past the plasma membrane into the dendrites of striatal MSNs, which contain sGCs [32]. As a result, modulating various forms of short and long-term corticostriatalsynaptic plasticity occurs. Nitric oxide synthase also inhibits dyskinesia in rats and mice, induced by the dopamine precursor L-DOPA (l-3, 4-dihydroxyphenylalanin [33].

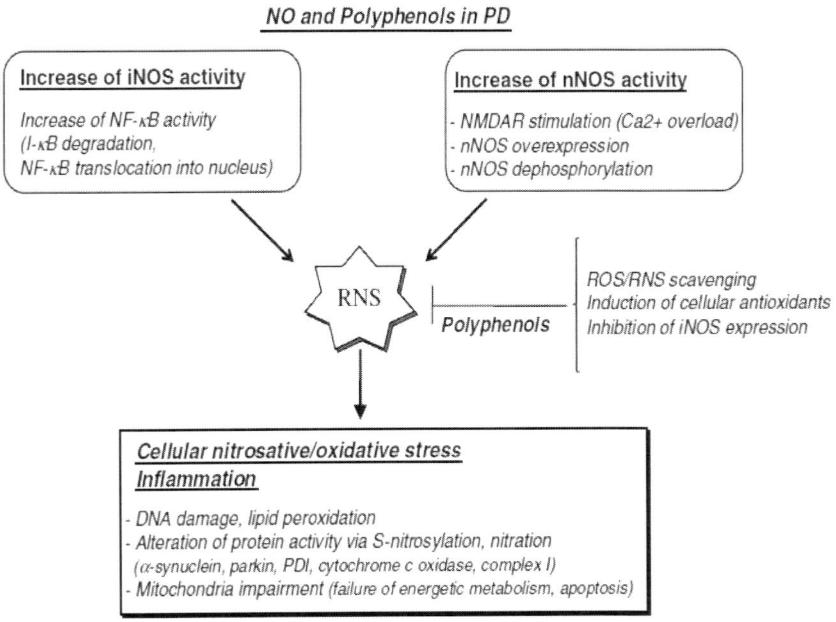

Figure 12.5. Diagram of the involvement of NO and the anti-inflammatory and anti-oxidant activity of polyphenols in PD. NMDAR is the *N*-methyl-D-aspartate receptor). is a glutamate receptor and ion channel protein f in nerve cells [from Springer Nature, 2008.

In a review [34], the following aspects of the cyclic GMP(cGMP) signaling pathway in the striatum PD were discussed: (1) behavioral effects of NO donors and nNOS inhibitors in the modulation of motor behavior; (2) disturbances in the nitrergic neurotransmission in PD and its 6-OHDA animal model; (3) contribution of both dopamine and glutamate to the regulation of NO biosynthesis in the striatum; (4) the role of NO in the tonic and phasic dopamine release; and (5) the regulation of striatal output pathways.

There is considerable evidence showing that cellular oxidative damage occurring in PD might result from the actions of altered reduction of NO. For example, high levels of nNOS and iNOS were found to produce cellular oxidative damage in PD [35, 36]. The involvement of NOS/NO in PD and exploration of anti-inflammatory and antioxidant actions of the neuroprotective activity of natural polyphenol compounds were evaluated [37]. It was found that catechins, oxyresveratrol, resveratrol, mangiferin, and quercetin appear to be the most efficient antioxidants to modulate neuroinflammation by inhibiting the expression of inflammatory genes and the level of intracellular antioxidants [38–40].

Pathogenic mechanism involved in PD suggested in [41] relates to NO increase, which induces DNA damage, protein modifications and cytotoxicity, and nitrosylation and nitration of proteins. For example, increased nitrotyrosine was detected in substantia nigra in *in vivo* models. In addition, an increased level of NO in PD leads to specific S-nitrosylation of protein-disulphide isomerase (PDI), which catalyses thiol-disulphide exchange, facilitating disulphide bond formation and the enzyme rearrangement. Another possible mechanism for NO toxicity is due to its high affinity for heme and implication in the release of iron from transferrin, enhancing Fenton reaction [42]. A diagram of the involvement of NO and the anti-inflammatory and antioxidant activity of polyphenols in PD is presented in Figure 12.5 [37].

Thus, NO and its reactive metabolites are important actors in the processes leading to neuronal cell death in PD both in terms of pro-oxidants and mediators of inflammatory responses and the blockage of NO synthesis or the scavenging of RNS by polyphenolic antioxidants [37].

Factors, including NO, stimulating the PD were discussed in [43]. In particular, neuroinflammation in PD was considered. Different genetic and/or environmental factors, such as parkin mutations or MTPT exposure, lead to the accumulation of α-synuclein aggregates in the brain, which triggers the activation of glial cells. The pro-inflammatory cytokines freed by astrocytes and microglia stimulate the release of several neuroinflammatory markers, including NO, IL-6, IL-1β, and TNF-α, which can promote neuronal death and aggravate the neurodegenerative process [43]. Similarity of the Parkinson's and Huntington's diseases were discussed [44-46].

12.5. Nitric Oxide and Schizophrenia

Schizophrenia is a severe mental disorder, which is characterized by thought disturbance, abnormal perception, impaired cognition, and bizarre behavior. Schizophrenia is associated with impaired antioxidant defense, including abnormal serum, plasma, and red blood cell (RBC) oxidative stress parameters. Nitric oxide as ubiquitous in the CNS and, therefore, is involved in pathophysiology of psychiatric disorders such as and schizophrenia, depression, and nevralgia. Therefore, NO and its metabolites have a significant influence on the development, progression, and treatment of schizophrenia [47–52]. Schizophrenia is a devastating psychiatric disorder with a broad range of behavioral and biologic manifestations. There are several clinical characteristics of the illness that have been consistently associated with poor premorbid adjustment, long duration of psychosis prior to treatment, and prominent negative symptoms.

In reviews [48, 53], the impact of NO metabolism on processes, which are disturbed in the neuropsychiatric disorder in schizophrenia, nerve cell migration, formation, and maintenance of synapses, N-methyl-D-aspartic acid receptor-mediated neurotransmission, adult hippocampal neurogenesis, membrane pathology, and cognitive abilities were considered. Special emphasis is given to aspects of genetic linkage between NO generating and modulating proteins and schizophrenia. Nitric oxide donors, such as sodium nitroprusside (SNP), glyceryl trinitrate, and methylene blue, were effective in treating depressive symptoms in patients and animal models with schizophrenia [51, 52] Sodium nitroprusside, a NO donor for novel treatment of schizophrenia, may also modulate dopaminergic systems and act through the glutamate N-methyl-D-aspartate (NMDA) receptor-NO-cyclic guanosine monophosphate (cGMP) pathway, which is altered in schizophrenia.

In schizophrenia, both decreases and increases in NOS activity, NOS protein, and mRNA content were found. For example, in blood samples of schizophrenics, the increased levels of NO metabolites were detected [54]. Nevertheless, according to [55], plasma levels of NO and its metabolites (NOx) are decreased in patients with schizophrenia. It is necessary take into consideration that the plasma NOx level is also influenced by many variables, including cardiovascular and metabolic diseases, obesity, lipids, plasma glucose, menopausal status, blood pressure, and smoking status. As evidenced by increased production of reactive oxygen or decreased antioxidant protection in schizophrenic patients, the excessive free radical production in ROS and most probably RNS may be involved in the pathophysiology of schizophrenia [56]. During schizophrenia, free radicals impair the antioxidant defense system, including antioxidant enzymes (superoxide dismutase, glutathione peroxidase, catalase), plasma antioxidant proteins (albumin, bilirubine, uric acid), and trace elements, and induce membrane pathology via lipid peroxidation. Experiments suggested that NO alters antioxidant enzyme activities and increases levels of lipid peroxidation, as well as altered levels of plasma antioxidants.

12.6. Nitric Oxide and Depression

Depression is a mental disorder accompanied by low self-esteem, loss of interest in normally enjoyable activities, low energy, and pain, and caused by a combination of genetic, environmental, and psychological effects. Recent studies demonstrated a potential role of NO in the relationship between major depressive disorder and effect of antidepressants on NO [57–66]. Nitric oxide contributes in the processes leading to depression, forming a biochemical and physiological basis for the disease.

Methylene blue (MB),

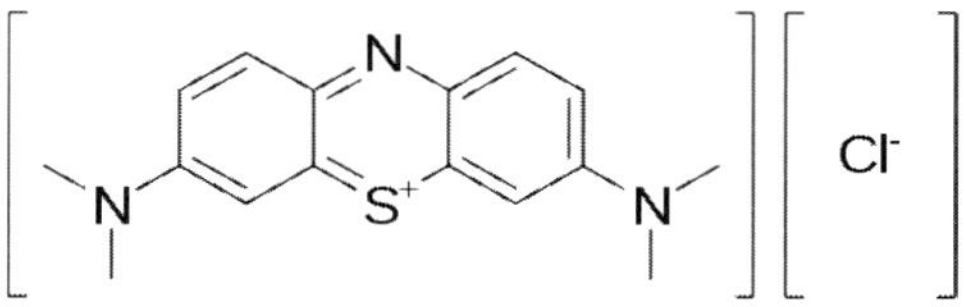

acting on both monoamine oxidase (MAO) and the NO-cGMP pathway, has antidepressant activity in rodents, reduces hippocampal nitrate levels, and affects the state of regional brain monoamines [57]. The antidepressant properties of MB and selected structural analogues and whether their actions involve MAO, NO synthase, and regional brain monoamines were evaluated.

Involvement of NO signaling pathway in the antidepressant action of the total flavonoids extracted from Xiaobuxin-Tang was reported [58]. It was revealed that NO controls the expression of aggressive behavior in a model organism [59]. For example, crickets treated with NO/cGMP pathway activators (SNAP and 8Br-cGMP) happened to be less aggressive, whereas treatment with inhibitors (LNAME, PTIO, and ODQ) led to more aggressive and longer contests. Aggression increases in male rodents after knockdown of neuronal NOS or treatment with the neuronal NOS inhibitor 7-nitroindazole. A study was carried out to investigate the effects of opioid and nitrergic systems on depression in an experimental model of cholestasis in mice [60]. Elevated levels of these substances in cholestatic subjects were also investigated. Studies suggested that naltrexone and L-NAME significantly reversed antidepressant-like effects of cholestasis and elevated levels of endogenous opioids, and NO in cholestatic mice induced an antidepressant-like effect. Effects of antidepressants such as melatonin, milnacipran, and paroxetine were also reported [62, 63].

To probe a suggestion that inflammation is associated with stress-induced depression and cardiovascular dysfunction, the effects of a tyrosine kinase inhibitor (tumor necrosis factor, TNF-α) on endothelium-dependent vascular reactivity, systemic blood pressure, and eNOS immunoreactivity in the chronic mild stress (UCMS) model of depression in rats were investigated [61]. Relaxation in response to the NO donor SNP and papaverine and KCl-induced contractile responses was observed. In UCMS, decreased expression of eNOS was also revealed. It was concluded that tumor necrosis factor-alpha (TNF-α) could be a major mediator of vascular dysfunction associated with UCMS, leading to decreased expression of eNOS.

A potential role of NO in the relationship between depression and coronary heart disease (CHD) risk and an effect of antidepressants on NO production were examined [63]. Experimentally detected decrease plasma NOx levels can be associated with the pathophysiology of depression. In depressed patients, these levels can be increased by treatment with milnacipran, a serotonin noradrenaline reuptake inhibitor. NO overproduction is involved in neurophysiologic processes of learning, memory, depression, and pain expression. For example, it was found that decreasing the levels or blocking the synthesis of NO in the brain induced antidepressant-like effects [64], and the level of plasma NO metabolites was significantly higher in suicidal patients than in nonsuicidal psychiatric patients or in normal control subjects [65]. In addition, nitric oxide synthase inhibitors, l-NAME or 7-nitroindazole, display antidepressant-like activity [66].

12.7. Nitric Oxide in Neuralgia, Migrane, and Mania

Neuralgia is pain in the distribution of a nerve or nerves, following nerve damage. Neuronal injury in the CNS typically leads to local degeneration of the nerve axon and myelin sheet. Considerable evidence implicates that impaired nitric oxide synthesis and inflammatory cytokines play a key role in the pathogenesis of persistent and exaggerated pain states [67–68].

Involvement nitric oxide in the development and maintenance of central sensitization and acute migraine and chronic tension type headache were well documented [68-75]. For example, nitroglycerin is capable of inducing a headache, and experimental vascular headaches causes the release of NO [67]. Chronic pain patients indicate a significant increase in plasma levels of NO in comparison to healthy controls, and pro-inflammatory cytokines (IL-1β, IL-2, IL-6, IFN-γ, TNF-α) in the plasma correlate with increasing pain intensity [68]. NO was produced by nNOS and eNOS expressed in the brain and spinal cord and vascular tissue, respectively, while iNOS was expressed in glial cells and immune cells during inflammation. Neuralgia-inducing cavitational osteonecrosis of the jaws (NICO) leads to the T-786C mutation and polymorphism of the eNOS gene affecting NO production [69].

Potential association of neuropathic pain with alterations of NOS immunoreactivity and catalytic activity in dorsal root ganglia and spinal dorsal horn was examined [70]. The results indicate that marked alterations of nNO in this system may contribute to spinal sensory processing as well as to the development of neuronal plasticity phenomena. To assess the possible roles of nNOS in spinal sensitization after nerve injury, NOS catalytic activity was determined by monitoring the conversion of [3H]arginine to [3H]citrulline in the lumbar (L4-L6) spinal cord segments and DRGs in rats [70]. Experiments for administration of inhibitors of NOSs strongly indicated a fundamental involvement of nitric oxide in biochemical and physiological processes leading to neuralgia. Mechanical allodynia, (central pain sensitization, increased response of neurons), induced by tightly ligating the left L5 and L6 spinal nerves, was suppressed by L-NAME, and cold allodynia and cold-stress exacerbated ongoing pain was also attenuated by L-NAME [71].

The administration of NNLA, a NOS inhibitor, prevented the hypoxia-induced phosphorylation of extracellular signal-related kinase (ERK) and c-Jun N-terminal kinase (JNK), which mediate signal transduction from cell surface receptors to the nucleus and phosphorylate anti-apoptotic proteins, thereby regulating programmed cell death [73]. This finding indicates that the hypoxia-induced activation of ERK and JNK in the cerebral cortical nuclei of newborn piglets is NO-mediated. Authors of work [74] investigated role of NOS in a mice spinal dorsal horn in herpetic and postherpetic pain, especially allodynia, which was induced by transdermal inoculation of the hind paw with herpes simplex virus type-1 (HSV-1). Herpetic allodynia was significantly inhibited by administration of the selective NOS2 inhibitor S-methylisothiourea, but not the selective NOS1 inhibitor 7-nitroindazole. NOS2 expression was observed around HSV-1 antigen-immunoreactive cells. On the other hand, postherpetic allodynia was significantly inhibited by administration of neuronal nitric oxide synthase inhibitor, 7-nitroindazole. Herpetic allodynia is also inhibited by administration of the selective NOS2 inhibitor S-methylisothiourea, while postherpetic allodynia was significantly inhibited by administration of 7-nitroindazole. The results suggest that both types of allodynia are mediated by nitric oxide and that NOS2 and NOS1 are responsible for herpetic and postherpetic allodynia, respectively [74]. Pain behavior and paw thickness after intraplantar injection of complete Freund's adjuvant in wild-type (WT) mice and in mice lacking either iNOS, eNOS, or nNOS were compared [75]. Results supported the proposed contribution of nNOS in sensitization of dorsal root ganglion (DRG), which is a cluster of nerve cell bodies in a dorsal root of a spinal nerve.

Nociception is the sensory nervous system's response to harmful stimuli. A major signaling mechanism of NO in spinal nociceptive processing is thought to be association with activation of NO-sensitive guanylyl cyclase (NO-GC) and subsequent cGMP production [76]. The distribution of NO-GC in the spinal cord and in DRG was investigated, and the nociceptive behavior of mice deficient in NO-GC (GC-KO mice) was characterized [77]. It was observed that GMP produced by NO-GC may activate signaling pathways different from cGMP-dependent protein kinase and essentially contributes to inflammatory and neuropathic pain. These findings, as well as the strongly inhibition of nociceptive behavior of GC-KO mice, justified that NO-GC-mediated cGMP production has a dominant role in the development of exaggerated pain sensitivity. It was suggested that three isoforms of NOS produce inflammatory pain via NO, involving pain transmission, hyperalgesia, chronic pain, inflammation, and central sensitization in a cyclic guanosinemono-phosphate (cGMP) dependent way. Nitric oxide is also implicated in the induction of a migraine attack in migraineurs [78].

Much evidence concerning a connection between oxidative and nitrosative stress and neuropathic pain exist. As an example, the involvement of the gluthatione system and NO, as a molecule related to nociception, (the sensory nervous system's response to harmful stimuli), in pain processes in a rat was experimentally found [79].

Migraine is a complex CNS-related headache disorder that involves multiple macromolecules and/or NOS [72, 78–82]. A series of 1,6-disubstituted indoline-based thiophene amidine compounds, including highly selective human neuronal nitric N-(1-(piperidin-4-yl)indolin-5-yl)thiophene-2-carboximidamide N-(1-(piperidin-4-yl)indolin-5-yl)thiophene-2-carboximidamide, were synthesized and employed as selective nNOS inhibitors to mitigate the cardiovascular liabilities and migraine pain [72]. The NOS inhibitor L-N(G) methyl arginine hydrochloride significantly reduced headaches and myofacial factors in patients with chronic tension-type headaches [81]. It was suggested that the analgesic effect of NOS inhibition in patients with chronic tension-type headaches is due to a reduction in the central sensitization at the level of the spinal dorsal horn, trigeminal nucleus, or both. Glyceryl trinitrate (GTN), a pro-drug for NO, causes headaches in normal volunteers, while a blockade of NOS by N^G-methyl-L-arginine acetate L-NMMA effectively treats attacks of migraines and chronic tension-type headaches and cluster headaches in sufferers [82]. Inhibition of the cGMP activity also provokes migraines, indicating that cGMP is the effector of NO-induced migraines. The established importance of NO as a potential initiator of the migraine attack paves the way for new directions for the pharmacological treatment of migraines and other vascular headaches.

Mania, manic syndrome, is a state of abnormally elevated arousal, affect, and energy level, or a state of heightened overall activation with enhanced affective expression together with lability of affect. Bipolar disorder, also known as manic depression, is a mental disorder that causes periods of depression and periods of elevated emotional state. A connection between NO and its derivatives and manic syndrome bipolar disorder has been well documented [82–90].

The nitric oxide level was measured in the serum samples obtained from the twenty-seven patients with bipolar disorder (BD) in euthymic phase, and twenty healthy volunteers [84]. The mean serum NO level in BD was found to be significantly higher than in controls. To evaluate the activity and levels of NO in bipolar I depressive episode (BD-DE), the serum levels of NO in thirty patients have been studied when admitted to hospital (first) and on the thirtieth days [85]. NO first-day levels were found to be significantly higher in patients, while NO levels markedly decreased and normalized on the thirtieth day. To evaluate the activity and levels of antioxidant superoxide dismutase (SOD), and NO in bipolar I depressive episode (BD-DE), the serum levels of NO and SOD in thirty BD-DE patients and thirty healthy volunteer controls were measured [86]. The experiments showed that NO first-day levels were significantly higher in patients, and SOD first-day activity was significantly low. Results on serum NOx levels of twenty patients with bipolar disorder, manic episodes (BD-ME), eighteen with major depressive disorder and depressive episodes (MDD-DE), and eighty healthy subjects suggested that increased levels of serum NOx are indicators in patients with BDME and MDD-DE after treatment [87].

Adrenomedullin (AM), a 52-amino-acid peptide hormone neurotransmitter in the brain, induces vasorelaxation by activating adenylate cyclase and by stimulating the release of NO [88]. The amount of AM and NO in the plasma of forty-four patients suffering from bipolar affective disorder (BPAD), type I, manic episodes, and twenty-one healthy control subjects were examined. AM levels of BPAD patients were approximately twofold higher than controls and were positively correlated with the duration of hospitalization for the current episode and negatively correlated with the total duration of illness. Thus, both NO and AM may have a pathophysiological role in BPAD. Similar findings associated with NO and AM were evaluated in autism. Asymmetric dimethylarginine (ADMA) is an endogenous competitive inhibitor of the nitricoxidesynthase. In studies on the function of NO in mood disorders of thirty patients, quantitative changes in the NO production have been observed that consequently showed a correlation between NO function and neuropsychiatric disorders [89]. Therefore, ADMA accumulation may have an important role in the regulation of signal transduction in the NO system. Arginase catalyzes the hydrolysis of L-arginine to urea and ornithine and competes with NOS for L-arginine [90]. Arginase activities, Mn, and total nitrite levels were measured in the plasma from forty-three patients with BPAD (Type one) and thirty-one healthy control subjects. Plasma arginase activities and Mn were found to be significantly lower, and the total nitrite levels were higher in patients with BPAD compared with controls. Results suggest that the arginine-NO pathway is involved in the pathogenesis of BPAD.

Potential dysfunction of the nitric oxide system in oxidative stress, inflammation, neurotransmission, and cerebrovascular tone regulation in bipolar disorder (BD) and suicide was examined [91]. By simultaneously analyzing variants of three isoforms of NOS, interindividual genetic liability to suicidal behavior in BD of 536 patients with BD (DSM-IV) and 160 healthy was explored. Results were evidenced in the potential involvement of eNOS gene variants in susceptibility to suicidal behavior.

REFERENCES

[1] Flint Beal M., Lang AE, Ludolph AC. *Neurodegenerative Diseases: Neurobiology, Pathogenesis and Therapeutics.* Cambridge University Press, 2010.

[2] Shamim A (Ed.) *Neurodegenerative Diseases*. Springer-Verlag New York 2102.

[3] Haun F, Nakamura T, Lipton SA. Dysfunctional mitochondrial dynamics in the pathophysiology of neurodegenerative diseases. *Journal of Cell Death* (2013), 6, 27-35.

[4] Hannibal L. Nitric Oxide Homeostasis in Neurodegenerative Diseases. *Curr Alzheimer Res.* 2016;13(2):135-149.

[5] Nakamura T, Shichun Tu S, Mohd Waseem Akhtar MW, Carmen R. Sunico CR, Okamoto S, Lipton SA. Aberrant Protein S-Nitrosylation in Neurodegenerative Diseases. *Neuron* 22, (2013) 78, 596-612.

[6] Alzheimer A, Stelzmann RA, Schnitzlein HN, Murtagh FR. An English translation of Alzheimer's 1907 paper, "Uber eine eigenartige Erkankung der Hirnrinde." *Clin Anat* (1995) 8, 429-431.

[7] de la Torre J, Toda N, Okamura T. Cerebral Blood Flow Regulation by Nitric Oxide in Alzheimer's Dsease. *Journal of Alzheimer's Disease* (2012) 32, 569-578.

[8] Fernandez AP, Pozo-Rodrigalvarez A, Serrano J, Martinez-Murillo R. Nitric oxide: Target for therapeutic strategies in Alzheimer's disease *Current Pharmaceutical Design* (2010) 16, 2837-2850.

[9] DiMarco LY, Venner A, Farkas E, Evans PC, Marzo A, Frangi AF. Vascular dysfunction in the pathogenesis of Alzheimer's disease—A review of endothelium-mediated mechanisms and ensuing vicious circles. *Neurobiology of Disease* (2015) 82, 593-606.

[10] Zhihui Q. Modulating nitric oxide signaling in the CNS for Alzheimer's disease therapy. *Future Med Chem.* (2013) 5, 1451-1468.

[11] Wang G, Moniri NH, Ozawa K, Stamler JS, Daaka Y. Nitric oxide regulates endocytosis by S-nitrosylation of dynamin. *Proc. Natl. Acad. Sci. USA* (2006) 103, 1295-1300.

[12] Westermann B. Nitric oxide links mitochondrial fission to Alzheimer's disease. *Science Signaling* (2009, 2, (69), pe29.

[13] Asiimwe N, Seung Geun Yeo, Min-Sik Kim, Junyang Jung, Na Young Jeong. Nitric Oxide: Exploring the Contextual Link with Alzheimer's Disease. *Oxidative Medicine and Cellular Longevity* Volume 2016 (2016), 2016 Article ID 7205747, 10 pages.

[14] Austin SA, Santhanam AV, Katusic ZS. Endothelial Nitric Oxide Modulates Expression and Processing of Amyloid Precursor Protein. *Circulation Research.* (2010)107, 1498-1502.

[15] Fernandez AP, Pozo-Rodrigalvarez A, Serrano J, Martinez-Murillo R. Nitric oxide: Target for therapeutic strategies in Alzheimer's disease. *Current Pharmaceutical Design* (2010) 16, 2837-2850.

[16] Ghosh C, Seal M, Mukherjee S, Ghosh Dey S Chandradeep Ghosh, Manas Seal, Soumya Mukherjee, Somdatta Ghosh Dey. Alzheimer's Disease: A Heme–Aβ Perspective. *Acc. Chem. Res.* (2015) 48, 2556–2564.

[17] Zhupp PP, Patterson A, Stadler J, Seeburg DP, Sheng M, Blackstone C. Intra-andintermolecular domain interactions of the C-terminal GTPase effector domain of the multimeric dynamin-like GTPase Drp1. *J. Biol. Chem.* (2004) 279, 35967-35974.

[18] Cho DH, Nakamura T, Fang J, Cieplak P, Godzik A, Gu Z, Lipton SA. S-nitrosylation of Drp1 mediates Aβ-related mitochondrial fission and neuronal injury. *Science*, (2009) 324, 102- 105,

[19] Wong D, Dorovini-Zis K, Vincent SR. Cytokines, nitric oxide, and cGMP modulate the permeability of an in vitro model of the human blood–brain barrier. *Experimental Neurology*. Volume 190, Issue 2, December 2004, Pages 446-455.

[20] Müller WF, Eckert A, Kurz C, Eckert GP, Leuner K. Mitochondrial Dysfunction: Common Final Pathway in Brain Aging and Alzheimer's Disease—Therapeutic Aspects. *Molecular Neurobiology (*2010) 41, 159-171.

[21] de la Torre J, Toda N, Okamura T. Cerebral Blood Flow Regulation by Nitric Oxide in Alzheimer's Disease *Journal of Alzheimer's Disease* (2012) 32, 569-578.

[22] Toda N, Ayajiki K, Okamura N. Obesity-Induced Cerebral Hypoperfusion Derived from Endothelial Dysfunction: One of the Risk Factors for Alzheimer's Disease. *Current Alzheimer Research* (2014), 11, 733-744.

[23] Walia V, Kansotia S. Nitric oxide-mediated neurodegeneration in Parkinson's Disease. *Asian Journal of Pharmaceutical and Clinical Research* (2016) 9, 9-13.

[24] Ramkumar Kavya, Rohit Saluja, Sarika Singh, Madhu Dikshit. Nitric oxide synthase regulation and diversity: Implications in Parkinson's disease. *Nitric Oxide* (2006) 15, 280-294.

[25] Jimenez-Jimenez FJ, Alonso-Navarro H, Herrero MT, Garcia-Martin E, Agundez, JAG. An Update on the Role of Nitric Oxide in the Neurodegenerative Processes of Parkinson's Disease. *Current Medicinal Chemistry* (2016) 3, 2, 2666-2679.

[26] Walia Vaibhav, Kansotia Santlal. Nitric oxide mediated neurodegeneration in Parkinson's disease. *Asian Journal of Pharmaceutical and Clinical Research* (2016), 9, 9-13.

[27] Tripathy D. Chakraborty J, Mohanakumar KP. Antagonistic pleiotropic effects of nitric oxide in the pathophysiology of Parkinson's disease. *Free Radical Research* (2015) 49, 1129.

[28] Lorenc-Koci E, Czarnecka A. Role of nitric oxide in the regulation of motor function. An overview of behavioral, biochemical and histological studies in animal models. *Pharmacological Reports* (2013), 65, 1043-1055.

[29] West AR, Tseng KY. Nitric oxide-soluble guanylyl cyclase-cyclic GMP signaling in the striatum: new targets for the treatment of Parkinson's disease*? Frontiers in Systems Neuroscience* (2011) 5, 55.

[30] Sammut S, Bray KE, West AR. Dopamine D2 receptor-dependent modulation of striatal NO synthase activity. *Psychopharmacology (Berl.)* (2007) 191, 793-803.

[31] DeLong MR, Wichmann T. Circuits and circuit disorders of the basal ganglia. *Arch. Neurol.* (2007) 64, 20-24.

[32] Sammut S, Threlfell S, West AR. Nitric oxide-soluble guanylyl cyclase signaling regulates corticostriatal transmission and short-term synaptic plasticity of striatal projection neurons recorded in vivo. *Neuropharmacology* (2007) 58, 624-631.

[33] Del-Bel E, Padovan-Neto FE, Raisman-Vozari R, Lazzarini M. Role of nitric oxide in motor control: Implications for Parkinson's disease pathophysiology and treatment. *Curr Pharm Des.* (2011) 17, 471-488.

[34] Lorenc-Koci E, Czarnecka A. Role of nitric oxide in the regulation of motor function. An overview of behavioral, biochemical and histological studies in animal models. *Pharmacological Reports* (2013) 65, 1043-1055.

[35] Mattson MP, Magnus T Ageing and neuronal vulnerability. *Nat. Rev. Neurosci.* (2006) 7, 278-294.

[36] Whitton PS. Inflammation as a causative factor in the aetiology of Parkinson's disease. *Br J Pharmacol* (2007) 150, 963-976.

[37] Aquilano K, Baldelli S, Rotilio G, Ciriolo MR. Role of Nitric Oxide Synthases in Parkinson's Disease: A Review on the Antioxidant and Anti-inflammatory Activity of Polyphenols. *Neurochemical Research* (2008), 33, 2416-2426.

[38] Mercer LD, Kelly BL, Horne MK, Beart PM. (2005). Dietary-polyphenols protect dopamine neurons from oxidative insults and apoptosis: Investigations in primary rat mesencephalic cultures *Biochemical Pharmacology* (2005) 69, 339-345.

[39] Jiménez-Jiménez FJ, Alonso-Navarro H, Herrero MT, García-Martín E, Agúndez JA. An Update on the Role of Nitric Oxide in the Neurodegenerative Processes of Parkinson's Disease. *Curr Med Chem.* (2016) 23:2666-2679

[40] Lorenz P, Roychowdhury S, Engelmann M, Wolf G, Horn TF. Oxyresveratrol and resveratrol are potent antioxidants and free radical scavengers: Effect on nitrosative and oxidative stress derived from microglial cells. *Nitric Oxide* (2003) 9, 64-76.

[41] Farooqui A, Farooqui T. Diet and Exercise in Cognitive Function and Neurological Diseases. *Neurochem Res.* (2008) 33, 2416-2426.

[42] Schapira AH. Mitochondrial dysfunction in Parkinson's disease. *Cell. Death. Differ.* (2007) 14:1261-1266.

[43] Yuste JE, Tarragon E, Campuzano CM, Ros-Bernal F. Implications of glial nitric oxide in neurodegenerative diseases. *Frontiers in Cellular Neuroscience* (2015) 9, 1-13.

[44] Walker FO. Huntington's disease. *The Lancet* (2007) 369, No. 9557, 218-228.

[45] Deckel A. Wallace Nitric oxide synthase in Huntington's disease. *Journal of Neuroscience Research* (2001) 64, 99-107.

[46] Bonelli RM, Hoedl AK, Kapfhammer H-P. Neuroprotection in Huntington's disease. Letters *in Drug Design & Discovery* (2005) 2, 143-147.

[47] Nasyrova RF, Ivashchenko DV. Ivanov MV, Nesnanov NG. Role of nitric oxide and related molecules in schizophrenia pathogenesis: biochemical, genetic and clinical aspects. *Front Physiol.* (2015) 6: 139.

[48] Bernstein HG, Bogerts B, Keilhoff G. The many faces of nitric oxide in schizophrenia. A review. *Schizophr. Res.* (2005) 78, 69-86.

[49] Nasyrova RF, Ivashchenko DV, Ivanov MV, Neznanov NG. Role of nitric oxide and related molecules in schizophrenia pathogenesis: Biochemical, genetic and clinical aspect. *Front Physiol.* (2015) 6, 1s.

[50] Flatow J, Buckley P, Miller BJ. Meta-analysis of oxidative stress in schizophrenia. *Biol. Psychiatry* (2013) 74, 400-409.

[51] Maia-de-Oliveira JP, Lobao-Soares B, Baker GB, Dursun SM, Hallak JE. Sodium nitroprusside, a nitric oxide donor for novel treatment of schizophrenia, may also modulate dopaminergic systems. *Schizophr. Res.* (2014) 159, 558-559.

[52] Kandratavicius L, Balista PA, Wolf DC' Abrao J, Evora PR, Rodrigues AJ, Chaves C, Maia-de-Oliveira JP, Leite JP, Dursun SM, Baker GB, Guimaraes FS, Hallak JE. Effects of nitric oxide-related compounds in the acute ketamine animal model of schizophrenia. *BMC Neuroscience* (2015) 16, 9.

[53] Bernstein HG, Keilhoff G, Steiner J, Dobrowolny H, Bogerts B. Nitric Oxide and Schizophrenia: Present Knowledge and Emerging Concepts of Therapy. *CNS Neurol Disord Drug Target*s (2011), 792-807.

[54] Herken H, Uz E, Ozyurt H, Akyol O. 2001. Red blood cell nitric oxide levels in patients with schizophrenia. *Schizophr. Res.* 52, 289-290.

[55] Nakano Y, Yoshimura R, Nakano H, Ikenouchi-Sugita A, Hori H, Umene-Nakano W, Ueda N, Nakamura J. Association between plasma nitric oxide metabolites levels and negative symptoms of schizophrenia: A pilot study. *Human psychopharmacology. Psychopharmacol. Clin. Exp.* (2010) 25, 139-144.

[56] Wu JQ, Kosten TR, Zhang XY. Free radicals, antioxidant defense systems, and schizophrenia. *Prog. Neuropsychopharmacol Biol. Psychiatry* (2013) 46, 200-206.

[57] Harvey BH, Duvenhage I, Viljoen F, Scheepers N, Malan SF, Wegener G, Brink CB, Petzer JP. Role of monoamine oxidase, nitric oxide synthase and regional brain monoamines in the anti depressant-like effects of methylene blue and selected structural analogues. *Biochem. Pharmacol.* (2010) 80, 1580-1591.

[58] Zhang LM, Wang HL, Zhao N, Chen HX, Li YF, Zhang YZ. . Involvement of nitric oxide (NO) signaling pathway in the antidepressant action of the total flavonoids extracted from Xiaobuxin-Tang *Neuroscience Letters* (2014) 575, 31-36.

[59] Stevenson PA, Rillich J. Adding up the odds—Nitric oxide signaling underlies the decision to flee and post-conflict depression of aggression. *Science Advances* (2015) 1, e1500060.

[60] Haj-Mirzaian A, Hamzeh N, Javadi-Paydar M, Abdollahzadeh Estakhri MR, Dehpour AR. Resistance to depression through interference of opioid and nitrergic systems in bile-ductligated mice. *European Journal of Pharmacology* (2013) 708, 38-43.

[61] Demirtaş T, Utkan T, Karson A, Yazır Y, Bayramgürler D, Gacar N. The Link between Unpredictable Chronic Mild Stress Model for Depression and Vascular Inflammation? *Inflammation* (2014) 37, 1432-1438.

[62] Wegener G, Joca SRL. Nitric Oxide Signaling in Depression and Antidepressant Action. In Venkataramanujam Srinivasan, Domenico de Berardis, Cecilio Álamo, Takahiro A. Kato (Eds.) *Melatonin, Neuroprotective Agents and Antidepressant Therapy*. Springer, 2016, pp 765-792.

[63] Ikenouchi-Sugita A, Yoshimura R, Hori H, Umene-Nakano W, Ueda N, Nakamura J. Effects of antidepressants on plasma metabolites of nitric oxide in major depressive disorder: Comparison between milnacipran and paroxetine. *Prog Neuropsychopharmacol Biol Psychiatry*. (2009) 33, 1451-1453.

[64] Chrapko WE, Jurasz P, Radomski MW. Lara N, Archer SL, LeMelledo JM, Decreased platelet nitric oxide synthase activity and plasma nitricoxide metabolites in major depressive disorder, *Biol Psychiatry* 56 (2004) 129-134.

[65] Moreno J, Gaspar E, Lopez-Bello G, Juarez E, Alcazar-Leyva S. Gonzalez-Trujano E, Pavon L, Alvarado-Vasquez N. Increase in nitric oxide levels andmitochondrial membrane potential in platelets of untreated patients with major depression. *Psychiatry Res*. (2013) 209, 447-452.

[66] Yildiz F, Erden BF, Ulak G, Utkan T, Gacar N. Antidepressant-like effect of 7-nitroindazole in the forced swimming test in rats. *Psychopharmacology* (Berl) (2000) 149, 41-44.

[67] . Choi JI, Kim WM, Lee HG, Kim YO, Yoon MH. Role of neuronal nitric oxide synthase in the antiallodynic effects of intrathecal EGCG in a neuropathic pain rat model. *Neurosci Lett*. (2012) 21:53-57.

[68] Koch A, Zacharowski K, Boehm O, Stevens M, Lipfert P, von Giesen HJ, Wolf A, Freynhagen R. Nitric oxide and pro-inflammatory cytokines correlate with pain intensity in chronic pain patients. *Inflamm Res*. (2007) 56, 32-37.

[69] Glueck CJ, McMahon RE, Bouquot JE, Khan NA, Wang P. T-786C polymorphism of the endothelial nitric oxide synthase gene and neuralgia-inducing cavitational osteonecrosis of the jaws. *Oral Surg Oral Med Oral Pathol Oral Radiol Endod*. (2010) 109, 548-553.

[70] Cízková D, Lukácová N, Marsala M, Marsala J. Neuropathic pain is associated with alterations of nitric oxide synthase immunoreactivity and catalytic activity in dorsal root ganglia and spinal dorsal horn. *Brain Res. Bull*. (2002) 161-171.

[71] Yoon YW, Sung B, Chung JM. Nitric oxide mediates behavioral signs of neuropathic pain in an experimental rat model. *Neuroreport*. (1998) 9, 367-372.

[72] Annedi SC, Maddaford SP, Ramnauth J, Renton P, Rybak T, Silverman S, Rakhit S, Mladenova G, Dove P, Andrews JS, Zhang D, Porreca F. Discovery of a potent, orally bioavailable and highly selective human neuronal nitric oxide synthase (nNOS) inhibitor, N-(1-(piperidin-4-yl)indolin-5-yl)thiophene-2-carboximidamide

as a pre-clinical development candidate for the treatment of migraine. *European Journal of Medicinal Chemistry* (2012) 55, 94e107.

[73] Mishra OP, Zubrow AB, Ashraf QM. Nitric oxide-mediated activation of extracellular signal-regulated kinase (ERK) and c-jun N-terminal kinase (JNK) during hypoxia in cerebral cortical nuclei of newborn piglets. *Neuroscience.* (2004) 123, 179-186.

[74] Sasaki A, Mabuchi N, Serizawa K, Takasaki I, Andoh T, Shiraki K, Ito S, Kuraishi Y. Different roles of nitric oxide synthase-1 and -2 between herpetic and postherpetic allodynia in mice. *Neuroscience* (2007) 150, 459-466.

[75] Boettger MK, Uceyler N, Zelenka M, Schmitt A, Reif A, Chen Y, Sommer C. Differences in inflammatory pain in nNOS-, iNOS- and eNOS-deficient mice. *Eur J Pain.* (2007) 11, 810-818.

[76] Tao YX, Johns RA. Activation and up-regulation of spinal cord nitric oxide receptor, soluble guanylate cyclase, after formalin injection into the rat hind paw. *Neuroscience* (2002) 112:439-446].

[77] Schmidtko A, Gao W, König P, Heine S, Motterlini R, Ruth P, Schlossmann J, Koesling D, Niederberger E, Tegeder I, Friebe A, Geisslinger G. GMP Produced by NO-Sensitive Guanylyl Cyclase Essentially Contributes to Inflammatory and Neuropathic Pain by Using Targets Different from cGMP-Dependent Protein Kinase. *The Journal of Neuroscience* (2008) 28, 8568 - 8576.

[78] Neeb L, Reuter U. Nitric oxide in migraine. *CNS Neurol Disord Drug Targets* (2007) 6, 258-264.

[79] Guedes RP, Dal Bosco L, Araújo AS, Belló-Klein A, Ribeiro MF, Partata WA. Sciatic nerve transection increases gluthatione antioxidant system activity and neuronal nitric oxide synthase expression in the spinal cord. *Brain Res Bull.* (2009) 80, 422-427.

[80] van der Kuy PH, Lohman JJ. The role of nitric oxide in vascular headache. *Pharm World Sci.* (2003) 25,146-151.

[81] Ashina M. Nitric oxide synthase inhibitors for the treatment of chronic tension-type headache. *Expert. Opin. Pharmacother* (2002) (4) 395-399.

[82] Olesen J. The role of nitric oxide (NO) in migraine, tension-type headache and cluster headache. *Pharmacol. Ther.* (2008) 120, 157-171.

[83] Berrios GE. (2004). "Of mania." *History of Psychiatry.* 15 (57 Pt 1), 105-124.

[84] Savas HA, Gergerlioglu HS, Armutcu F, Herken H, Yilmaz HR, Kocoglu E, Selek S, Tutkun H, Zoroglu SS, Akyol O. Elevated serum nitric oxide and superoxide dismutase in euthymic bipolar patients: Impact of past episodes. *World J Biol Psychiatry* (2006) 7, 51-55.

[85] Selek S, Savas HA, Gergerlioglu HS, Bulbul F, Uz E, Yumru M. The course of nitric oxide and superoxide dismutase during treatment of bipolar depressive episode. *J. Affect Disord.* 2008 Apr; 107, 89-94.

[86] Gergerlioglu HS, Savas HA, Bulbul F, Selek S, Uz E, Yumru M. Changes in nitric oxide level and superoxide dismutase activity during antimanic treatment. *Prog Neuropsychopharmacol. Biol Psychiatry* (2007) 31,697-702.

[87 Hsu C-w, Huang T-l Tsai M-C. Increased levels of serum nitrogen oxides are indicators of post-treatment response in mood disorder patients with acute episodes. *Neuropsychiatry (London)* (2016) 6, 417-423.

[88] Zoroğlu SS, Yürekli M, Meram I, Söğüt S, Tutkun H, Yetkin O, Sivasli E, Savaş HA, Yanik M, Herken H, Akyol O. Pathophysiological role of nitric oxide and adrenomedullin in autism. *Cell Biochemistry and Function* (2003), 21, 55-60.

[89] Demet Sağlam Aykut, Ahmet Tiryaki, Evrim ײzkorumak, Caner Karahan. Nitric Oxide and Asymmetrical Dimethylarginine Levels in Acute Mania. *Bulletin of Clinical Psychopharmacology* (2012) 22, 10-16.

[90] Yanik M, Vural H, Tutkun H, Zoroğlu SS, Savaş HA, Herken H, Koçyiğit A, Keleş H, Akyol O. The role of the arginine-nitric oxide pathwayin the pathogenesis of bipolar affective disorder. *Eur Arch Psychiatry Clin Neurosci* (2004) 254, 43-47.

[91] Bennabi M, Hamdani N, Lajnef M, Bengoufa D, Fortier C, Boukouaci W, Bellivier F, Kahn J-P, Henry C, Charron D, Krishnamoorthy R, Leboyer M, Tamouza R. Violent suicidal behaviour in bipolar disorder is associated with nitric oxide synthase 3 gene polymorphism. *Acta Psychiatrica Scandinavica* (2015) 132, 3218-

Chapter 13

MISCELLANEOUS DISEASES

ANNOTATION

Nitric oxide is deeply involved in various diseases, such as endocrine dysfunction, diabetes mellitus, thrombosis, asthma, gastronomic tract, and immune response dysfunction. Although these diseases are quite different in their medical symptoms, they are characterized by similar biochemical and physiological mechanisms. In most cases, these processes include the radical storm or/and the NO–cGMP pathway.

13.1. NITRIC OXIDE AND ENDOCRINOLOGY

13.1.1. Endocrine Dysfunction

Endocrinology is dealing with the endocrine system, its diseases, and hormones. Erectile dysfunction (ED) is one of the serious problems in medicine. A significant participation of NO in biochemical and physiological processes in the endocrine system are well accepted [1–9].

13.1.1.1. Sexual Activity

Nitric oxide has been proven to play a key role in male and female sexual activity, involving hormonal release, both as an activator and an inhibitor [9]. For example, NO is a physiological mediator of male erectile function and mechanism of NO deficiency in erectile dysfunction is in focus of attention. NO may also participate in the regulation of testicular functions via the NO–cGMP pathway, which stimulates muscle activity (Figure 13.1) [1]. Nitric oxide in the corpus cavernosum of the penis binds to guanylate cyclase receptors, which results in increased levels of cGMP, leading to vasodilation (smooth muscle relaxation causing widening of blood vessels) and increase inflow of blood into the penis, causing an erection. cGMP is converted into GMP by a specific phosphodiesterase (PDE.) Low levels of NO, a transient stimulatory effect, that correlates with increased cGMP levels, was observed. In contrast, high levels of NO were associated with testosterone inhibition with no change in cGMP production [1].

NO formation or bioavailability is decreased by oxidative stress [1]. During severe oxidative stress, the heme iron on sGC can be oxidized, rendering the enzyme unresponsive to NO or sGC stimulators. In normal conditions, sGC stimulators bind to the reduced NO-sensitive form of sGC to increase cGMP formation and promote erection. The sGC stimulators produce normal erectile responses when NO formation is inhibited and the nerves innervating the corpora cavernosa are damaged. During severe oxidative stress, the heme iron on sGC can be oxidized, rendering the enzyme unresponsive to NO or sGC stimulators.

The role of NO in endothelial relaxation and erectile function is well documented. The involvement of H_2S and CO in sexual function and dysfunction is also extensively investigated. Review [1] focused on the role of these three "sister" gasotransmitters in the physiology, pharmacology, and pathophysiology of sexual function in man, specifically erectile function and of sGC/cGMP pathway as a common target of this transmitter. Contribution of NO as the principal mediator of penile erection was stressed. Modulation of sGC for the treatment of ED using PDE-5 inhibitors was reported [12].

Synthesis and mechanisms of gaseous neurotransmitters in the relaxation of penile or other vascular tissues include the following steps [1]: (1) CO is synthesized from hemine by constitutive (HO-2 and HO-3) and inducible (HO-1) haem oxygenases; (2) caveolin interacts and inactivates both eNOS and HO-1. 5: Hsp90 (HSP90) activates eNOS; (3)ROS decreases the availability of NO to act on sGC; and (4)7 NO induces relaxation via inhibition of Rho-kinase (ROCK) signaling in the penile tissue horse of penile resistance arteries.

NO/CO/H_2S/sGC pathways in vascular tissues including the penis are [1]: (1) CO inhibits eNOS in the presence of higher amounts of NO and activates eNOS when there is a low amount of NO; (2) high levels of CO inhibit NOS activity and NO generation, lower concentrations of CO induce the release of NO. 3: NO donors activate HO-1; (3) NO donors upregulate the expression and activity of cystathionine gamma-lyase (CSE) in vascular tissues and cultured aortic smooth muscle cells; (4) H_2S causes eNOS activation in the aorta through Akt and directly increases the expression of eNOS in CC; (5) CO inhibits the CBS sensor; 7: (6) CO modulates NO-stimulated sGC activation in the presence of low concentrations of NO and inhibit sGC activation; (7) cGMP causes an increase in H_2S production in vasculature; and (8) H_2S acts as an endogenous inhibitor of PDE activity [1].

Nitric oxide generated by neuronal NO synthase (nNOS) initiates penile erection but does not participate in the sustained erection required for normal sexual performance [13]. It was found that cAMP-dependent phosphorylation of nNOS mediates erectile physiology, including sustained erection. nNOS is phosphorylated by cAMP-dependent protein kinase (PKA) at serine(S)1412. Stimulation of cAMP formation by forskolin, which is commonly used as a tool in biochemistry to raise levels of cyclic AMP, activates nNOS phosphorylation.

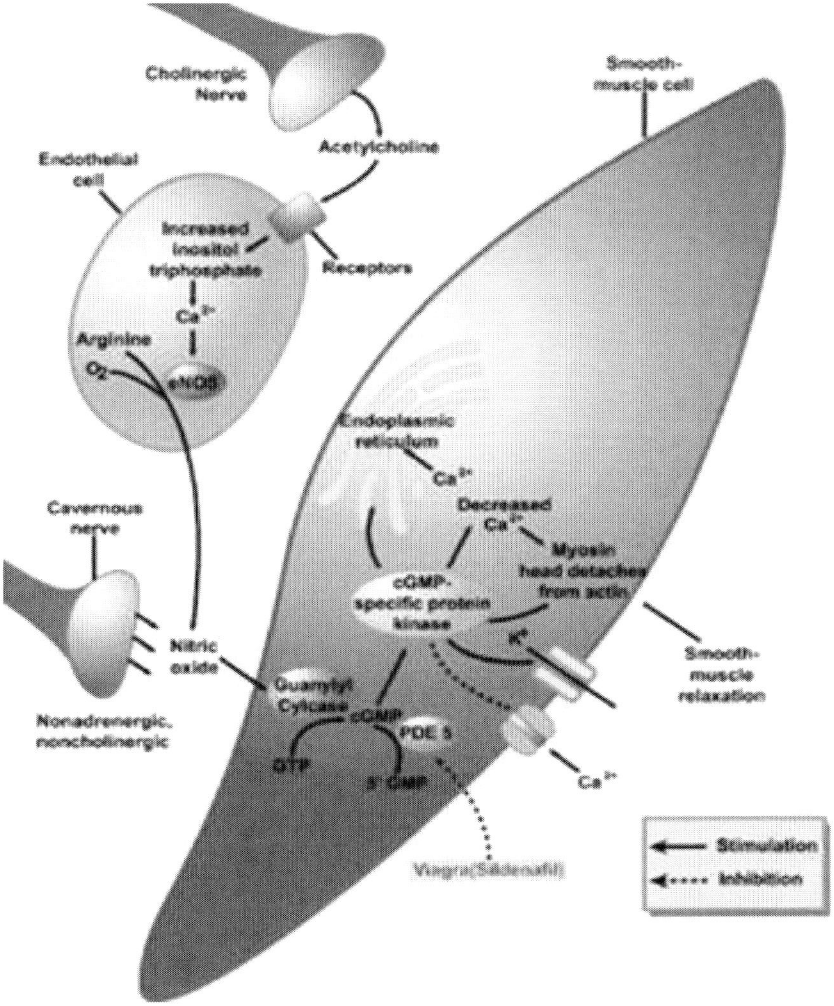

Figure 13.1. Stimulation and inhibition of muscles by NO. www.wiley.com/legacy/college/boyer/0470003790/cutting.../viagra.htm.

13.1.1.2. Steroidogenesis

NO has been shown to impact steroidogenesis in ovaries, testes, and adrenal glands in which it inhibits steroid production [9]. Effects NO on steroidogenesis and other processes in follicle development and oocyte competence were demonstrated.

Nitric oxide's role in the sexual hormonal pathway has been discussed in a series of publications [9–21]. The NO inhibitory effect on steroidogenesis can occur by both direct and indirect mechanisms [14, 15]. In the frame of the first mechanism introduced by Ignaro, NO activates guanylate cyclase by altering its conformation via binding to the heme group of this enzyme, leading to elevation of intracellular cyclic guanosine monophosphate (cGMP) with subsequent activation of PKG. According to the second mechanism, NO competitively interacts with the heme-oxygen binding site of key steroidogenic enzymes such as Cytochrome P450 17A1. The heme-oxygen complex attacks the steroid substrate. Nitric oxide can also act via two distinct post-translational modifications of proteins mediated by NO and NO-derived species: S-nitrosylation of cysteine residues and nitration of tyrosine residues.

Examples are given below indicate a connection of nitroxide with steroidogenesis.

An inverse relationship between ovarian NOS activity and synthesis of steroidogenic acute regulatory protein (StAR) was reported [16]. The endocrine system can protect against nitric-oxide-mediated tissue damage by producing corticosteroids, growth factors, and cytokines that can be inhibitors of NO overproduction. 17β-estradiol regulates NO amount produced by NOS-2 (inducible NOS) and NOS-1 (brain or neuronal NOS), which mediate cytotoxic effects and participate in inflammatory reactions. In the female of several animal species, NO displays positive effects on follicle development and selection related to angiogenic events and plays a modulatory role in steroidogenesis in ovarian cells [2]. The NO/NO synthase system is also involved in the control of meiotic maturation of cumulus-oocyte complexes.

Endogenously generated NO can be transported to remote tissues and exerts physiological activity [5–7]. For example, NO overexpressed in one organ (e.g., heart), could modulate I/R injury at a distant site (e.g., liver) [7]. Nitric oxide synthesized within a specified location (a mouse cardiomyocytes) may be transported in the blood, stored in distant tissues, and protects against ischemic injury in nitrosothiol-dependent fashion This finding demonstrated that endogenously generated NO is capable of exerting endocrine activity and, therefore, exhibits the properties typical of the classical hormones.

Increase of oxidative stress and endothelial dysfunction in the penis is attributed to age-related ED. Endothelial nitric oxide synthase uncoupling in the aged rat penis, as a contributing mechanism was evaluated [18]. The effect of replacement with eNOS cofactor tetrahydrobiopterin (BH(4)) on erectile function in the aged rats was examined. eNOS uncoupling and oxidative stress (thiobarbituric acid reactive substances, TBARS) were measured by conducting western blot in penes samples. Erectile response was found to be significantly reduced in the aged rat penis compared with young rats. These findings suggest that aging induces eNOS uncoupling in the penis, resulting in increased oxidative stress and ED.

Results of work [19] suggested that eNOS S-nitrosylation/denitrosylation is an important mechanism regulating eNOS activity during erectile function, and S-nitrosoglutathione reductase (GSNO-R) is a key enzyme involved in the denitrosylation of eNOS. The increase in eNOS S-nitrosylation (inactivation) of the enzyme observed with tumescence (a state of being swollen) may begin a cycle leading to detumescence.

In sixty Sprague-Dawley rats, changes in gene expression and the enzymatic activity of iNOS, heme oxygenase isoenzymes 1 and 2 (HO-1 and HO-2), and cGMP tissue levels in the corpora cavernosa (CC) were investigated [20]. The use of either NOS or HO inducers can equally enhance erectile function via upregulation of gene expression of the two signaling genes involved in erection as well as through upregulation of the tissue levels of cGMP, the mediator of vasodilatation.

The contribution of nNOS and eNOS in the synthesis of NO as a mediator of penile erection, at the levels of both the penile corpora cavernosa and the hypothalamic regions that control the erectile response, was well established [21] Transcriptional induction of NOS was postulated to be a key factor in two opposite related pathological processes, namely neurotoxicity in critical related regions of the hypothalamus during senescence and as a defense mechanism against the aging or injury-associated fibrosis [21]. iNOS may protect the erectile tissue by counteracting fibrosis that impairs cavernosal smooth muscle compliance.

It is well established that sildenafil citrate (Viagra)

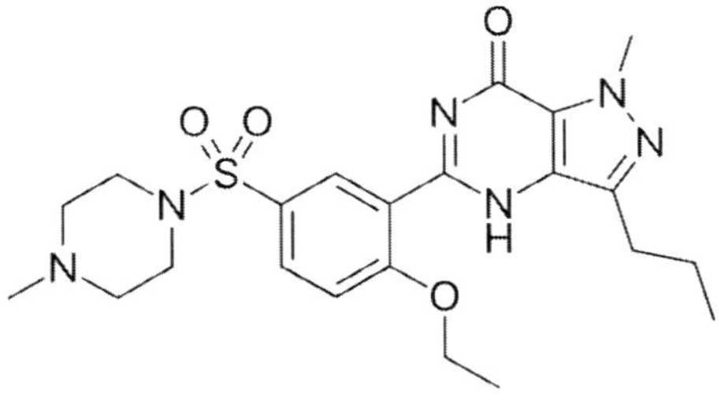

and other similar drugs for the treatment of ED act by inhibiting cGMP-specific phosphodiesterase type 5 (phosphodiesterase 5, PDE5), an enzyme that promotes degradation of cGMP, which regulates blood flow in the penis (Figure 13.1) [22–24]. The molecular structure of sildenafil is similar to that of cGMP and acts as a competitive binding agent of PDE5. In fact, sildenafil protects cGMP from degradation by PDE5. By blocking the breakdown of cGMP, sildenafil acts to prolong the effects of cGMP and, hence, the erection. As essential regulators of cyclic nucleotide signaling with diverse physiological functions, PDEs are also drug targets for the treatment of various diseases, including heart failure, depression, asthma, and inflammation.

In treatment for ED, sildenafil acts as a specific inhibitor for PDE 5 that enhances NO-mediated vasodilation in the CC by inhibiting cGMP breakdown and improves endothelial function of healthy men [23]. Experiments indicated that sildenafil increased sensitivity to nitroglycerin, an exogenous NO donor, approximately fourfold.

The crystal structure of human PDE5 with bound sildenafil was established [22]. The action mechanism and specificity of PDE was discussed on the base of atomic structure of PDE4 [24]. The catalytic domain of the PDE5 molecule can be divided into three subdomains: an N-terminal cyclin-fold domain (residues 537–678), a linker helical domain (residues 679–725), and a C-terminal helical bundle domain (residues 726–860). The amide moiety of the pyrazolopyrimidinone group of sildenafil forms a bidentate hydrogen bond with the gamide group of Gln 817, which is well ordered by a hydrogen bond relay involving Gln 817 to Gln 775, Gln 775 to Ala 767, and Gln 775 to Trp 853 (see details in [22].

Cialis or Nitric Oxide Tadalafil

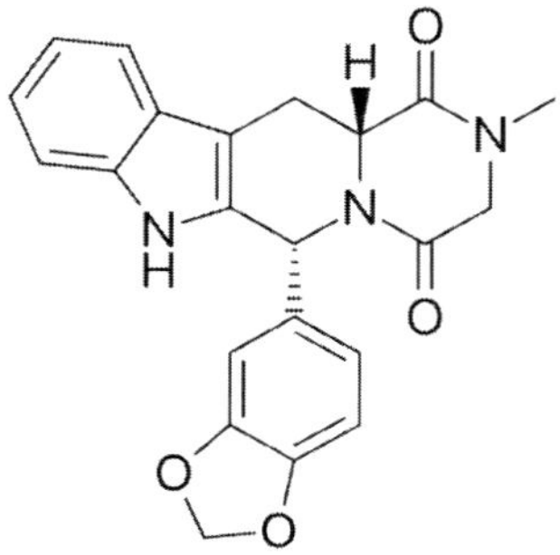

is a PDE5 inhibitor marketed in pill form for treating ED [25].

Cialis is one of a class of drugs called PDE inhibitors and acts by molecular mechanisms similar to one for sildenafil [26]. Cialis decreases the activity of cGMP, which allows more NO to stay in the blood vessels of the penis. The NO relaxes those blood vessels, allowing them to expand and cause an erection. Since Cialis also inhibits cAMP, it relaxes blood vessels in other parts of the body as well. The three-dimensional structures of the catalytic domain (residues 537–860) of human PDE5 complexed with the three drug molecules sildenafil, tadalafil (Cialis), and vardenafil (Levitra) were presented in [27].

13.2. Diabetes Mellitus

Diabetes mellitus (DM), diabetes, is a group of metabolic disorders in which there are high blood sugar levels over a prolonged period [29]. Type 1 DM results from the pancreas's failure to produce enough insulin, which is a hormone that helps move sugar, or glucose, into your body's tissues. Type 1 diabetes happens when the immune system destroys beta cells in the pancreas. Type 2 DM begins with insulin resistance, a condition in which cells fail to respond to insulin properly.

The integrative model of hypoglycemia detection suggested a number biochemical events [29]: (1) a falling glucose is detected by peripheral and central glucose-sensing cells/neurons; (2) peripheral glucose sensors signal back to glucose-sensing regions of the hindbrain, in turn, activating efferent pathways that initiate a counterregulatory response; (3) at the same time, glucose sensors in the hypothalamus can detect falling glucose, also activating efferent pathways that initiate a counterregulatory response; (4) integrative pathways between the hindbrain, hypothalamic, and other forebrain regions are reciprocally connected and can modulate responses to the hypoglycemic signal; and (5) glucose-sensing regions in the brain, such as the VMH, contain GE and GI neurons and an astrocytic support structure. [29].

Nitric oxide, as a key metabolic and vascular regulator, is significantly involved in the principle biochemical and physiological processes of diabetes mellitus. A connection between NO production and DM was evaluated in a series of studies. Serum nitrite and nitrate concentrations were assessed as an index of NO production in thirty adolescents and young adults with type 1 diabetes [28]. In young diabetic patients with early nephropathy, chronic hyperglycemia is associated with an increased NO biosynthesis and action that contributes to generating glomerular hyperfiltration and persistent on inclusion.

The mechanisms involved in the destruction of insulin-producing β cells of pancreatic islets of Langerhan human type 1 diabetes were examined [30]. Cell-mediated immune reactivity against islet constituents is assumed to play a major role in the development of this disease. In an attempt to identify cellular effector mechanisms involved in β-cell destruction, it was established that NO released from activated macrophages is able to exert cytotoxic activity against islet cells. NO was identified as a major regulatory and cytotoxic mediator during islet inflammation in system under investigation. In their report [31], the authors summarized data and conclusions on a main culprit for β-cell dysfunction and death in type 1 diabetes in humans and animal models: (1) type 1 diabetes (T1DM) is characterized by strong expression of the pro-inflammatory cytokines IL-1β and TNFα; (2) IL-1β alone with its dominant effect on NO generation causes β-cell dysfunction; (3) cytokine-treated β-cells show high manganese superoxide dismutase expression in the mitochondria; (4) peroxynitrite is not a main culprit for β-cell dysfunction and death in type 1 diabetes; and (5) the pathogenesis of β-cell death in T1DM and therefore conclusion that NO is the primary mediator of cellular toxicity is inadequate.

Polymorphisms in the endothelial nitric oxide synthase gene (eNOS) was thought to be implicated in the development of nephropathy in patients with type 1 or insulin-dependent diabetes mellitus (IDDM) [32]. Studies demonstrate clearly that DNA sequence differences in eNOS influence the risk of advanced nephropathy in type 1 diabetes.

Both cyclooxygenase (COX) and NOS contribute to sweating of patients, whereas NOS alone contributes to cutaneous vasodilation during exercise in the heat [33]. In individuals with T1DM, performing moderate intensity exercise in the heat, NOS-dependent sweating but not cutaneous vasodilation is attenuated, whereas COX inhibition increases sweating.

Nitric oxide production is stimulated by insulin. A reduced urinary excretion of NO products (NOx) is frequently found in type 2 diabetes, particularly in association with nephropathy, which is damage or disease of a kidney [34]. NOx fractional and absolute synthesis rates in type 2 diabetic patients with diabetic nephropathy and in control subjects, after infusion L-[$^{15}N_2$-guanidino]-arginine (NOS inhibitor), and use of precursor–product relationships, were measured. It was found that whole-blood NOx production from arginine and its response to insulin are decreased in type 2 diabetes with nephropathy under both basal and hyper-insulinemic states.

The effects of insulin on production of nitrites and nitrates after ^{15}N-arginine intravenous infusion and on asymmetric dimethylarginine (ADMA) and symmetric dimethylarginine (SDMA) concentrations was tested [35]. In the experiment condition, i.e., aging, hypertension, hypercholesterolemia, and type 2 diabetes mellitus (T2DM), NO availability was alternated [35]. The authors concluded that whole-body NOx production is decreased in aging and T2DM. ADMA concentration, and T2DM, but not insulin resistance, appear as negative regulators of whole-body NOx production. ADMA elevation in patients with type 2 diabetes mellitus was found in [36]. A glucose-induced impairment of dimethylarginine dimethylaminohydrolase (DDAH) causes ADMA accumulation and may contribute to endothelial vasodilator dysfunction in DM.

- The association between NO production and insulin sensitivity (Si) in obese subjects with and without metabolic syndrome (MetSyn) and the relationship between NO production and ADMA was evaluated [37]. Circulating levels of ADMA were significantly higher in the obese group with MetSyn. The association between Si and NO production suggests a close mechanistic link between endothelial function and insulin signaling. Uncoupling of NOS secondary to redox signaling is a central mechanism in endothelial and macrophage activation [38]. Evidences that recoupling eNOS by increasing the eNOS cofactor tetrahydrobiopterin (BH_4) could restore endothelial function and prevent kidney injury in experimental kidney transplantation were presented in [38]. To investigate the effects of diabetes on renal tissue NO bioavailability, a diabetic mouse model *in vivo*, NO trapping, followed by electron paramagnetic resonance spectroscopy was employed. Experimental findings suggested coupling of NOS by supplying the cofactor BH_4 could restore glomerular endothelial barrier function.
- The role of NO in glucose leg blood flow (LBF) uptake during exercise in individuals with type 2 diabetes was evaluated [39]. After the N^G-monomethyl-L-arginine (L-NMMA) administration, LBF was lower and arterial plasma glucose and insulin levels were higher in individuals with diabetes. L-NMMA had no effect on LBF or arterial plasma glucose and insulin concentrations during exercise in

both groups. In the diabetic group L-NMMA reduced leg glucose uptake significantly greater as compared with healthy individuals. These data suggest a greater reliance on NO for glucose uptake during exercise in individuals with type 2 diabetes compared with control subjects.

- The association between hyperglycemia and serum NO levels in T2DM patients and T2DM with cardiovascular complication was established [40]. It was concluded that glucose levels observed in in serum of South Indian diabetic patients might be responsible for activation of endothelial cells to enhance NO levels.
- It was reported that increased glucose levels are associated with higher serum nitric NO levels in fructose-fed insulin resistant rats [41]. Obtained higher glucose levels in serum was proposed to be responsible for activation of endothelial cells to enhance NO levels. Nitric oxide more effectively inhibits neointimal hyperplasia in type 2 diabetic *versus* nondiabetic and type 1 diabetic rodents. In addition, NO decreases the ubiquitin-conjugating enzyme UbcH10, which is critical to cell-cycle regulation [41]. These data suggest a disconnect between UbcH10 levels and neointimal hyperplasia formation in type 2 diabetic models.

13.3. Nitric Oxide and Thrombosis

A thrombus is the product of the blood coagulation consisting of aggregated platelets and red blood cells and cross-linked fibrinprotein [42–50]. Platelets are a component of the blood whose function is to stop bleeding. Thrombus forms as platelets aggregate via the binding of bivalent fibrinogen to glycoprotein IIb/IIIa.

Nitric oxide released by the endothelium prevents platelet adhesion to the vessel wall and inhibits further recruitment of platelets to a growing thrombus. These processes provide a negative feedback mechanism for the propagation of thrombus formation. Nitric oxide can also inhibit platelet function by stimulating platelet adenylyl and guanylyl cyclases, which synthesize cyclic AMP (cAMP) and cyclic GMP (cGMP), respectively [45–47]. These products stimulate cAMP-dependent protein kinase (protein kinase A [PKA]I and PKAII) and cGMP-dependent protein kinase (protein kinase G [PKG]I) to phosphorylate proteins. The protein phosphorylation results in the inactivation of small G-proteins inhibition of the release of contribute to platelet hyperreactivity in cardiovascular disease.

The following cascade reactions in the platelet-derived NO process was suggested [47]: (1). Thrombin stimulation of the protease-activated receptor -(PAR1) increases intracellular Ca^{2+} concentration both by releasing Ca^{2+} from intracellular stores and allowing external Ca^{2+} to enter the cytosol; (2) The binding of the Ca^{2+}-calmodulin complex in addition to post-translational modifications (including phosphorylation, dephosphorylation, myristoilation, and palmitoylation) of cNOS activates the enzyme to catalyze the conversion of L-arginine to L-citrulline and NO; (3) Nitric oxide subsequently binds the heme moiety of sGC activity of this enzyme to increase continuous-glucose-monitoring (Cgm).

Inactivation of NO occurs mainly through its interaction with ROS and can be favored by a deficiency of antioxidant enzymes such as glutathione peroxidase. Therefore, the antioxidant status is important in normal platelet function and the prevention of thrombosis. Antioxidants, vitamin E (a-tocopherol) in particular, may alter platelet function, for example by inhibiting its aggregation. In addition, platelet aggregation might be associated with a burst of oxygen consumption with consequent appearance of the reactive oxygen and nitrogen species [45, 47, 49]. Involving ROS in the thrombus formation and inactivation of NO is depicted in Figure 13.9. Inactivation of NO occurs mainly through its interaction with ROS and can be favored by a deficiency of antioxidant enzymes such as glutathione peroxidase.

In the intact vessel, the production of prostacyclin and NO by the endothelium prevents adherence and activation of platelets to the vessel surface [47]. In the disrupted endothelium, platelets adhere to the subendothelial matrix, become activated, and release the prothrombotic substances thromboxane A2, serotonin, and ADP. This process supports the recruitment of additional platelets to the growing thrombus and may by limited by the release of NO by the activated platelet. Inhibition of platelet activation/aggregation in NO/cGMP signaling by cGMP-dependent protein kinase I (cGKI), as a central regulatory pathway in mutant mice, was examined [48]. The inositol-1,4,5-trisphosphate receptor-associated cGMP kinase substrate (IRAG) was found to be expressed in platelets and assembled in a macrocomplex together with protein kinase, cGMP-dependent, type I (cGKIbeta), and the inositol-1,4,5-trisphosphate receptor type I (InsP3RI).

Drugs that chemically link a nonsteroidal anti-inflammatory drugs NSAID with a NO-donating moiety (cyclo-oxygenase-inhibiting NO-donating drugs [CINODs]) were synthesized and exploited as a vasodilator and an inhibitor of platelet aggregation, and having anti-inflammatory properties [50]. Effects of CINODs in animal studies include inhibition of vasopressor responses, blood pressure reduction in hypertensive rats, and inhibition of platelet aggregation. In addition, CINODs may reduce ischemic damage to compromised myocardial tissue.

13.4. Nitric Oxide and Asthma

Asthma is a disorder that is characterized by variable airflow obstruction, airway inflammation, and hyperresponsiveness [51–55]. Main factors underlying asthma exacerbations include allergen exposure, viral infections, exercise, irritants, and ingestion of nonsteroidal anti-inflammatory agents. Figure 13.1, which demonstrates schematically biochemical and physiological processes causing inflammatory and remodeling responses in asthma, were discussed.

In the acute inflammatory in asthma, the cytokine network including IL-3, IL-4, IL-5, IL-9, and IL-13mast cells, eosinophils, neutrophils, and lymphocytes are contributors to the initiation of asthma with the release of acute-phase mediators [52, 53]. It was assumed that the factors affecting the inflammation will be also in force for corresponding processes in asthma. Exhaled NO measured in the expired air of asthmatic subjects is increased during an asthma exacerbation [51, 54, 55].

Nitric oxide is released by several pulmonary cells, including epithelial cells, eosinophils, and macrophages, and NO has been shown to be increased in conditions associated with airway inflammation and correspondent increase of synthesis of NO and ROS [54]. Pro-inflammatory cytokines such as TNFalpha and IL-1beta are secreted in asthma and result in inflammatory cell recruitment, but also induce calcium- and calmodulin-independent nitric oxide synthases (iNOS) and perpetuate the inflammatory response within the airways.

13.5. Nitric Oxide and the Gastrointestinal Tract

13.5.1. Nitric Oxide Contribution in Normal Gastric Processes

In the gastrointestinal tract, NO critically contributes in motility, secretion, digestion, absorption, and elimination and, therefore, is involved in normal physiology and the pathophysiology of many serious diseases [56–82]. Nitric oxide takes part in processes of intestinal injury, repair, carcinogenesis, and apoptosis in the tract.

The three NOS isoenzymes, nNOS, eNOS, and iNOS, play double face role to the function and diseases of the gastrointestinal tract [56]: (1) absence of eNOS-derived NO results in an increased susceptibility of the gastrointestinal tract to injury; (2) small (nanomolar) quantities of NO, produced by calcium-dependent nNOS, can have an effect on peristalsis (a radially symmetrical contraction and relaxation of muscles) and circular muscle (sphincter) function of the intestine; (3) decreased nNOS function can result in peristalsis and obstructive sphincters; (4) large amounts of NO can increase gut permeability, induce apoptosis, and stimulate intestinal secretion; and (5) NO can also kill bacteria, block apoptosis, and reduce inflammation by inhibiting activation of nuclear factor-κB (NF κB).

On the physiological level, NO: (1) maintains mucosal integrity providing a continuous supply of blood to the gastrointestinal mucosa; (2) is involved in the calcitonin gene-related peptide mediation, in hyperaemic response (the increase of blood flow) to mucosal injury; (3) inhibits intestinal muscular contraction; (4) is essential in maintaining mucosal blood flow; (5) is important to non-adrenergic, non-cholinergic (NANC) inhibitory neurotransmitters in the gut; and (6) is involved in the intestinal water transport by acting directly on the epithelium and blood flow or indirectly by stimulating neuronal reflexes and releases of, or interactions, with other agents [56]. On the molecular level, NO (1) activates sGC resulting in cGMP generation, a potent activator of intestinal secretion; (2) induces vasoactive intestinal polypeptide, an important neurotransmitter in secretomotor neurons; (3) causes an increase of prostaglandin E2 production, a secretory molecule; and (4) exerts direct secretory effects by the opening of chloride channels.

NO regulates processes in both the barrier and immune function NO which maintainsan intact mucosal barrier mediated by peroxynitrite (PN). However, high amounts of NO can increase the intestinal epithelial cells permeability induced by the NO donor, S-nitroso-N-acetylpenicillamine (SNAP), and the PN generator, 3-morpholinosydnonimine (SIN-1) [57]. The induction of iNOS is mediated by the nuclear transcription factor κB (NF-κB), which can be activated by physical and chemical stress initiated in turn by viruses, microbial products, pro-inflammatory cytokines, and T-and B-cell mitogens [58].

NO, NO-derived aspirin, and glyceryl trinitrate have a protective effect of NO on the gastrointestinal tract [59–62]. Nitric oxide has various effects on the gastrointestinal tract: (1) it helps maintain gastric mucosal integrity, (2) it inhibits leukocyte adherence to the endothelium, and (3) it repairs NSAID-induced damage [62]. In addition, the use of NO-donating agents with NSAIDs or aspirin resulted in reduced risk for gastrointestinal bleeding and cyclo-oxygenase inhibiting NO-donating drugs (CINODs), effective anti-inflammatory agents.

Gastritis is the inflammation of the lining of the stomach, and NO can be involved in processes developing and healing this disease. The protective effect of NO on intestinal epithelial cells was confirmed by finding of the reduced gastrointestinal toxicity of NSAID [59].

According to review [63], the GI tract nitric oxide is involved in the following important events: (1) participates in the modulation of the smooth musculature tone, namely, the regulation of intestinal peristalsis, gastric emptying, and antral motor activity; (2) regulates acid and gastric mucus secretion, alkaline production; (3) involved in the maintenance of mucosal blood flow; and (4) acts as an endogenous mediator modulating both the repairing and integrity of the tissues and demonstrates gastroprotective properties against different types of aggressive agents. High concentrations of NO are related to numerous pathological processes of the GI tract [63].

13.5.2. Nitric Oxide and Gastrointestinal Tract Diseases

Stated below, results illustrate the involvement of NO in key biochemical processes in diseases of the GI tract [64–82]. The enhanced expression of iNOS and nitrotyrosine in gastric mucosa (the mucous membrane layer of the stomach), which may contribute to the carcinogenesis, was examined [64]. Evaluation of the participation of NO in the chronic gastritis of children led to the finding that NO has got a great value in the evaluation of the inflammatory process and its activity in gastritis [65]. It was shown that berberine, a quaternary ammonium salt from the protoberberine group of benzylisoquinoline alkaloids, significantly protects gastric mucosa from damage by ethanol by the increasing expression of eNOS mRNA and inhibiting expression of iNOS mRNA [66]. Treating the rat gastric ulcers induced by acetic acid with NOS inhibitor, N(G)-nitro-L-arginine methyl ester (L-NAME), significantly increased the number of inflammatory cells with endogenous peroxidase, reduced the number of apoptotic inflammatory cells, and delayed ulcer healing [67]. Thus, iNOS, as "a double sword" agent, not only participates in ulcer formation but also plays a beneficial role in ulcer healing, in part by the exclusion of iNOS-positive inflammatory cells from the regenerating mucosa.

It was found that peritoneal injection of sodium nitroprusside (SNP), a NO donor, decreased the ulcer area, inflammatory cell infiltration, and MPO degree in acetic acid-induced gastric ulcer in rats [68]. This effect was abolished by a transient receptor potential cation channel (TRPV1) antagonist. Other reported important activities of SNP are an increase in the jejunal mesenteric afferent discharge, which was largely diminished by pretreatment of S-nitrosylation blocker N-ethylmaleimide, depolarization of the resting membrane potential of NG neurons, and enhanced capsaicin-induced inward current. These results suggest that NO donor SNP alleviates acetic acid-induced gastric ulcers in rats via the vagus nerve, via S-nitrosylation of a transient receptor.

- Bacterial infection is an important factor in the occurrence of gastritis. Bacteria could contribute to this production either by stimulating the mucosa to produce NO or generating NO themselves [64, 69–74]. Finding that the *Helicobacter pylori* (HP) bacterium is capable of interfering with the recurrent peptic ulcer healing process has been well established [69]. HP infection, associated with gastric cancer, can lead to a sustained production of RNS that may contribute to DNA damage [70]. Local expression of iNOS and nitration of tyrosine as an indicator of peroxynitrite formation were revealed in patients with (HP)-associated gastric ulcers [71]. In HP-positive gastric ulcers, iNOS and nitrotyrosine immunoreactivity was observed at active ulcer margins, in surface epithelial, and in the lamina propria. In cases of HP-positive ulcer and HP-related gastritis, iNOS and nitrotyrosine reactivity were found in areas remote from the lesion. Thus, NO and peroxynitrite formation is increased in HP-infected gastric mucosa, suggesting that HP promotes nitric oxide stress.
- Results of work [72] suggest that high production of iNOS and nitrotyrosine in the gastric mucosa, mucous membrane layer of the stomach, infected with *H. pylori*, may contribute to the carcinogenesis. The relation between high levels expression

of iNOS and pro-inflammatory cytokines in HP-infected gastric mucosa and serum markers of gastritis was examined [73]. It was found that iNOS and nitrotyrosine may contribute to development of gastric cancer and iNOS-producing gastritis and is correlated with high levels of interleukin-6, a protein that acts as both a pro-inflammatory cytokine and an anti-inflammatory myokine.

- Besides NO, molecular hydrogen, carbon monoxide, and hydrogen sulfide have been proved to be signaling molecules in the GI tract enhancing, for example, ulcer healing in rats [75–82]. CO, as a ligand of the NOS heme, inhibits the enzyme activity NOS activity *in vitro* [76]. This molecule, derived from heme oxygenase 1, also acts as a tonic regulator of NO-dependent vasodilation [77]. NO and H_2S compete for site recognition of cysteine residues for nitrosylation and sulfhydration, respectively, and both are required for the regulation of angiogenesis and endothelium-dependent vasorelaxation for certain physiological actions [82].

13.6. NITRIC OXIDE AND IMMUNO RESPONSE

The immune system is a defense system that protects against disease. Immune response is a response originating from immune system activation by antigens, including immunity to pathogenic microorganisms. In this process, the main cells involved are the T cells, B cells of lymphocytes, and macrophages https://en.wikipedia.org/wiki/Immune_system). B cells produce immunoglobulins, or antibodies, that react with antigens. The activation of a resting helper T cell causes it to release cytokines and other stimulatory signals that stimulate the activity of macrophages, killer T cells, and B cells, the latter producing antibodies. The stimulation of B cells and macrophages succeed a proliferation of T helpers.

Review [83] summarized studies on NO contribution in the immune system published by 2000, focusing on following subjects: (1) NO production in the immune system, (2) mechanisms of regulation of NO production, (3) overview of immune-system NO function, (4) regulation of NO production by iNOS, (5) NO signaling in leukocyte adhesion and chemotaxis, (6) NO and the thymus, (7) NO and tumor growth, (8) NO and infectious disease, and (9) NO and transplantation, inflammation, and autoimmunity. The chemistry of NO and ROS was discussed in the context of antipathogen activity and immune regulation. Nitric oxide and ROS are key mediators of immunity to regulate immune responses. These species trigger the eradication of pathogens and modulate immune-suppression during tissue-restoration and wound-healing processes. Fluctuations in the levels of these reactive intermediates induce other phases of the immune response and activates specific signal transduction pathways in tumor cells, endothelial cells, and monocytes. The NO/ROS balance is also important during Th1 to Th2 transition. The Th cells (T helper cells) are a type of T cell that play an important role in the immune system, particularly in the adaptive immune system. Levels of NO produced by iNOS in the microenvironment of the cell can range from as low as 10 nM to μM amounts for days [84]. Thus, there are two major roles for these reactive small molecules in the immune response: direct participation in the eradication of pathogens or regulation of immune pathways. Specific roles of NOS isoforms during acute and chronic inflammation to illustrate the temporal and concentration dependence of NO with respect to outcome were compared in [83]. The difference between murine models and humans with respect to regulation of these molecules was also discussed.

A diagrammatic scheme of interactive pathways intracellular pathways that regulate levels of NO/RNS and ROS that are important to the immune response was provided.

According to the diagram provided in [83], metabolic pathways that control arginine and BH_4 is involved in determining the NO/superoxide balance, which can be shifted by altering the effect of NO on its targets. Commonly accepted mechanisms of the macrophages activation in a course of pathogen eradication is associated with the production and release of pro-inflammatory cytokines, such as TNF-α, IL-6, and IL-1β of proteases, and of NO/RNS and O_2^-/ROS. In addition, the early increase in ROS, followed by increasing production of NO, results in a rapidly changing NO/RNS profile within the immune cell.

The role of NO in immune response and inflammation and its mechanisms of action in these processes was also considered in reviews [85-87]. Macrophages activated by pathogens and a large number of other immune system cells produce and respond to nitric oxide. NO is not only important as a toxic defense molecule against infectious organisms but regulates the functional activity, growth, and death of many immune and inflammatory cell types, including macrophages, T lymphocytes, antigen-presenting cells, mast cells, neutrophils, and natural killer (NK) cells, which are a type of cytotoxic lymphocyte critical to the innate immune system.

Nuclear factor-kappa B (NF-κB) is thought to be critical in orchestrating the innate immune response outcomes [86]. As an example, the effects of NO on MAP kinase signaling and NF-kB activation in lipopolysaccharide (LPS)-stimulated RAW 264.7 macrophages were reported. These effects correlate to the induction target genes, including interferon-beta (IFN-beta) and IkaB-ras. Simultaneous treatment with LPS and the NO donor, diethylamine NONOate (DEA/NO), enhances and prolongs proteins JNK and p38 phosphorylation and the LPS-induced degradation of the NF-kB inhibitory subunit, IkB-alpha. The enhancement of IkB-alpha degradation by DEA/NO correlates with an increase in the nuclear levels of the p50 and p65 subunits and DNA-binding activity.

NO is also involved in immunosuppression by regulating circulating immune cells. Combinations of the three pro-inflammatory cytokines, TNFalpha, IL-1alpha, or IL-1beta, provoke the expression of high levels of several chemokines and iNOS by mesenchymal stem cells (MSCs) [88]. Chemokines drive T cell migration into proximity with MSCs, where T cell responsiveness is suppressed by NO. These findings suggest that pro-inflammatory cytokines are required to induce immunosuppression by MSCs through the concerted action of chemokines and NO. In another example, myeloid-derived suppressor cells (MDSC), activated by NO-mediated increases in cGMP, reduce activated T-cell number and inhibit their function by multiple mechanisms [89]. These mechanisms include depletion of l-arginine by arginase-1 (ARG1) production of NO, ROS, and reactive nitrogen oxide species by iNOS. The resulting increase in chemokines and iNOS leads to the attenuation of T cell responsiveness.

Besides regulating immune signaling pathways, NO/RNS and ROS participate in tissue restoration mediated by macrophages by helping to regulating the cleanup of debris, the restructuring of extracellular matrix ECM, cell proliferation, and the formation of new blood vessels [90]. To fight pathogens within the phagosome, NOX2 is involved in two processes, namely, the nitrite accumulation through scavenging mechanisms and providing peroxide as a source of ROS [91]. Killing by nitrosative and oxidative modification of critical bacterial macromolecules and providing a multifaceted pathogen eradication program are two independent mechanisms of action of the variety of chemical species generated from NO/RNS and ROS.

It was reported that NO provides significant protection to mammalian cells from the cytotoxic effects of hydrogen peroxide (H_2O_2) [92]. Treating bacteria (*Escherichia coli*) with NO resulted in minimal toxicity, but greatly enhanced (up to 1,000-fold) in H_2O_2-mediated killing, the combination of NO/H_2O_2-induced DNA double-strand breaks in the bacterial genome, and DNA damage correlated with cell killing. The requirement for the combination of RNS and ROS, rather than NO or superoxide/H_2O_2 by themselves, was confirmed in work [93]. Viral replication (the formation of biological viruses during the infection process), inhibited by the induction of iNOS and the subsequent production of NO, caused the eradication of viruses, such as HIV-1, coxsackievirus, influenza A and B, rhino virus, CMV, vaccinia virus, ectromelia virus, human herpesvirus-1, and human parainfluenza virus type 3 [94].

References

[1] Yetik-Anacak G, Sorrentino R, Linder AE, Murat N. Gas what: NO is not the only answer to sexual function. *British Journal of Pharmacology* (2014) 172, 1434-1454.

[2] Basini G, Grasselli F. Nitric oxide in follicle development and oocyte competence. *Reproduction.* (2015) 150, R1-9.

[3] Laskin JD, Diane E. Heck DE, Debra L, Laskin DL. Multifunctional role of nitric oxide in inflammation. *Trends in Endocrinology and Metabolism* (1994) 5, 9377-9382.

[4] Kauser K, Rubanyi GM. Estrogen and nitric oxide in vasculature. *Current Opinion in Endocrinology & Diabetes* (1999) 6, 230.

[5] Schechter AN, Gladwin MT. Hemoglobin and the paracrine and endocrine functions of nitric oxide. *N. Engl. J. Med.* (2003) 348, 1483-148.

[6] Elrod JW, Calvert JW, Gundewar S, Bryan NS, Lefer DJ. Nitric oxide promotes distant organ protection: evidence for an endocrine role of nitric oxide. *Proc. Natl. Acad. Sci. USA* (2008) 105,11430-11435.

[7] Ghasemi A, Zahedias S.. Is Nitric Oxide a Hormone? *Iran Biomed J.* (2011) 15, 59-65.

[8] Chen Z, Yuhanna IS, Galcheva-Gargova Z, Karas RH, Mendelsohn ME, Shaul PW. Estrogen receptor α mediates the nongenomic activation of endothelial nitric oxide synthase by estrogen. *J. Clin. Invest.* (1999) 103, 401-406.

[9] Kevil CG, Lefer DJ. Nitric Oxide's Role in Sexual Hormonal Pathway. *Cardiovasc Res.* (2011) 89, 489-491.

[10] di Villa Bianca Rd, Sorrentino R, Sorrentino R, Imbimbo C, Palmieri A, Fusco F, Maggi M, De Palma R, Cirino G, Mirone V. Endogenous Urotensin II Selectively Modulates Erectile Function through eNOS. *J. Pharmacol. Exp. Ther.* (2006) 316, 703-708.

[11] Valenti S, Cuttica CM, Fazzuoli L, Giordano G & Giusti M. Biphasic effect of nitricoxide on testosterone and cyclic GMP production by purified rat Leydig cells cultured in vitro. *International Journal of Andrology* (1999) 22, 336-341.

[12] Lasker GF, Pankey EA, Kadowitz PJ. Modulation of soluble guanylate cyclase for the treatment of erectile dysfunction. *Physiology (Bethesda)* (2013) 28, 262-269.

[13] Hurt KJ, Sezen SF, Lagoda GF, Musicki B, Rameau GA, Snyder SH, Burnett AL. Cyclic AMP-dependent phosphorylation of neuronal nitric oxide synthase mediates penile erection. *Proc. Natl. Acad. Sci. USA* (2012) 109, 16624-16629.

[14] Ignarro LJ. Biosynthesis and metabolism of endothelium-derived nitric oxide. *Annual Review of Pharmacology and Toxicology* (1990) 30, 535-560.

[15] Ignarro LJ. Nitric oxide as a unique signaling molecule in the vascular system: A historical overview. *Journal of Physiology and Pharmacology* (2002) 53, 503-514.

[16] Srivastava VK, Dissen GA, Ojeda SR, Hiney JK, Pine MD, Dees WL. Effects of alcohol on intraovarian nitric oxide synthase and steroidogenic acute regulatory proteinin the prepubertal female rhesus monkey. *Journal of Studies on Alcohol and Drugs* (2007) 68, 182-191.

[17] Musicki B, Ross AE, Champion HC, Burnett AL, Bivalacqua TJ. Posttranslational modification of constitutive nitric oxide synthase in the penis. *J. Androl.* (2009) 30, 352-362.

[18] Johnson JM, Bivalacqua TJ, Lagoda GA, Burnett AL, Musicki B. eNOS-uncoupling in age-related erectile dysfunction. *Int. J. Impot. Res.* (2011) 23, 43-48.

[19] Palmer L, Kavoussi P, Lysiak J. S-Nitrosylation of endothelial nitric oxide synthase alters erectile function. *Nitric Oxide.* (2012) 27(Suppl), S22–

[20] Abdel Aziz MT, El-Asmar MF, Mostafa T, Atta H, Wassef MAA, Fouad HH, et al. Effects of nitric oxide synthase and heme oxygenase inducers and inhibitors on molecular signaling of erectile function. *J. Clin. Biochem.* (2005) 37, 103-111.

[21] Gonzalez-Cadavid NF, Rajfer J. The pleiotropic effects of inducible nitric oxide synthase (iNOS) on the physiology and pathology of penile erection. *Curr. Pharm.* (2005) 11, 4041-4046.

[22] Sung B-J, Hwang KY, Jeon YH, Lee J-Il, Heo YS, Kim JH, Moon J, Yoon JM, Hyun Y-L, Kim E, Eum SJ, Park S-Y, Lee J-O, Lee TG, Ro S, Cho JM. Structure of the catalytic domain of human phosphodiesterase 5 with bound drug molecules. *Nature* 425 (2003) 98.

[23] Dishy V, Sofowora G, Harris PA, Kandcer M, Zhan F, Wood AJ, Stein CM. The effect of sildenafil on nitric oxide-mediated vasodilation in healthy men. *Clin Pharmacol Ther.* 2001 Sep;70(3):270-279.

[24] Xu RX, Hassell AM, Vanderwall D, Lambert MH, Holmes WD, Luther MA, Rocque WJ, Milburn MV, Zhao Y, Ke H, Nolte RT. Atomic structure of PDE4: Insights into phosphodiesterase mechanism and specificity. Insights into Phosphodiesterase Mechanism and Specificity. *Science* (2000) 288, 1822 -18225.

[25] Daugan A, Grondin P, Ruault C, Le Monnier de Gouville AC, Coste H, Kirilovsky J, Hyafil F, Labaudinière R. "The discovery of tadalafil: a novel and highly selective PDE5 inhibitor. 1: 5,6,11,11a-tetrahydro-1H-imidazo[1',5':1,6]pyrido[3,4-b]indole-1,3(2H)-dione analogues." *Journal of Medicinal Chemistry* (2003) 46, 4525-4532.

[26] Lee HJ, Feliers D, Mariappan MM, Sataranatarajan K, Choudhury GG, Gorin Y, Kasinath BS. Tadalafil Integrates Nitric Oxide-Hydrogen Sulfide Signaling to Inhibit High Glucose-induced Matrix Protein Synthesis in Podocytes. *J. Biol. Chem.* (2015) 290, 12014-12026.

[27] Sung BJ, Hwang KY, Jeon YH, Lee JI, Heo YS, Kim JH, Moon J, Yoon JM, Hyun YL, Kim E, Eum SJ, Park SY, Lee JO, Lee TG, Ro S, Cho JM. Structure of the catalytic domain of human phosphodiesterase 5 with bound drug molecules. *Nature* (2003) 425, 98-102.

[28] Chiarelli F, Cipollone F, Romano F, Tumini S, Costantini F, di Ricco L, Pomilio M, Pierdomenico SD, Marini M, Cuccurullo F, Mezzetti A. Increased Circulating Nitric Oxide in Young Patients with Type 1 Diabetes and Persistent Microalbuminuria Relation to Glomerular Hyperfiltration. *Diabetes* (2000) 49, 1258-1263.

[29] Rory J. McCrimmon RJ, Sherwin RS. Hypoglycemia in Type 1 Diabetes. *Diabetes* (2010) 59, 2333-2339.

[30] Burkart V, Kolb H. Nitric Oxide in the Immunopathogenesis of Type 1 Diabetes. *Nitric Oxide* Part of the Handbook of Experimental Pharmacology book series (HEP, volume 143), 2000, pp 525-544.

[31] Gurgul-Convey E, LenzenIs S. Nitric Oxide Really the Primary Mediator of Pancreatic β-Cell Death in Type 1 Diabetes? *The Journal of Biological Chemistry* (2015) 290, 10570.

[32] Zanchi A, Moczulski D, Hanna LS, Andrzej S, Krolewski AS. Risk of advanced diabetic nephropathy in type 1 diabetes is associated with endothelial nitric oxide synthase gene polymorphism. *Kidney International* (2000) 57, 405-413.

[33] Fujii N, Dervis S, Sigal RJ, Kenny GP. Type 1 diabetes modulates cyclooxygenase- and nitric oxide-dependent mechanisms governing sweating but not cutaneous vasodilation during exercise in the heat. *American Journal of Physiology - Regulatory, Integrative and Comparative Physiology* (2016) 311, R1076-R1084.

[34] Tessari P, Cecchet D, Cosma A, Vettore M, Coracina A, Millioni R, Iori E, Puricelli L, Avogaro A, Vedovato M. Nitric Oxide Synthesis Is Reduced in Subjects with Type 2. Diabetes and Nephropathy. *Diabetes* (2010) 59, 2152-2159.

[35] Tessari P, Cecchet D, Artusi C, Vettore M, Millioni R, Plebani M, Puricelli L, Vedovato M. Roles of insulin, age, and asymmetric dimethylarginine. *Diabetes* (2013) 62, 2699-2708.

[36] Lin KY, Ito A, Asagami T, Tsao PS, Adimoolam S, Kimoto M, Tsuji H, Reaven GM, Cooke JP. Impaired Nitric Oxide Synthase Pathway in Diabetes Mellitus. Role of Asymmetric Dimethylarginine and Dimethylarginine Dimethyl-aminohydrolase. *Circulation* (2002) 106, 987-992.

[37] Siervo M, Bluck LJ. *In vivo* nitric oxide synthesis, insulin sensitivity, and asymmetric dimethylarginine in obese subjects without and with metabolic syndrome. *Metabolism* (2012) 61, 680-688.

[38] Boels MGS, van Faassen EEH, Avramu MC, van der Vlag J, van den Berg BM, Rabelin TJ. Direct Observation of Enhanced Nitric Oxide in a Murine Model of Diabetic Nephropathy *PLoS ONE* (2017) 12, e0170065.

[39] Kingwell BA, Formosa M, Muhlmann M, Bradley SJ, McConell G. Nitric Oxide Synthase Inhibition Reduces Glucose Uptake During Exercise in Individuals with Type 2 Diabetes More Than in Control Subjects. *Diabetes* (2002) 51, 2572-2580.

[40] Adela R, Nethi SK, Bagul PK, Barui AK, Mattapally S, Kuncha M, Patra CR, Reddy PN, Banerjee SK Hyperglycaemia Enhances Nitric Oxide Production in Diabetes: A Study from South Indian Patients. *PLoS One* (2015) 10, e0125270.

[41] Rodriguez MP, Tsihlis ND, Emond ZM, Wang Z, Varu VN, Jiang Q, Vercammen JM, Kibbe MR. Nitric oxide affects UbcH10 levels differently in type 1 and type 2 diabetic rats. *J. Surg. Res.* (2015) 196, 180-189.

[42] 1. Ruggeri ZM. Platelets in atherothrombosis. *Nat. Med.* (2002) 8, 1227-1234. 2. Walter U, Gambaryan S. Roles of cGMP/cGMPdependent protein kinase in platelet activation [letter]. *Blood* (2004) 104, 2609.
[43] Schwarz UR, Walter U, Eigenthaler M. Taming platelets with cyclic nucleotides. *Biochem. Pharmacol.* (2001) 62, 1153-1161.
[44] Jackson SP, Nesbitt WS, Kulkarni S. Signaling events underlying thrombus. *J. Thromb. Haemost.* (2003) 1,1602-1612.
[45] Smolenski AJ. Novel roles of cAMP/cGMP-dependent signaling in platelets. *Thromb Haemost*. (2012) 10, 167-176.
[46] Walter U, Gambaryan S. cGMP and cGMP-dependent protein kinase in platelets and blood cells. Handb. *Exp. Pharmacol.* (2009) (191), 533-548.
[47] Freedman JE, Loscalzo J. Nitric oxide and its relationship to thrombotic disorders. *Journal of Thrombosis and Haemostasis* (2003) 1, 1183-1188.
[48] Antl M, von Brühl ML, Eiglsperger C, Werner M, Konrad I, Kocher T, Wilm M, Hofmann F, Massberg S, Schlossmann J. IRAG mediates NO/cGMP-dependent inhibition of platelet aggregation and thrombus formation. *BLOOD* (2007) 109, 552-559.
[49] Voetsch B, Jin RC, Loscalzo J. Nioxide insufficiency and atherothrombosis. *J. Histochem. Cell Biol.* (2004) 353-367.
[50] Mackenzie IS, Rutherford D, MacDonald TM. Nitric oxide and cardiovascular effects: New insights in the role of nitric oxide for the management of osteoarthritis. *Arthritis Research & Therapy* (2008) 10, (Suppl 2).
[51] Robert F. Lemanske RF, Jr., William W. Busse WW. Asthma: Clinical Expression and Molecular Mechanisms. J. *Allergy Clin. Immunol.* (2010) 125, (2 Suppl 2), S95-102.
[52] Holgate ST, Polosa R. The mechanisms, diagnosis, and management of severe asthma in adults. *Lancet.* (2006) 368, 780-793.
[53] Barnes PJ. Immunology of asthma and chronic obstructive pulmonary disease. *Nat. Rev. Immunol.* (2008) 8,183-192.
[54] Yates DH. Role of exhaled nitric oxide in asthma. *Immunology and Cell Biology* (2001) 79, 178-190.
[55] LaForce C, Brooks E, Herje N, Dorinsky P, Rickard K. Impact of exhaled nitric oxide measurements on treatment decisions in an asthma specialty clinic. *Ann. Allergy Asthma Immunol.* (2014) 113, 619e623.
[56] Dijkstra G, van Goor H, Jansen PL, Moshage H. Targeting Nitric Oxide in the Gastrointestinal Tract. *Current Opinion in Investigational Drugs* (2004) 5, 529-536.
[57] Menconi MJ, Unno N, Smith M et al. Nitric oxide donor-induced hyperpermeability of cultured intestinal epithelial monolayers: role of superoxide radical, hydroxyl radical, and peroxynitrite. *Biochim Biophys Acta* (1998)1425:189-203.

[58] Xie QW, Kashiwabara Y, Nathan C. Role of transcription factor NF-kappa B/Rel in induction of nitric oxide synthase. *J. Biol. Chem.* (1994) 269, 4705-4708.

[59] Jobin C, Sartor RB. The I kappa B/NF-kappa B system: A key determinant, of mucosal inflammation and protection. *Am. J. Physiol. Cell Physiol.* (2000) 278C451-C462.

[60] Fiorucci S, Antonelli E, Santucci L et al. Gastrointestinal safety of nitric oxide-derived aspirin is related to inhibition of ICE-like cysteine proteases in rats. *Gastroenterology.* (1999) 116, 1089-1106.

[61] Konturek SJ, Brzozowski T, Majka J, Pytko-Polonczyk J, Stachura J. Inhibition of nitric oxide synthase delays healing of chronic gastric ulcers. *Eur J Pharmacol.* (1993) 23, 215-217.

[62] Lanas A. Role of nitric oxide in the gastrointestinal tract. *Arthritis Research & Therapy (*2008) 10, (Suppl 2), S4.

[63] Kochar NI, Chandewal AV, Bakal RL, Kochar PN. Nitric Oxide and the Gastrointestinal Tract. *International Journal of Pharmacolog,* (2011) 7, 31-39.

[64] Goto T, Haruma K, Kitadai Y, Ito M, Yoshihara M, Sumii K, Hayakawa N, Kajiyama G. Enhanced Expression of Inducible Nitric Oxide Synthase and Nitrotyrosine in Gastric of Gastric Cancer Patients. *Clinical Cancer Research* (1999) 5, 1411-1415.

[65] Ignyś I, Krauss H. Nitric oxide in children with chronic gastritis. *Pediatria Wspolczesna* (2004) 6, 241-248. ·

[66] Pan LR, Tang Q, Fu Q, Hu BR, Xiang JZ, Qian JQ. Roles of nitric oxide in protective effect of berberine in ethanol-induced gastric ulcer mice. *Acta Pharmacol. Sin.* (2005) 26, 1334-1338.

[67] Akiba Y, Nakamura M, Mori M, Suzuki H, Oda M, Kimura H, Miura S, Tsuchiya M, Ishii H. Inhibition of inducible nitric oxide synthase delays gastric ulcer healing in the rat. *Clin Gastroenterol.* (1998) 27, Suppl 1, S64-73.

[68] Han T, Tang Y, Li J, Xue B, Gong L, Li J, Yu X, Liu C. Nitric oxide donor protects against acetic acid-induced gastric ulcer in rats via S-nitrosylation of TRPV1 on vagus nerve. *Sci Rep.* (2017) 7, (2063).

[69] Elliott SN, Wallace JL. Nitric oxide, bacteria and ulcer healing. In Hunt RH. (ed.) *Kluwer Academic Publisher*, 1996, pp. 132-138.

[70] Komoto K, Haruma K, Kamada T, Tanaka S, Yoshihara M, Sumii K, Kajiyama G, Talley NJ.· *Helicobacter pylori* Infection and Gastric Neoplasia: Correlations with Histological Gastritis and Tumor Histology. *Am. J. Gastroenterol.* (1998) *93,* 1271-1276.

[71] Sakaguchi AA, Miura S, Takeuchi T, Hokari R, Mizumori M., Yoshida H., Ishii H. Expression of inducible nitric oxide synthase and peroxynitrite in *Helicobacter pylori* gastric ulcer. *Free Radical Biology and Medicine* (1999) *27,* 781-789.

[72] Goto T, Haruma K, Kitadai Y, Ito M, Yoshihara M, Sumii K, Hayakawa N, Kajiyama G. Enhanced Expression of Inducible Nitric Oxide Synthase and Nitrotyrosine in Gastric Mucosa of Gastric Cancer Patients. *Clinical Cancer Research* (1999) 5, 1411-1415.

[73] Kai H, Ito M, Kitadai Y, Tanaka S, Haruma K, Chayama K. Chronic gastritis with expression of inducible nitric oxide synthase is associated with high expression of interleukin-6 and hypergastrinaemia. *Alimentary Pharmacology & Therapeutics* (2004) 19, 1309-1314.

[74] Sobko T, Reinders C, Norin E, Midtvedt T, Gustafsson LE, Lundberg JO. Gastrointestinal nitric oxide generation in germ-free and conventional rats. *American Journal of Physiology - Gastrointestinal and Liver Physiology* (2004) 287, 5, G993-G997.

[75] Farrugia G, Szurszewski JH. Carbon Monoxide, Hydrogen Sulfide, and Nitric Oxide as Signaling Molecules in the Gastrointestinal Tract. *Gastroenterology* (2014) 147, 303-313.

[76] Dallas ML, Yang Z, Boyle JP, Boycott HE, Scragg JL, Milligan CJ, Elies J, Duke A, Thireau J, Reboul C, Richard S, Bernus O, Steele DS, Peers C. Carbon monoxide induces cardiac arrhythmia via induction of the late Na current. *Am. J. Respir. Crit. Care Med.* (2012)186, 648-656.

[77] Ishikawa M, Kajimura M, Adachi T, Maruyama K, Makino N, Goda N, Yamaguchi T, Sekizuka E, Suematsu M. Carbon monoxide from heme oxygenase-2 is a tonic regulator against NO-dependent vasodilatation in the adult rat cerebral microcirculation. *Circ Res* (2005) 97, e104-e114.

[78] Wallace JL, Dicay M, McKnight W, Martin GR. Hydrogen sulfide enhances ulcer healing in rats. *FASEB J* (2007) 21, 4070-4076.

[79] Bhatia M, Wong FL, Fu D, Bhatia M, Wong FL, Fu D, Lau HY, Moochhala SM, Moore PK. Role of hydrogen sulfide in acute pancreatitis and associated lung injury. *FASEB J* 2005;19:623-625.

[80] Ondrias K, Stasko A, Cacanyiova S, Sulova Z, Krizanova O, Kristek F, Malekova L, Knezl V, Breier A. H_2S and HS^-donor NaHS releases nitric oxide from nitrosothiols, metal nitrosyl complex, brain homogenate and murine L1210 leukaemia cells. *Pflugers Arch.* (2008) 457, 271-2791313.

[81] Sha L, Linden DR, Farrugia G, Szurszewski JH. Effect of endogenous hydrogen sulfide on the transwall gradient of the mouse colon circular smooth muscle. *J. Physiol.* (2014) 59, 1077-1089.

[82] Coletta C, Papapetropoulos A, Erdelyi K, Olah G, Módis K, Panopoulos P, Asimakopoulou A, Gerö D, Sharina I, Martin E, Szabo C. Hydrogen sulfide and nitric oxide are mutually dependent in the regulation of angiogenesis and endothelium-dependent vasorelaxation. *Proc. Natl. Acad. Sci. USA* (2012), 109, 9161-9166.

[83] Wink DA, Hines HB, Cheng RYS, Switzer CH, Flores-Santana W, Vitek MP, Ridnour LA, Carol A. Colton CA. Nitric oxide and redox mechanisms in the immune response. *J. Leukoc. Biol.* (2011) 89, 873-891.

[84] Thomas DD, Ridnour LA, Isenberg JS, Flores-Santana W, Switzer CH, Donzelli S, Hussain P, Vecoli C, Paolocci N, Ambs S, Colton CA, Harris CC, Roberts DD, Wink DA. The chemical biology of nitric oxide: Implications in cellular signaling. *Free Radic. Biol. Med.* (2008) 45,18- 31.

[85] Von Knethen A, Brüne B. Activation of peroxisome proliferator-activated receptor gamma by nitric oxide in monocytes/macrophages down-regulates p47phox and attenuates the respiratory burst. *J. Immunol.* (2002) 169, 2619-2626.

[86] Jacobs AT, Ignarro LJ. Nuclear factor-κ B and mitogen-activated protein kinases mediate nitric oxide-enhanced transcriptional expression of interferon-β. *J. Biol. Chem.* (2003) 278, 8018-8027.

[87] Colton CA, Wilcock DM. Assessing activation states in microglia. *CNS Neurol. Disord. Drug Targets* (2010) 9, 174-191.

[88] Ren G, Zhang L, Zhao X, Xu G, Zhang Y, Roberts AI, Zhao RC, Shi Y. Mesenchymal stem cell-mediated immunosuppression occurs via concerted action of chemokines and nitric oxide. *Cell Stem Cell* 2, (2008) 141-150.

[89] Serafini P, Meckel K, Kelso M, Noonan K, Califano J, Koch W, Dolcetti L, Bronte V, Borrello I. Phosphodiesterase-5 inhibition augments endogenous antitumor immunity by reducing myeloid-derived suppressor cell function. *J Exp Med.* (2006) 203, 2691-2702.

[90] Martinez FO, Sica A, Mantovani A, Locati M. Macrophage activation and polarization. *Front. Biosci.* (2008) 13, 453-446.

[91] Delledonne M, Polverari A, Murgia I. (2003) The functions of nitric oxide-mediated signaling and changes in gene expression during the hypersensitive response. *Antioxid. Redox Signal*. 5, 33-41.

[92] Pacelli R, Wink DA, Cook JA, Krishna MC, DeGraff W, Friedman N, Tsokos M, Samuni A, Mitchell JB. Nitric oxide potentiates hydrogen peroxide-induced killing of *Escherichia coli*. *(J. Exp. Med.* 1995) 182,1469-1479.

[93] Manchado M, Michan C, Pueyo C. Hydrogen peroxide activates the SoxRS regulon *in vivo*. *J. Bacteriol.* (2000) 182, 6842-6844.

[94] Xu W, Zheng S, Dweik RA, Erzurum SC. (2006) Role of epithelial nitric oxide in airway viral infection. *Free Radic. Biol. Med.* 41, 19-28.

AUTHOR CONTACT INFORMATION

Professor Gertz I. Likhtenshtein
Emeritus Professor
Department of Chemistry
Ben-Gurion University of the Negev
Beer-Sheva, Israel
Email: gertz@bgu.ac.il
and
Institute of Problems of Chemical Physics, RAS,
Chernogolovka, Russia
E-mail: likht@cp.ac.ru

INDEX

#

A

B

C

D

E

F

G

H

I

L

M

N

O

P

Q

R